Music Theory and Mathematics

Eastman Studies in Music

Ralph P. Locke, Senior Editor
Eastman School of Music

Additional Titles in Music Theory and in Music since 1900

Analyzing Wagner's Operas: Alfred Lorenz and German Nationalist Ideology
Stephen McClatchie

Aspects of Unity in J. S. Bach's Partitas and Suites: An Analytical Study
David W. Beach

CageTalk: Dialogues with and about John Cage
Edited by Peter Dickinson

Concert Music, Rock, and Jazz since 1945: Essays and Analytical Studies
Edited by Elizabeth West Marvin and Richard Hermann

Elliott Carter: Collected Essays and Lectures, 1937–1995
Edited by Jonathan W. Bernard

Explaining Tonality: Schenkerian Theory and Beyond
Matthew Brown

The Music of Luigi Dallapiccola
Raymond Fearn

Opera and Ideology in Prague: Polemics and Practice at the National Theater, 1900–1938
Brian S. Locke

Pentatonicism from the Eighteenth Century to Debussy
Jeremy Day-O'Connell

The Pleasure of Modernist Music: Listening, Meaning, Intention, Ideology
Edited by Arved Ashby

Representing Non-Western Music in Nineteenth-Century Britain
Bennett Zon

Ruth Crawford Seeger's Worlds: Innovation and Tradition in Twentieth-Century American Music
Edited by Ray Allen and Ellie M. Hisama

Schumann's Piano Cycles and the Novels of Jean Paul
Erika Reiman

The Sea on Fire: Jean Barraqué
Paul Griffiths

The Substance of Things Heard: Writings about Music
Paul Griffiths

Theories of Fugue from the Age of Josquin to the Age of Bach
Paul Mark Walker

A complete list of titles in the Eastman Studies in Music Series, in order of publication, may be found at the end of this book.

Music Theory and Mathematics

Chords, Collections, and Transformations

EDITED BY JACK DOUTHETT, MARTHA M. HYDE, AND CHARLES J. SMITH

UNIVERSITY OF ROCHESTER PRESS

First published 2008
Transferred to digital printing 2015

University of Rochester Press
668 Mt. Hope Avenue, Rochester, NY 14620, USA
www.urpress.com
and Boydell & Brewer Limited
PO Box 9, Woodbridge, Suffolk IP12 3DF, UK
www.boydellandbrewer.com

ISBN-13: 978-1-58046-266-2
ISBN-10: 1-58046-266-9
ISSN: 1071-9989

Library of Congress Cataloging-in-Publication Data
Music theory and mathematics : chords, collections, and transformations / edited by Jack Douthett, Martha M. Hyde, and Charles J. Smith.
p. cm. — (Eastman studies in music, ISSN 1071–9989 ; v. 50)
Memorial volume for John Clough.
Includes bibliographical references (p.) and index.
ISBN-13: 978-1-58046-266-2 (hardcover : alk. paper)
ISBN-10: 1-58046-266-9 (hardcover : alk. paper) 1. Music theory—Mathematics. I. Douthett, Jack M. (Jack Moser) II. Hyde, Martha M. III. Smith, Charles J., 1950– IV. Clough, John (John L.)
MT6.M96204 2008
781.2—dc22

2007033160

A catalogue record for this title is available from the British Library.

This publication is printed on acid-free paper.

Cover image and design: Geri McCormick

Contents

Preface

In the fall of 2001, John Clough announced his intention to retire from his faculty position at the University at Buffalo. As a celebration of John's central role in the field of music theory, a collection of essays by scholars who had worked and studied with him seemed timely and appropriate. Martha Hyde and I (John's music theory colleagues at UB) first discussed this idea at the Society for Music Theory annual meeting that fall in Philadelphia, and then began to put together a roster of possible contributors and to issue invitations.

Sadly, John's severe health problems appeared at the beginning of the next semester, in January 2002. What had been planned as an orderly transformation of the UB theory program became a mad scramble to cover John's teaching, advising, and departmental duties. After his January surgery, he seemed well enough; nonetheless, a slow, inexorable process of diminishment had begun. The dissertation defenses of Nora Engebretsen and Jonathan Kochavi, both in the spring of 2002, were the last he participated in, and virtually his last public appearances in the department. By the summer, the decline had become impossible to ignore. John's children had to manage the difficult task of looking after him in Buffalo, which they did with courage and determination. His UB colleagues continued to visit, but every time we saw him, there seemed to be less and less of him. Finally, John was moved to be nearer his family, and died in the late summer of 2003—less than two years after making those retirement plans.

What was originally conceived as a collection honoring one of the great lights in the field of music theory, on the occasion of his retirement, was first reconceived as a get-well offering, and now is appearing as a post-mortem tribute. In this context, it seemed more appropriate to Martha Hyde and me that we present a collection of papers in the areas of John's specialization, rather than a more general offering from friends and colleagues. Since neither of us has much expertise in mathematical music theory, we decided to withdraw our own planned contributions, and invited Jack Douthett, a professional mathematician and one of John's best friends, as well as his frequent co-author, to share the editorial duties. Jack has done most of the work getting John's own piece, his last music theoretical paper, ready for publication. He did the same for David Lewin's paper, since, by tragic coincidence this great scholar has also left us, shortly after submitting his piece for this collection.

Martha compiled the master bibliography for the entire collection, negotiated with publishers, and co-wrote the introduction along with Jack and Norman Carey. I made most of the initial contacts with contributors, and formatted the manuscripts according to the publisher's specifications. It has also fallen to me to preface the collection—perhaps because I had known John Clough since 1972, the longest of any of the co-editors.

I am left with an almost impossible task—to try to communicate some sense of a remarkable person to those of you not privileged to have known him, and (much harder!) to do justice to a central figure in the lives of almost everyone who did. It's relatively easy to honor John professionally—he transformed music theory, though not single-handedly (because none of us works alone). His contributions to the theory of diatonicism in all its manifestations have revolutionized the way we all think about the basic elements of music. Expressions like "maximally even" and "cardinality equals variety," popularized by John, are now indispensable to theorists grappling with tonality and post-tonality. The three Buffalo Symposia on neo-Riemannian transformations, with their many distinguished invited participants, were John's, both in conception and execution. They reflected his commitment to and excitement about this emerging field. David Lewin and Richard Cohn may have been the actual parents of transformation theory, but John Clough was its midwife.

One of the central themes of John's long music-theoretical career, from its leading-tone beginnings (the first article in the first issue of *Journal of Music Theory!*) to its neo-Riemannian cadence, was his refusal to espouse any theoretical orthodoxies. All his life, he remained a skeptic about clubs of all sorts—especially those whose main purpose seemed to be that of excluding the uninitiated from their ranks. He never could stand the idea of doing something just because everyone else did it; his advice was always to think things out on their own terms. I well remember expressing dismay at finding myself in some theoretical quandary (a frequent occurrence!), at which John's response would be "Perhaps you're asking the wrong question," or "Perhaps you haven't thought through your assumptions carefully enough."

Also notable was his commitment to precision and rigor—not least in his lifelong fascination with the interaction between music and mathematics. John turned to math not to *avoid* talking about the details of music; rather, he saw how certain important things about music could be said precisely and effectively *only* with the help of mathematics. In the process, he created theoretical structures of great elegance, fascination, and beauty—objects that are deserving of aesthetic appreciation and study in their own right.

My primary duty here, however, is to honor John Clough personally, because it's the person we miss more than the theorist. I had known John since my days in the graduate program at the University of Michigan, where he was teacher, supervisor, and mentor for so many of us. John became the Slee Chair of Music Theory at the State University of New York at Buffalo (as it was then called), and I was later privileged to join him on that faculty, where I was a Clough colleague for over fifteen years. John pulled his professional and collegial weight; he was, in fact, one of the most respected senior faculty figures in Buffalo's history. The academic landscape is littered with his theoretical progeny—Buffalo and Michigan students he advised, others he helped, and colleagues he encouraged and inspired.

As a music theorist, John's work was primarily motivated by his ready love of music—music of all flavors, from every place and group. He had a sophisticated but firm taste in jazz, and played the old established standards on the piano with resolute and solid enthusiasm. Another important backdrop for his life was recreational mathematics and puzzles of all kinds (not to mention his love for blackjack, a game in which he could adjust the odds by his considerable skill at card-counting)—passions that would help him to congeal many of the most important friendships of his life.[1]

Finally, though, John Clough's real importance to those who knew him well was his all-encompassing integrity. I don't think I've ever known anyone with as fine and deep a natural sense of fair play. If I had to pick a literary character to capture John's essence, my choice would be Atticus Finch, in Harper Lee's *To Kill a Mockingbird*—especially in Gregory Peck's screen realization of that character.[2] John never played favorites, and never advocated his own interests above those of the department or whatever community he was working within. He was unfailingly generous with his time and his encouragement, but had no tolerance for empty rhetoric. If he had something critical or unpleasant to say to you, he simply said it to your face, without rancor. In short, who he was in public was exactly the same as who he was in private; there was no role-playing, no hypocrisy, no attempt to escape from the standards he had set for himself. John never sought to impress anyone; I don't think he ever felt the need to do so. I can't think of a single serious scholar in the world of music theory who has a bad word to say about him. His UB students knew this better than any of us, and loved him for it; to them, he was simply "Papa Clough."

I knew John as my colleague for many years, and as my teacher for many years before that. I'm sure that I'll never know another like him: father figure, the most civilized of colleagues, dear friend. This, then, is the dedication of this book: to Papa Clough.

Charles J. Smith
Coordinator of Music Theory and Chair of the Music Department
The University at Buffalo

Notes

1. One of many world-puzzles I showed John over the years particularly delighted him as a diatonic set theorist. As a tribute to his sense of play and for those who care about such things, I have embedded that puzzle within this paragraph. . . .

Introduction

The essays in this collection celebrate the work of the late John Clough, a revolutionary musical thinker and a pioneer in enquiry into the nature of diatonic systems. Clough, who held the Slee Chair of Music Theory at University at Buffalo (SUNY) for many years, brought to music theory new perspectives in four roughly chronological phases.

The first phase of Clough's work, in the late 1970s, focused on the definition of diatonic sets and an enquiry into interval cycles and sequences; he became interested in extending Allen Forte's atonal methodology to the diatonic system. Milton Babbitt and Carlton Gamer, among others, had noticed intriguing structural properties of the diatonic system when considered as a subset of the equal-tempered chromatic scale.[1] Clough conjectured that, just as arithmetic modulo 12 serves to formalize chromatic pitch-class space, arithmetic modulo 7 would be the appropriate tool for heptatonic systems. This conjecture was not as obvious as it seems now. The elements in the 12-tone chromatic scale have an obvious logarithmic equality of step sizes. Clough realized that a more abstract but no less powerful equivalence rules the diatonic system, in which steps of different sizes are perceived as identical. That is, steps or seconds define an equivalence class in the set of diatonic intervals, as do thirds, fourths, and so on. Using this methodology, Clough established the mod-7 diatonic sets, and defined diatonic set classes under transpositional equivalence only, rather than through the standard T/I invariance of atonal theory.[2] The insight behind this choice led to important and powerful results in his later work.

Clough's earliest work in scale theory grew out of his interest in interval cycles in the mod-12 and mod-7 universes. These interval cycles provided a framework for a hierarchical non-Schenkerian tonal theory. His work on interval cycles also intersected with a lifelong fascination with diatonic sequences. Clough's essay in this collection contains more comprehensive thoughts on the topic—returning to his starting point with the benefit of a lifetime of study.[3]

The second period of Clough's work undertook a deeper study of the nature of diatonic systems. In the mid-1980s, Clough worked with mathematician Gerald Myerson and published seminal works that reveal three unexpectedly interrelated properties of the diatonic scale.[4] The most fundamental of these properties, "Myhill's Property" (MP), is named after mathematician John Myhill, who was instrumental in bringing John Clough and Gerald Myerson together and advanced a conjecture about the property that bears his name. A set has MP if every *generic* interval in the set comes in two *specific* sizes—for example, among diatonic intervals the *generic* second comes in 1 and 2 half-steps, the third in 3 and 4 half-steps, and so forth. This property may account, at least in part, for the ubiquity of the diatonic scale. Clough and Myerson went on to show Myhill's Property to be the basis of two other properties: "Cardinality Equals Variety" (CV), and "Structure Yields Multiplicity" (SM).[5]

Under CV, any diatonic sequence with *n* distinct members from a diatonic set comes in exactly *n* species. For example, the sequence < *C, E, G* > has three distinct members, all of which are in the C Major scale. When Clough and Myerson considered all diatonic transpositions of this sequence, they found that the sequences come in three species of triads—major, minor, and diminished. In fact, the diatonic transpositions of any sequence of three distinct members from a diatonic set come in three species, and Clough and Myerson demonstrated that the diatonic transpositions of any sequence with *n* distinct members from a diatonic set come in *n* species.

Clough and Myerson's second concept, "Structure Yields Multiplicity" (SM) provides an algorithm to determine how the set of diatonic transpositions of a sequence is partitioned into subsets of species. They then showed that a set has MP if and only if it has CV, and that a set has CV if and only if it has SM.[6]

The third period of Clough's work, beginning in the 1990s, was marked by his collaboration with mathematician and music theorist Jack Douthett. Greatly broadening the earlier work with Myerson, their 1991 article advances the idea of a *maximally even set*, which is both simple to describe and powerful in implication. In a maximally even set, the pitch classes are distributed as evenly as possible among the notes of the underlying chromatic universe. Clough and Douthett discovered that as long as the number of pitch classes in the set is less than the number of chromatic pitch classes, every chromatic universe has one maximally even set for each cardinality. For example, the diatonic scale is the unique seven-note maximally even set in the twelve-tone universe. Similarly, the augmented triad is the maximally even three-note set in that same universe. Clough and Douthett further described what they call *second-order maximally even sets*, which account for structures such as the triads and seventh chords embedded in a diatonic scale. The account of three- and four-note maximally even sets in the "7-in-12" system helps explain the basic harmonic materials of tonal music. A further generalization led them to posit *hyperdiatonic* systems that share key features with the ordinary diatonic system. They later collaborated with Lewis Rowell and N. Ramanathan on a study that used second-order maximally even sets to model the structure of ancient Indian theoretical systems.[7] Later, in collaboration with John Cuciurean, they generalized the concept of second-order maximal evenness and defined n^{th}-*order maximally even sets*.[8]

The fourth and last phase of Clough's work was cataloging and organizing recent scale theory. In collaboration with Nora Engebretsen, and Jonathan Kochavi, Clough mapped this new scalar territory, proposing *distributionally even* (DE) sets as a further generalization of maximally even ones.[9] The DE sets provide a framework in which to view the scale theories of Eytan Agmon, Gerald Balzano, Norman Carey and David Clampitt, Clough and Myerson, Clough and Douthett, and Carlton Gamer.[10]

Clough's work set the stage for advances in scale theory, as well as in transformational theory and voice-leading studies, including neo-Riemannian theory. The contributors to this collection have made some of the most significant and recognized contributions to these fields. Richard Cohn, for instance, is the only two-time recipient of the Society for Music Theory (SMT) Outstanding Publication Award and is one of the most prolific music theorists today.[11] Clough and Douthett won the SMT Outstanding Publication Award in 1993, and both separately and together have published several other important articles in diatonic theory and related topics.[12] Timothy Johnson has developed pedagogical approaches for relating diatonic and neo-Riemannian theory directly to the study of music fundamentals.[13] David Clampitt, in collaboration with Norman Carey, has investigated tuning properties related to the work of Clough and Myerson, and has discovered properties of diatonicism that expand the historical scope of diatonic set theory by clarifying the work of medieval theorists.[14] Clough and others have significantly broadened the application of scale theory to explore systems of tone relations and scale systems from ancient and medieval India.[15] This expansion into the realm of ethnomusicology has been particularly promising, for it isolates certain fundamental properties inherent both in Western scales and in scales of other cultures. More recently, theorists have found useful applications of scale theory in the analysis of non-serial music by such composers as Stravinsky, Barber, and Prokofiev.[16] Clough, Douthett and others have explored correlations between scale theory and the physical properties of magnetic ordering and phase diagrams.[17]

One of the most exciting recent developments in music theory has been the emergence of neo-Riemannian and transformational theory, a seemingly unrelated field that arose independently beginning in the 1980s with the work of David Lewin and, later, Richard Cohn. Neo-Riemannian theory arose in response to analytical problems posed by later nineteenth-century music that is triadic, but not fully unified tonally—that is, music typical of Wagner, Brahms, Liszt, for example. Because highly chromatic music uses the harmonies and cadences of diatonic tonality, it has attracted analytical models designed for purely diatonic music. Yet it frustrates these models, a problem unsuccessfully confronted by theorists throughout the twentieth century.[18] Not quite what we normally label as "atonal," this music resides somewhere between diatonic tonality and atonality. Persistent failure to explain how this music works led nineteenth-century theorists to explore other possible principles, and Cohn isolates six of these principles: triadic transformations, common-tone maximization, voice-leading parsimony, "mirror" or "dual" inversion, enharmonic equivalence, and the "Table of Tonal Relations" (parsimony and the "Table of Tonal Relations" are discussed below in more detail).[19] He then argues: "Neo-Riemannian theory strips these concepts of their tonally centric and dualist residues, integrates them, and binds them within a framework already erected for the study of the atonal repertoires of our own century."[20]

Neo-Riemannian theory originated in David Lewin's transformational approach to triadic relations, in which he explores classes of contextual transformations that, following Riemann and Hauptmann, act upon consonant triads whose constituent pitches are arranged in a line of alternating minor and major thirds (transformations that largely comprise incremental shifts along this line that change one triad into another).[21] Although triads and diatonic collections are among the objects produced, Lewin's transformations are conceptually independent of the system of diatonic tonality. The exploration of musical transformations culminates in Lewin's *Generalized Musical Intervals and Transformations*, which identifies three areas for further exploration: geometric representations of pitch space, group structures, and compositional logic.[22]

Henry Klumpenhouwer, Brian Hyer, Julian Hook, and others have extended Lewin's group-theoretic approach.[23] Theorists began exploring geometric representations of pitch space based on the *Tonnetz*, a two-dimensional lattice used by many nineteenth-century theorists; but unlike the nineteenth-century *Tonnetz*, the objects and relations are conceived as equally tempered. Thus, instead of a two-dimensional lattice of pitch space that extended indefinitely, axes of tempered perfect fifths and major and minor thirds intersect to form a triangular lattice on the surface of a torus in which vertices represent pitch classes and triangles represent consonant triads. The appropriation of the *Tonnetz* showed the relationship between triadic transformational theory and nineteenth-century harmonic theory.[24] Such graphical interpretations enhance the understanding of group structures that model what one might call "quasi-tonality" (or "quasi-atonality") within the chromatic twelve-tone universe.

Clough was a contributor to transformational theory as well. One of the important developments in neo-Riemannian and transformational theory began in a Kansas City café where Clough, Cohn, and Douthett met in 1992 and started a discussion that is still going on, with many more participants. Cohn observed that, in the usual mod-12 universe, set classes in which non-trivial cycles (cycles of length 3 or more) of pitch-class sets within a given set class can be constructed with the following property: any pitch-class set in the cycle can be obtained from either of its adjacent pitch-class sets by moving a single note by a half-step. These cycles are known as *maximally smooth cycles* or *Cohn cycles*. The most familiar Cohn cycle is the circle of fifths, in which adjacent sets are closely related keys. But Cohn observed that the diatonic set class is not the only set class in the modulo 12 universe that contains Cohn cycles. Others are

Forte's set class (SC) 5-35 (pentatonic set), SC 3-11 (consonant triads), and the complement of SC 3-11, SC 9-11. Cohn had also explored set classes in other chromatic universes capable of supporting Cohn cycles and gave these observations to Clough and Douthett in Kansas City.[25] Over the next six months, Douthett circulated a series of letters that formalized and extended many of Cohn's observations.[26] Clough then observed important similarities between consonant triads and half-diminished and dominant-seventh chords. Led by Clough, an organizing committee of four people (Clampitt, Clough, Cohn, and Douthett) was formed, and the first of three symposiums on these issues was held in 1993 at the University at Buffalo.

The first two important papers that resulted from the 1993 Buffalo symposium were Lewin's work on "Cohn functions" and Cohn's award winning article on maximally smooth cycles.[27] A third paper, also by Cohn, introduced the term *parsimony* in connection with consonant triads: two triads are parsimonious if they have precisely two pitch classes in common.[28] The triads are of opposite mode, and the pitch classes which are not in common differ by a diatonic step (a half-step or a whole-step).[29] Following the lead of Clough, Douthett, and others, Cohn then searched for "parsimonious set classes" in chromatic universes with other than twelve pitch classes, in order to create a broader platform from which to explore parsimony in microtonal music. Further discussions of these topics led to the second Buffalo symposium, titled "Neo-Riemannian Transformations: Mathematics and Applications." The twelve papers delivered at this symposium subsequently appeared in a special issue of the *Journal of Music Theory*, which has become one of the most frequently cited collections in transformational and neo-Riemannian theory.[30] The third Buffalo symposium, "On Neo-Riemannian Theory," was held in 2001. The ten essays in this collection—eight of which are authored by contributors to the *Journal of Music Theory* collection—further explore the interaction of diatonic theory and transformational theory.

In his contribution to this collection, Clampitt distinguishes between two types of properties investigated by diatonic theory or scale theory, "operational" and "systemic," while stressing that they are not absolutely separable, as exemplified in our paradigmatic case: "the 'transformational' property of the diatonic that allows one to modulate in a 'maximally smooth' way between fifth-related sets . . . becomes an essential aspect of a fully developed and comprehensible tonal system."[31] The mathematical methods prominent in both Clough and Lewin's work exemplify how both scale theory and transformational theory lend themselves to mathematical methods (the former emphasizing combinatorial and number theories; the latter, group- and graph-theoretic approaches). In both, however, the mathematics serves a better understanding of the musical elements.

The essays in this collection constitute a close-knit body of work—a family in the sense that they trace their descent from a few key breakthroughs by John Clough, David Lewin, and Richard Cohn in the 1980s and 1990s. For example, both Clough and Hook exploit mathematical group theory, the former to investigate a particular relationship among Riemannian transformations, the latter to reconsider important elements of tonal theory; some parts of Douthett's modeling of scales and chords with dynamical systems are related to Hook's essay, others to Clough's. And Hook's essay has important connections with Johnson's pedagogical approach to diatonic and neo-Riemannian theory; the geometric devices used by Johnson can be related to the dynamical approach taken by Douthett; and so forth, through the entire volume. Because of this dialogue, and the resistance of this work to easy categorization, the editors have chosen to organize the essays in alphabetical order by author.

The collection begins with Clampitt's "Cardinality Equals Variety for Chords." Clampitt's *chord CV* differs from Clough's CV in that Clampitt considers pitch-class sets rather than sequences. This difference leads Clampitt to a connection between chord CV and an unsolved problem in mathematics known as the "Twin Primes conjecture."

Clough's essay, "Flip-Flop Circles and their Groups," deals with cycles of major and minor triads. The essay asks: what is the space of all cycles formed by alternating major and minor triads formed in a uniform manner? These cycles are called *uniform flip-flop circles* (UFFCs).

In the next essay, Cohn explores the conflict between two classes of rhythmic pattern in the first movement of Bartók's *Sonata for Two Pianos and Percussion* using what he calls *pitch-time analogies* and *open versus closed generators*. The structure of the thematic material in this movement mimics the "open and closed" design of the metric domain.

Douthett's essay deals with dynamical systems of moving concentric circles. The innermost circle emits beams of light that pass through holes in the other circles. The outermost circle's twelve holes represent pitch classes. Douthett shows that many well-known cycles of scales and chords can be dynamically generated by choosing circle configurations and control parameters (frequencies and phases of the cycles). Linked to his earlier work with Clough, Douthett's dynamical systems visually represent iterated maximally even sets.

In "The Over-Determined Triad as a Source of Discord," Nora Engebretsen studies the group-combinatorial approaches and related concepts of tonality implicit in the works of Moritz Hauptmann, Arthur von Oettingen, and Hugo Riemann. The shift from Hauptmann's combinatorial system, which favored generative relationships defined on the basis of voice-leading parsimony, to Riemann's, which favored those defined on the basis of acoustically pure root-interval relations, involved a shift toward a chromatic conception of tonality, under which certain diatonic relationships were accounted for through combinations of chromatic generators. Engebretsen points out, however, that Riemann's functional system (which co-existed with the combinatorial *Schritt/Wechsel* system), is conceptually similar to Hauptmann's diatonically predicated common-tone approach. She explores how Riemann attempted to reconcile these two competing concepts.

In "Signature Transformations," Hook constructs a group with the familiar mod-12 transposition operators as well as diatonic mod-7 transposition operators. Both types of operators are expressible in terms of *signature transformations,* which reinterpret any diatonic object in the context of a key signature. Using his group of order 84, Hook provides fresh insight to Schubert's *Valse sentimentale* in A major, D. 799, no. 13, Torke's *Yellow Pages,* and the first movement of Debussy's Violin Sonata.

Johnson's essay, "Some Pedagogical Implications of Diatonic and Neo-Riemannian Theory," presents a pedagogy for introducing aspects of diatonic and neo-Riemannian theory at the beginning stages of college-level music theory instruction. Johnson describes student-driven methodologies based on special features related to interval cycles such as well-formedness, Myhill property, distributional and maximal evenness, deep scales, Balzano's microtonal systems, and cycles of alternating major and minor triads, as well as more general cycles of paired musical objects.

In "A Parsimony Metric for Diatonic Sequences," Kochavi uses the notions of voice-leading parsimony and structural representation, both implicit in neo-Riemannian theory. A formal definition of parsimony generates a natural classification of diatonic sequences. Structural representations of diatonic sequences mimic the dualistic design inherent in neo-Riemannian transformations to bring the essay full circle.

Lewin, in "Transformational Considerations in Schoenberg's *Piano Pieces,* op. 23, no. 3," defines three contextual transformations that model Schoenberg's use of motivic material. With this collection of operators, it is easy to design networks that model the appearance of different forms of the subject, which remain audibly recognizable. Lewin traces his subject through the score's interaction between hand assignments (left and right), identifying different forms of the subject and the relevant group of operators.

Finally, Stephen Soderberg explores six basic interval-string transformations in "Basic Principles and Applications of Interval String Theory." These transformations—*rotation, retrograde, sum, split, scalar multiplication,* and *configure*—allow Soderberg to newly interpret Scriabin's "mystic chord" as well as some of Stravinsky's musical structures.

Norman Carey
Jack Douthett
Martha M. Hyde

Notes

1. Babbitt 1960, 1965; Gamer 1967.
2. Clough 1979, 1979–80.
3. Clough 1994.

4. Clough and Myerson 1985, 1986.
5. Ibid.
6. It should be noted, however, that recently a slight error has been discovered in Clough and Myerson's work. This error was corrected by Silverman and Wiseman (2006) and requires some minor changes to a couple of Clough and Myerson's proofs, as well as slight modifications to the statements in several of their corollaries.
7. Clough, Douthett, Ramanathan, and Rowell 1993.
8. Clough, Cuciurean, and Douthett 1997.
9. Clough, Engebretsen, and Kochavi 1999.
10. Agmon 1996; Balzano 1980; Carey and Clampitt 1989; Clough and Meyerson 1985; Clough and Douthett 1991; Gamer 1967.
11. Cohn 1996, 1997.
12. Clough and Douthett 1991.
13. Johnson 2003a, 2003b.
14. Clampitt and Carey 1996a; Clough and Meyerson 1985, 1986.
15. Clough, Douthett, Ramanathan, and Rowell 1993.
16. Santa 2000.
17. Douthett and Krantz 1996; Douthett and Steinbach 1998; Krantz, Douthett and Clough 2000.
18. For a historical survey of neo-Riemannian theory, see Cohn 1998a. Our discussion summarizes Cohn's survey.
19. Cohn 1998a.
20. Ibid., 169.
21. Lewin 1982. For an excellent account of the development of transformational theory, see Kopp 2002.
22. Cohn 1998a, 171.
23. Klumpenhower 1994; Hyer 1995; Hook 2002b. A variety of these group-theoretic approaches can be found in the *Journal of Music Theory* 42 (1998), a special issue on "Neo-Riemannian Theory."
24. Extending this geometric approach, Douthett and Steinbach (1998) explored musical interpretations of cubes, tetrahedra, and other shapes.
25. Cohn (1997) subsequently discusses a variety of trichordal cycles including a particular class of Cohn cycles. It is here that Cohn determines what chromatic universes are necessary to support particular families of cycles.
26. Cohn 1998a, 177.
27. Lewin 1996; Cohn 1996. Lewin's "Cohn functions" are mathematical tools that allowed Lewin to investigate the structure of Cohn cycles.
28. Cohn 1997.
29. Since then, the definition of parsimony has been extended in many different ways (see the *Journal of Music Theory* 42 [1998]), but these extensions generally require voice-leading motion to be restricted to diatonic steps.
30. *Journal of Music Theory* 42 (1998).
31. See chapter 1, pp. 9–22.

Chapter One

"Cardinality Equals Variety for Chords" in Well-Formed Scales, with a Note on the Twin Primes Conjecture

David Clampitt

1.1 Scope, Method, and Aim of Scale Theory

Researchers have, in the past several decades, used formal approaches to diatonic theory in an attempt to show why the features of certain pitch collections have had such appeal for composers.[1] The results relate either to what musicians have discovered they can do with a given collection—through moves, routines, or processes within the collection, or through manipulation of the collection itself—or to how a given collection functions cognitively, based upon measures of symmetry versus asymmetry, simplicity versus complexity, or information versus redundancy.[2]

Investigations of the first type use transformational theory and analysis. For example, harmonic triads, and the usual pentatonic and diatonic sets, all participate in maximally smooth cycles, the starting-point for neo-Riemannian analysis. Triads in an octatonic setting, or dominant and half-diminished seventh chords in non-maximally smooth settings, also suggest a neo-Riemannian approach; certain pitch-class sets in an atonal setting allow for analogous procedures using transformational analysis.[3]

Investigations of the second type tend to be too general to apply to analysis, but help explain the popularity of certain systems—always within particular cultural contexts, be it understood. Such explanations are by no means the exclusive province of self-described diatonicists and other scale theorists. Carl

Dahlhaus offers a diatonicist distinction between Guido's hexachord, on the one hand, and the usual pentatonic and diatonic, on the other: the latter are "systems," the former "is not, in contrast to the heptatonic and pentatonic scales, a self-significant system of tones," but is "a mere auxiliary construction."[4] What Dahlhaus means is that in the pentatonic and diatonic one always understands which intervals are steps and leaps, relative to the system: "minor thirds" in the pentatonic are always "steps," while in the diatonic they are always "leaps," as opposed to the situation in the hexachord, where "the listener would have to alternate between . . . the idea of the minor third as a 'step' and as a 'leap'."[5] He labels this an "absurd consequence." The vehemence of Dahlhaus's language here must be understood within the context of a particular historical/theoretical discussion, on the development of the tonal system. But the implicit assumption, that a simpler relationship between intervals and their description in terms of scale steps provides for smoother cognitive processing, has more general applications. Taking Dahlhaus's observation and considering it closely (nowhere does he expressly say that all three of these entities are generated by the perfect fifth, though this is the case), and formulating it in mathematical terms, one is led to the concept of a "well-formed scale," defined by Norman Carey and myself and discussed later in this article.

These two types of properties—we might call them "operational" and "systemic"—are not entirely separable: the transformational property of the diatonic that allows one to modulate in a maximally smooth way between fifth-related sets, by moving a single note by a minimal distance (a chromatic semitone), is an essential aspect of a fully developed tonal system. Both types of properties lend themselves to mathematical investigation, one of the hallmarks of scale theory. The aim of the mathematical formalization and generalization of such properties is to understand how they work and how they relate to each other: for example, what is the relationship between the fact that a given scale admits a generating interval, and the number and arrangement of the scale's various step intervals? A mathematically derived result is that in such cases step intervals come in at most three sizes (as exemplified in the three entities—pentachordal, hexachordal, and heptachordal—discussed by Dahlhaus).[6]

The subject of this paper, the property "cardinality equals variety," to be defined, is a systemic property that assumes that describing some musical objects both in terms of intervals and in terms of scale steps is useful. In particular, that the number of varieties of such objects, such as chords or pitch-class lines, matches the number of distinct constituents of these objects, is hypothesized to be advantageous. Along the axis of information and redundancy, "cardinality equals variety" entails a system that is not too exciting, not too boring, and that has a high degree of organization. "Cardinality equals variety" also forms the starting point for Johnson's textbook, which attempts to be at least a partial answer to Cohn's call for a "new pedagogibility" in music theory.[7]

1.2 Cardinality Equals Variety for Chords

In their 1985 *Journal of Music Theory* article "Variety and Multiplicity in Diatonic Systems," Clough and Myerson showed that MP ⇔ CV for lines; that is, any pitch-class set with Myhill's Property (in chromatic universes of any cardinality) has cardinality equals variety for lines, and (trivially) vice versa.[8] The basic notion these concepts depend upon is the distinction between generic and specific interval measures (diatonic and chromatic lengths, respectively). Generic measure is simply the measure musicians commonly use when they take an assumed underlying diatonic set as a ruler, and speak of "seconds" or "thirds" or "sixths," without regard to quality. The appropriate mathematics for discussions of generic measure in the usual diatonic set is arithmetic mod 7.[9] Specific measure, on the other hand, deals with the actual quality of the interval. In Clough and Myerson's approach, the diatonic is assumed to be embedded within a mod 12 chromatic universe; thus, specific interval measure is just the usual mod 12 measure of musical set theory. More generally, they apply generic and specific interval measures in chromatic universes of arbitrary cardinality c, and in pitch-class subsets of cardinality $d < c$.

Clough and Myerson also categorize unordered and ordered pitch-class sets, i.e., chords and lines, according to the generic/specific distinction. Unlike the practice in traditional musical set theory, however, Clough and Myerson's equivalence classes use transposition only, not inversion. Thus, the unordered pitch-class subsets (chords) in the C-major diatonic set {C, D, F} and {D, F, G} are considered to be non-equivalent, even though they map onto each other under inversion under both mod 7 and mod 12 measures. Chord types are displayed using interval normal form.[10] The trichord {C, D, F} is a member of generic class (124) (i.e., step, third, fifth), and specific class (237), (i.e., major second, minor third, perfect fifth), while the trichord {D, F, G} is a member of generic class (214) and specific class (327). The C-major subset {F, A, B}, on the other hand, is also a member of generic class (214), but with respect to specific measure is of type (426). As a line, the ordered set ⟨F A B⟩ is described generically as ⟨2 1 4⟩, and specifically as ⟨4 2 6⟩.

A set is said to have Myhill's Property (MP) if every non-zero generic interval comes in two specific varieties (with "non-zero" understood to include "non-octave equivalent"). The usual diatonic is an example of a set with MP. A set has cardinality equals variety (CV) for lines if any line of a given generic description comes in k specific varieties, where k is the number of distinct pitch classes in the line. Again, the usual diatonic is an example, as is the complementary pentatonic. Diatonic seventh chords, for instance, come in four varieties: major, minor, dominant, and half diminished; arpeggiated, these form four different types of pitch-class lines. Since an interval may be construed as a two-note line, if a set has CV for lines, it automatically has MP. Clough and Myerson proved the non-trivial converse, that any set that has MP exhibits CV for lines.

As Clough and Myerson pointed out, however, the situation regarding CV for chords is more delicate. First of all, if c is the cardinality of the chromatic universe and d is that of a pitch-class subset that has MP, while CV for lines holds for lines with d distinct pitch classes, there is clearly only one chord containing all d pitch classes. In the usual diatonic, for example, the seven line species corresponding to the generic line class ⟨1 1 1 1 1 1 1⟩ are the seven diatonic modes, whereas the 7-pc chord is just the diatonic aggregate, of specific description (1 2 2 1 2 2 2). Thus, one must exclude the set as a whole from the discussion, and say that a set exhibits CV for chords if for all integers e such that $1 \leq e < d$, every chord of cardinality e of a given generic description comes in e specific varieties. In the usual diatonic and pentatonic scales, CV for chords holds, but Clough and Myerson gave a counterexample of an MP set where CV for chords fails. This occurred in a case where the cardinality of the chord is not coprime with d, the cardinality of the MP set. Clough and Myerson concluded with the following conjecture: "It seems that the best one can say is that, if a scale has MP, then CV holds except for certain chords of cardinality not coprime with d."[11]

The purpose of this essay is to determine precisely under what conditions CV for chords holds. I will consider a slightly more general setting: while all the MP sets in Clough and Myerson are embedded (that is, subsets of a finite chromatic universe), I will also consider non-embedded sets, wherein the sizes of the constituent intervals may be incommensurable with the size of the octave or other modular unit. For example, in the diatonic scale in Pythagorean tuning or in quarter-comma mean-tone temperament, the sizes of the non-zero intervals (those other than unison or its octave equivalents) are irrational values if the octave itself is taken as the unit of measure. Hence, it is impossible to express these scales as subsets of an equal division of the octave. Nonetheless, it is still possible to invoke the generic/specific distinction, and to say that the diatonic scale in either of these two tunings exhibits a slightly generalized Myhill's Property. To do so, we simply adopt the notion of generic measure already given, and generalize the notion of specific measure to cover the cases of irrational divisions of the octave.

As Carey and I have shown, sets with generalized MP are equivalent to nondegenerate well-formed sets.[12] A set is well-formed if it admits a generator, where that generator is spanned by a constant number of scale-step intervals. For example, the usual diatonic is generated by the perfect fifth (or perfect fourth), and in the diatonic set, the perfect fifth is always spanned by 4 step intervals (and the perfect fourth is always spanned by 3 step intervals). While it may sound tautological to say that a fifth in the diatonic set is always a fifth, that is only because we customarily use diatonic nomenclature. If we consider the hexachord generated by perfect fifths or fourths, we see that there are some "perfect fifths" that are spanned by 4 steps, and others that are spanned by 3 steps (within the hexachord).

Any equal division of the octave trivially satisfies the definition of a well-formed scale: if U_c is a chromatic universe of cardinality c, then any interval represented by the integer h where h is coprime with c will generate the whole set and will be spanned by a constant number of steps. Setting $h = 1$ and generating all c elements will always work, for example. Carey and I call these *degenerate* well-formed sets, and they obviously do not have MP. Hereinafter, whenever I write "well-formed" I intend "*non-degenerate* well-formed," and given the equivalence mentioned above, "generalized MP" may always be substituted for "well-formed."

The results on CV for chords in well-formed sets, to be demonstrated below, are as follows: CV for chords holds in D if and only if D is a non-degenerate well-formed set of cardinality d where d is prime and where if D is embedded in a chromatic universe of cardinality c, $d \leq \lfloor c/2 \rfloor + 1$.[13] Furthermore, if $d > \lfloor c/2 \rfloor + 1$, CV fails for a chord of cardinality $e = 2d - c$, and if d is composite, CV fails for chords of cardinality e where the greatest common divisor of d and e is $n > 1$. The proofs of these statements are given later in this paper.

Before considering the mathematical treatment of CV for chords in general, it may be useful to observe the situation in a concrete case, involving a familiar musical object of small cardinality, the usual pentatonic scale.

The pentatonic case is displayed in table 1.1. For now, we are just looking at the first three columns. The third column shows the literal unordered subsets of the pentatonic scale with pitch classes C D F G A, except for the bottom box, which gives the five different scalar orderings of the pentatonic scale (the five "modes"). The first column displays the pentatonic genera, up to pitch-class equivalence; the second column partitions each of the genera into species, each of which is measured by the semitone of 12-tone equal temperament. The top two boxes give the dyads, or two-note literal subsets: they come in two genera, each of which contains two species. Observe that genera are given by partitions of 5 and species by partitions of 12—this is the Clough-Regener notation introduced above in the case of the usual diatonic set. Once we have checked CV for the dyads and find that Myhill's Property holds, we know from Clough and Myerson's result that CV for lines holds for the pentatonic. The rest of the table's first two columns verifies this for CV for chords (i.e., for unordered subsets of cardinalities 2–4), a stronger property than CV for lines (even though there are an infinite number of instantiations in pitch space of CV for lines).

Columns 4 and 5 use a tool from the foundations of well-formed scale theory: the generating interval of constant span. This defining characteristic implies that one can use the generating interval as a measure of both specific and generic intervals. Consider again the familiar diatonic case. All specific diatonic intervals modulo the octave may be represented by some number, positive or negative, of perfect fifths: thus, +2 fifths (reduced by an octave) is the ascending whole step. That is, +2 represents all directed pitch-class intervals of the type C–D, D–E, F♯–G♯, etc. The rising semitone E–F, on the other hand, is represented in perfect fifths measure by −5 (moving counterclockwise on the diatonic circle of fifths, if

Table 1.1 Table of pentatonic chords of 2 to 4 notes (and the pentatonic modes)

Genus (steps)	Species (semitones)	literal chords	Species (fifths)	Genus (fifths)
		{F, G}		
	(2 10)	{C, D}	(2 −2)	
(1 4)		{G, A}		(2 3)
	(3 9)	{D, F}	(−3 3)	
		{A, C}		
	(4 8)	{F, A}	(−4 4)	
		{C, F}		
(2 3)		{G, C}		(1 4)
	(5 7)	{D, G}	(1 −1)	
		{A, D}		
		{F, G, C}		
	(2 5 5)	{C, D, G}	(1 1 −2)	
		{G, A, D}		
(1 2 2)	(3 4 5)	{D, F, A}	(1 −4 3)	(1 1 3)
	(3 5 4)	{A, C, F}	(−4 1 3)	
	(2 2 8)	{F, G, A}	(−4 2 2)	
		{C, D, F}		
(1 1 3)	(2 3 7)	{G, A, C}	(1 2 −3)	(1 2 2)
		{D, F, G}		
	(3 2 7)	{A, C, D}	(1 −3 2)	
	(2 2 3 5)	{F, G, A, C}	(−4 1 1 2)	
		{C, D, F, G}		
(1 1 1 2)	(2 3 2 5)	{G, A, C, D}	(1 1 1 −3)	(1 1 1 2)
	(3 2 2 5)	{D, F, G, A}	(1 1 −4 2)	
	(3 2 3 4)	{A, C, D, F}	(1 −4 1 2)	
(lines)	⟨2 2 3 2 3⟩	⟨F G A C D⟩	⟨1 1 1 1 −4⟩	
	⟨2 3 2 2 3⟩	⟨C D F G A⟩	⟨1 1 1 −4 1⟩	
⟨1 1 1 1 1⟩	⟨2 3 2 3 2⟩	⟨G A C D F⟩	⟨1 1 −4 1 1⟩	⟨1 1 1 1 1⟩
	⟨3 2 2 3 2⟩	⟨D F G A C⟩	⟨1 −4 1 1 1⟩	
	⟨3 2 3 2 2⟩	⟨A C D F G⟩	⟨−4 1 1 1 1⟩	

you will, to avoid the diminished fifth F–B). In both cases, the generic ascending step interval is represented by 2, since all generic diatonic steps are equivalent to two generic diatonic fifths.

The fourth and fifth columns of table 1.1 show the pentatonic subsets, but using the generating interval of constant span to specify the chord types. The pentatonic scale is well-formed, because it admits a generating interval of constant span. This interval is again the perfect fifth, using diatonic nomenclature, but that interval has a span of 3 in a pentatonic context. The "bump," equivalent to the diatonic scale's "imperfect fifth" B–F, in the F C G D A pentatonic is the A–F interval, also spanning 3 pentatonic steps. The rightmost column again shows the genera, as measured by generic intervals of span 3, while the fourth column shows the chords, with specific intervals measured by perfect fifths ("perfect fifth," in the conventional diatonic nomenclature, but again, these intervals span 3 pentatonic step intervals). For the purposes of these descriptions, the elements of the (unordered) sets are arranged in circle-of-fifths order, just as when, in computing their descriptions in terms of scale measurement, they are placed in scale order. In all cases, note that the numerical entries sum to zero: in the first and last columns, under mod 5 addition; in the second column, under mod 12 addition; in the fourth column, under ordinary addition. For example, the genus (1 1 3) comprises all those trichords with notes separated by one generic interval of span 3, followed by another interval of span 3, and completed by three intervals of span 3. The three specific varieties are determined by the location of the "imperfect fifth," represented by the negative integer in the specific description. The negative integer appears because in the specific description we are using only the "perfect fifth" intervals of span 3 to count, so to avoid the "bump" of the A–F interval, we have to turn around and use "perfect fifth" intervals of span 3 in the opposite direction. Thus, in the trichord type with multiplicity 3, the specific description (1 1 –2) means up a perfect fifth, up a perfect fifth, then down two perfect fifths to return to the starting point. For example, F to C, C to G, then back from G to F. The two "singleton" forms are of specific descriptions (1 –4 3) and (1 3 –4): the set {D, F, A} is measured by the perfect fifth D to A, then down four perfect fifths A to F, with 3 perfect fifths to return to D; the set {F, A, C} is up a perfect fifth F to C, up three perfect fifths C to A, then down 4 perfect fifths from A to return to F. There are three locations for the imperfect fifth, yielding three non-rotationally equivalent descriptions, hence there are three possible varieties of (1 1 3) pentatonic trichord. Note that when we are talking about chords, we are considering all inversions of the chord to be equivalent (inversion here in the sense of triadic inversions, not set theory inversion). Thus, all rotations of the descriptions, both generic and specific, are considered to be the same: (1 1 3) is equivalent to (1 3 1) and to (3 1 1). On the other hand, the specific varieties (1 –4 3) and (1 3 –4) are distinct, because they are not rotations of each other, but retrogrades; which is to say, they are inversions of each other in musical set theory terms. In the case of the five pentatonic modes, as ordered subsets, they are distinct, but they collapse into a single unordered set. This is represented in the fourth column by the fact that the descriptions of the modal ordering in terms of generating intervals are rotations of each other: starting on a different note merely changes the location of the unique –4 element.

The usual pentatonic—along with its complement, the usual diatonic—exhibits CV for chords as well as lines, and perhaps the discussion of the generalized circle of fifths in the pentatonic case affords some insight into why CV for lines holds in the general case of non-degenerate well-formed scales. But can we learn anything about CV for chord failures from this example?

The description of pentatonic chords in terms of generating intervals frees us from assuming that the pentatonic is embedded in the 12-note chromatic. The generating interval might, for example, be a pure perfect fifth (of frequency ratio 3:2), or a quarter-comma mean-tone fifth (of frequency ratio $\sqrt[4]{5}$). The notations still carry the same structural information, in terms of generic and specific types. But what if the pentatonic were embedded in a 7-note chromatic? Then the specific descriptions of column 4 would be in terms of arithmetic mod 7 rather than ordinary arithmetic. Consider the (1 –4 3) and (1 3 –4) species. Mod 7, –4 is equivalent to +3, so in this case, the two species collapse into a single (1 3 3) species, and CV for chords fails. If the pentatonic were embedded in a 6-note chromatic, we would have –4 equivalent to +2, and the four-note chord genus would collapse into three species. But if it were embedded in an 8-note chromatic (or any larger chromatic space, including of course the 12-note space), the problem disappears. It turns out that a necessary condition for CV for chords to hold is that if the scale is embedded in a chromatic, it must be no larger than the next integer up from one-half the cardinality of the chromatic. The usual diatonic, then, is maximal in the 12-tone chromatic for CV for chords, since 7 is one more than half of 12.

This is a necessary, but not sufficient, condition. The other condition, whether the scale is embedded in a chromatic or not, is that it be of prime cardinality. One can see in the case of the pentatonic modes, with generic description ⟨1 1 1 1 1⟩, that as lines they are distinguished by the position of the –4 marker, but these descriptions are all rotations of each other, and there is just one 5-note chord. Of course, we have excluded the case of the whole set from the question of CV for chords. If the cardinality d is composite, though, then there is a similar situation for proper subsets. If *d* is composite, there is an integer *k* strictly between 1 and *d* that divides *d*; thus, there is a generic chord of the form $(d/k\ d/k \ldots d/k)$ for some $k > 1$. The specific chords collapse into a single species.

Based upon the analysis above, what follows is a mathematical treatment of CV for chords, with proofs of the previously stated results. That is, the results are generalized and formalized, but the ideas are just those above. As before, pitch-class sets are designated by curly brackets enclosing elements separated by commas, while chord types are designated by interval numbers enclosed in parentheses, and lines by interval numbers enclosed in angle brackets.

Let D be a well-formed set of cardinality *d*. If it is embedded, the cardinality of its chromatic universe U_c is *c*.

Result 1: If CV for chords holds for D, then d is prime, and if CV holds and D is embedded, we also have $d \leq \lfloor c/2 \rfloor + 1$. In contrapositive form: CV fails if d is composite, or $d > \lfloor c/2 \rfloor + 1$. Furthermore, if $d > \lfloor c/2 \rfloor + 1$, then CV fails for a chord of cardinality e where $e = 2d - c$.

Proof: If d is composite, CV fails irrespective of the universe, chromatic or infinite: Suppose d is composite; then there exists $e < d$ such that e and d have a common factor greater than 1, i.e., gcd $(d,e) = n > 1$. The generic chords of cardinality e are determined by partitions of d into e parts, that is, e numbers summing to d. Rotations of partitions are considered indistinguishable, but other permutations of a partition determine different generic chords. Since D is well-formed, we may consider the partitions to measure numbers of generalized fifths (generators). The specific chords are uniquely determined by the location of the remainder fifth. This is the method of the "generalized circle of fifths" used by Clough and Myerson.[14] That is, each element of the partition is a number s_i representing an interval spanning s_i fifths. For exactly one s_j the specific interval contains $s_j - 1$ generalized fifths and the unique remainder fifth. While $\sum_{i=1}^{e} s_i \equiv 0_{\text{mod } d}$ by definition, the sum of the specific intervals must actually equal 0. So the specific chords are represented by the partition elements in order, but with s_j replaced with $s_j = -\sum_{i \neq j} s_i$. We now construct a generic chord E of cardinality e where CV fails. Since gcd $(d,e) = n > 1$, d may be partitioned into n identical subpartitions of d/n, each of e/n parts. There are now at most e/n distinguishable specific chords, corresponding to the e/n positions in the subpartition. For example, where $d = 10$, $e = 8$, the gcd $(10,8) = 2$, and generic chord E = (1 1 1 2 1 1 1 2) comes only in the specific varieties (1 1 1 2 1 1 1 −8), (1 1 1 2 1 1 −10 2), (1 1 1 2 1 −10 1 2) and (1 1 1 2 −10 1 1 2); the other four replacements of the elements s_j yield only rotations of these four, so CV fails.

Next we drop the assumption that d is composite, but assume that D is embedded in U_c and $d > \lfloor c/2 \rfloor + 1$. We exhibit a chord E of cardinality $e = 2d - c$ for which CV fails:

Note that $e \geq 3$. Also, since $e = 2d - c$, $d - 1 = c + e - d - 1$. Then $d - 1 \equiv (e - d - 1)_{\text{mod } c}$. The generic chord E = $(1\ 1\ 1 \ldots d - e + 1)$ is a CV failure. This description of E is a partition of d into $e - 1$ ones, followed by $d - e + 1$. Each entry represents a number of generalized fifths. The specific varieties are represented by the following types, all entries reduced mod c:

$E_1 = (-(d-1)\ 1\ 1 \ldots d - e + 1)$, $E_2 = (1\ -(d-1)\ 1 \ldots d - e + 1)$, … $E_e = (1\ 1\ 1 \ldots -(e-1))$.

However, recalling that $d - e + 1 = -(e - d - 1)$ is congruent to $-(d-1)$ mod c, it is clear that for all i from 1 to $e - 1$, E_i and E_{e-i} are of the same variety:

$E_1 = (-(d-1)\ 1\ 1 \ldots -(d-1))$ is a rotation of $E_{e-1} = (1\ 1\ 1 \ldots (d-1)\ -(d-1))$, for example. Thus, CV fails, and instead of e varieties, there are at most $\lfloor e/2 \rfloor + 1$ varieties.

For a concrete example, let $c = 13$, $d = 11$. Then $e = 9$, and E = (1 1 1 1 1 1 1 1 3). Since 3 is congruent to −10 mod 13, we may replace −10 by 3 in the representations of the specific chords, and it becomes apparent that E_1 and E_8 are rotations of each other, as are E_2 and E_7, etc., reducing the number of specific varieties to 5:

$$E_1 = (3\ 1\ 1\ 1\ 1\ 1\ 1\ 1\ 3)$$

$$E_2 = (1\ 3\ 1\ 1\ 1\ 1\ 1\ 1\ 3)$$

$$E_3 = (1\ 1\ 3\ 1\ 1\ 1\ 1\ 1\ 3)$$

$$E_4 = (1\ 1\ 1\ 3\ 1\ 1\ 1\ 1\ 3)$$

$$E_5 = (1\ 1\ 1\ 1\ 3\ 1\ 1\ 1\ 3)$$

$$E_6 = (1\ 1\ 1\ 1\ 1\ 3\ 1\ 1\ 3)$$

$$E_7 = (1\ 1\ 1\ 1\ 1\ 1\ 3\ 1\ 3)$$

$$E_8 = (1\ 1\ 1\ 1\ 1\ 1\ 1\ 3\ 3)$$

$$E_9 = (1\ 1\ 1\ 1\ 1\ 1\ 1\ 1\ 5)$$

Result 2: If D is a non-degenerate well-formed set where d is a prime and D is either non-embedded or embedded in a chromatic with $d \leq \lfloor c/2 \rfloor + 1$, then CV holds for chords.

Proof: Assume D is non-embedded with d a prime, and suppose there exists a chord E such that CV fails. CV holds for lines, as can be seen if one considers the generic description of E as a line: $\langle s_1\ s_2 \ldots s_e \rangle$ where the s_i represent measures by generalized fifths.[15] Then the specific varieties of the line are given by the following descriptions, numbering e in all:
$\langle -\sum_{i=2}^{e} s_i\ s_2 \ldots s_e \rangle, \langle s_1 - \sum_{i \neq 2} s_i\ s_3 \ldots s_e \rangle, \ldots \langle s_1\ s_2 \ldots - \sum_{i \neq j} s_i \ldots s_e \rangle, \ldots \langle s_1 \ldots s_{e-1} - \sum_{i=1}^{e-1} s_i \rangle$. While the generic elements must represent a partition of d, thus sum up to 0 mod d, in the specific cases they must sum to zero under ordinary arithmetic, with a single negative element representing the interval containing the unique generalized tritone. If CV fails for the chord E, it must be that there is a rotation that takes some specific E_j onto E_k for $j \neq k$. Since there is exactly one negative element for each E_i, a rotation of some E_j onto E_k is possible only if E is composed of a repeated sequence of elements. But then the number of repetitions divides d, contradicting the assumption.

Next, assume the set D is embedded in a chromatic with $d \leq \lfloor c/2 \rfloor + 1$, d prime. Let $E = (s_1\ s_2 \ldots s_e)$ be a chord with CV failure. The specific varieties E_j of the chord are determined by changing a single element s_j in the generic description of E to $-\sum_{i \neq j} s_{i_{\bmod c}}$. With a little reflection, one can see that for CV to fail, at a minimum the replacement entry of a specific E_j equals one of the values s_i

in E's generic description. This is not sufficient, but it would be necessary for CV to fail for E. That is, we know that for some s_j its replacement in the specific chord E_j must equal some s_k, $k \neq j$. By definition, in the generic description of E, $s_j = d - \sum_{i \neq j} s_i$. $E_j = (s_1, \ldots s_{j-1}, c - \sum_{i \neq j} s_i, \ldots s_e)$, since the sum of the entries must equal c, i.e., 0 mod c. Since $\sum_{i \neq j} s_i = d - s_j$, we have $c - \sum_{i \neq j} s_i = c - d + s_j = s_k$. Since $d \leq \lfloor c/2 \rfloor + 1$, $c - d \geq \lfloor c/2 \rfloor - 1$, then $c - d + s_j = s_k \geq \lfloor c/2 \rfloor$. But since $s_k < d$, this can only happen if $d = \lfloor c/2 \rfloor + 1$ and $s_k = d - 1$, so E would have to be the two-note chord with generic description $(1 \; d-1)$. The varieties of E would be described by $(1 \; c-1)$ and $(-(d-1) \bmod c \; (d-1)) = (\lfloor c/2 \rfloor + 1 \; \lfloor c/2 \rfloor)$ for c odd and $(c/2 \; c/2)$ for c even. This contradicts the assumption of CV failure, since $c-1$ is different from $\lfloor c/2 \rfloor$, unless $c = 3$ and $d = 2$ and CV holds, contrary to assumption.

Since CV implies the Myhill Property, which in turn implies well-formedness, we have the following theorem summarizing the two results demonstrated above:

THEOREM: CV holds for chords if and only if D is a non-degenerate well-formed set of cardinality d where d is prime, and where if D is embedded in a chromatic universe of cardinality c, then $d \leq \lfloor c/2 \rfloor + 1$.

1.3 Twin Primes and Dual CV

When these results were communicated to Jack Douthett in 1993, he used them to define a concept he called dual CV, and in two papers he asserted a relationship between this formulation and a longstanding conjecture in number theory, the Twin Primes conjecture.[16]

Douthett is only concerned with scales that are embedded in chromatic universes. He also excludes chromatic clusters, i.e., the cases of well-formed scales in which both the generator and all but one of the step intervals have a chromatic length of 1. Under his definition, a set (that is not a chromatic cluster) has dual CV if it has CV for chords and its complement within the given chromatic universe also has CV for chords. Thus, either both a set and its complement have dual CV, or neither do. For example, the complementary diatonic and pentatonic sets in the usual 12-note universe have dual CV.

It has long been conjectured but not proven that there are an infinite number of primes that differ by two (such as 5 and 7). Such primes are called twin primes.[17] Douthett asserted that if there were a finite number of twin primes, the number of chromatic universes that support dual CV would be one more than the number of pairs of twin primes (and if the Twin Primes conjecture is true, i.e., there are an infinite number of such pairs, there is still a natural one-to-one correspondence between the pairs and the dual CV universes, excluding the case where $c = 5$). Douthett applies the conditions for chord CV (D must be a well-formed set of cardinality d where d is prime and $d \leq \lfloor c/2 \rfloor + 1$) to conclude that a chromatic universe of cardinality c where $c > 5$ and is odd cannot support dual CV sets, since the condition on the size of d forces the complementary sets to have cardinalities $(c-1)/2$ and $(c+1)/2$, which differ by 1, so one is odd and one is even. Since 2 is the only even prime, the only universe of odd cardinality which supports dual CV sets has cardinality 5 (where the sets with dual CV have cardinalities 2 and 3). If c is even, it is easy to see that the only well-formed sets of cardinality $c/2$ are the chromatic clusters, such as the pitch classes from 0 to $(c/2) - 1$. Douthett excludes these cases. The only other complementary well-formed sets with CV for chords that could exist would be in universes of even cardinality $c > 5$, with cardinalities $(c/2) + 1$ and $(c/2) - 1$, where both of these are primes. Given these values, such well-formed sets do always exist, and are moreover *maximally even* or ME sets, as proved by Clough and Douthett.[18] Thus, other than the unique case where $c = 5$, the only chromatic universes that support dual CV sets are of even cardinality c where c is the sum of twin primes.

It turns out, then, that loosely speaking, the number of pairs of twin primes is one less than the number of chromatic universes that support complementary ME sets that have CV for chords ("loosely speaking" because this would be strictly true only if there are in fact a finite number of twin primes; otherwise both the number of pairs of twin primes and chromatic universes of the required type are countably infinite).

I doubt very much that this advances the problem of the existence of an infinitude of twin primes, but it is an interesting case of a combinatorial statement in music theory that is equivalent to a long-standing conjecture in number theory. It is also curious that the Norwegian mathematician Viggo Brun, who approached the twin primes problem with a sieve argument, wrote (separately) about music, concerning tuning questions.[19]

The musical relevance of CV for chords is primarily as yet another indicator of the special status of the diatonic and pentatonic sets. If we do not confine dual CV to maximally even sets, then in the usual chromatic the 5- and 7-note chromatic clusters are also dual CV sets, but the chromatic clusters are, for one thing, rife with "contradiction" in the sense of Rahn.[20] To understand why CV for chords has cognitive significance for music that is essentially diatonic, one has to perform a contrary-to-fact thought experiment and imagine how our music would be different if, say, two different diatonic seventh chords were aurally indistinguishable. It may be useful to translate the case of the diminished seventh chord into the language of this paper. Consider registrally ordered diminished seventh chords with the same note in the bass, for example, the arpeggiated chords (or lines) ⟨C E♭ G♭ B𝄫⟩, ⟨C E♭ G♭ A⟩, ⟨C E♭ F♯ A⟩, ⟨C D♯ F♯ A⟩. Expressed in terms of perfect fourths, the specific diminished seventh line types are ⟨3 3 3 −9⟩, ⟨3 3 −9 3⟩, ⟨3 −9 3 3⟩, ⟨−9 3 3 3⟩, where "−9" represents the position of the notated augmented second. Since $-9 \equiv 3 \bmod 12$, within the 12-note equal-tempered universe the four varieties of diminished seventh collapse into the single type ⟨3 3 3 3⟩. Of course, this ambiguity is itself an important resource for tonal music. The tension between the functioning of the diminished seventh chord as a (chromatically altered) diatonic entity, and as a symmetrical entity in the twelve-note chromatic universe is a fruitful one, but one that highlights the very different geometries of the asymmetrical (though highly ordered) diatonic system and of its symmetrical equal-tempered chromatic background.

Beyond the musical meaning and significance of these results—and of those of Clough and Myerson, to which this chapter is merely a long footnote—they provide an instance of mathematical music theory where the two disciplines are brought into close conjunction. The method of generalization, which is proper to mathematics, is essential to uncovering what is really going on, whatever the value of insights into possible or impossible microtonal universes. Part of the beauty of music is the beauty of ideas. Usually by this we mean purely musical ideas. Clough's work, with his various co-authors, permits us to contemplate mathematical beauty reflected by musical objects.

Notes

1. A selective bibliography includes Babbitt 1965, Gamer 1967, Rahn 1973, Balzano 1980, Browne 1981, Gauldin 1983, Clough and Myerson 1985, Agmon 1989, Carey and Clampitt 1989, Clough and Douthett 1991, Carey and Clampitt 1996a and 1996b, Tymoczko 1997, and Clough, Engebretsen, and Kochavi 1999. A survey of mathematical music theory is found in Noll 2005.

2. Misunderstanding or ignoring the distinction between these two types of explanations leads Dmitri Tymoczko to criticize—after praising, with qualifications—Rahn 1991: "Furthermore, it is doubtful that these properties would have been valued by composers of the late nineteenth and early twentieth centuries" (Tymoczko 1997, 136 n. 3). The properties Rahn isolates (e.g., minimization of contradiction between intervals' diatonic and chromatic lengths) are of the second type, not available for compositional application except insofar as they are intrinsic features of a given collection.

3. Maximally smooth cycles are discussed in Cohn 1996, which provides an introduction to neo-Riemannian theory and analysis. For extensions of neo-Riemannian theory, see Cohn 1997 and 1998a, and see Childs 1998 for extensions to seventh chords. Shimbo 2001 provides a transformational framework for triads of the 3-11 and seventh chords of the 4-27 set class or set classes taken together. For post-tonal analogs, see Clampitt 1997 and 1999, and especially Lewin's essay in the present collection.

4. Dahlhaus [1968] 1990, 172.

5. Ibid.

6. The Three Gap Theorem or Three Lengths Theorem, formerly the Steinhaus Conjecture, states that the distances between adjacent points $n\alpha$ mod 1, for a positive real number α and integers n, $0 \leq n < N$, come in at most three sizes. See Sós 1958 for one of the first comprehensive treatments. For musical applications and further references, see Clampitt 1995.

7. Johnson 2003 and Cohn 1998b.

8. Clough and Myerson 1985. The following year they published a version of this article for a mathematical audience, "Musical Scales and the Generalized Circle of Fifths," in *American Mathematical Monthly*. Silverman and Wiseman 2006 note a gap in one of Clough and Myerson's proofs and provide a patch for it.

9. The generic perspective was first explored in a thorough way in Clough 1979.

10. Interval normal form is the cyclic interval representation derived from the normal form of the mod 7 (diatonic) pitch-class set. Clough and Myerson 1985 use this convention, from Clough 1979, which is in turn a variant of the notation introduced in Regener 1974.

11. In their counterexample, $d = 9$, $c = 17$, and the MP set is {0, 2, 4, 6, 8, 10, 12, 14, 16}. Then the trichord type of generic description (333) has just one specific variety (Clough and Myerson 1985, 265–66).

12. Carey and Clampitt 1989 and 1996b. We sometimes speak of "scales" rather than "sets" to emphasize the role that scale order plays in the definition of well-formedness, but I will use the terms interchangeably.

13. The bracket denotes the floor or integral part of a number. Thus, $\lfloor c/2 \rfloor$ is the greatest integer less than or equal to $c/2$. If c is even, $\lfloor c/2 \rfloor = c/2$, and if c is odd, $\lfloor c/2 \rfloor = (c-1)/2$.

14. Clough and Myerson 1985, 258–64.

15. From the correction to Clough and Myerson 1986 in Silverman and Wiseman 2006, we may assume that E is in circle-of-fifths order.

16. Douthett 1993a and 1993b.

17. There has been recent work pointing toward an affirmative answer to the conjecture. See Peterson 2005 for an account of the efforts of Daniel Goldston and his colleagues Motohashi, Printz, and Yildrim.

18. Clough and Douthett 1991, 99.

19. Among his papers on music is Brun 1961. He used his sieve method to prove that the sum of the reciprocals of the twin primes is convergent (Brun 1919), in contrast to Euler's result of 1737 that the sum of the reciprocals of all the primes is divergent.

20. Contradiction (where the specific size of an interval of generic span k is greater than the specific size of an interval of generic span greater than k) is discussed in Rahn 1991. See also discussions in Agmon 1989, Carey 2002 and Carey 2007.

Chapter Two

Flip-Flop Circles and Their Groups

John Clough

2.1 Introduction

In neo-Riemannian theory, an essential construct is the cycle (or circle) of triads, alternately major and minor. Figure 2.1 shows four such cycles, called "hexatonic systems" by Cohn, which partition the set class (sc) of consonant triads into four subclasses of six triads each.[1] The four systems are transpositions of each other, and they embody relations traceable to Riemann: *Parallel* (P) (equivalent to Riemann's *Quintwechsel*) and *Leittonwechsel* (L) apply alternately as we make our way around any of the circles.[2] The meanings of these terms, based on double common-tone retention, are evident from the context.

Cohn also describes three *octatonic systems*—cycles of eight triads each, which also partition the sc of major/minor triads.[3] These are shown in figure 2.2. Like their hexatonic cousins, the octatonic cycles, again all transpositions of one another, arise from alternation of two *Wechsel,* in this case P and *Relative* (R) (equivalent to Riemann's *Terzwechsel*). The remaining pair from P, L, and R (L and R), when applied alternately, generates the circle of figure 2.3—a circle running through the full set class of 24 consonant triads, put forth by Werckmeister in 1698, and placed in historical context in the work of Joel Lester.[4] P, L, and R show *parsimonious* voice leading: they are the only transformations that change one consonant triad into another by replacing a single pitch class. They are also *exchange* operations: they transform a major or minor triad to a triad of the opposite mode. There are 12 *Wechsel* in the Riemannian group, including P, L, and R.[5]

It seems natural to ask: what is the space of all circles formed by alternating a pair of *Wechsel*? It is this question that first motivated the present investigation. In the next few sections of the paper, I will enumerate and characterize the circles that comprise the space and show how they relate to subgroups of the *Schritt/Wechsel* group. In the final section of the paper I will expand the field of inquiry to other kinds of circles, including diatonic sequences, and study other groups

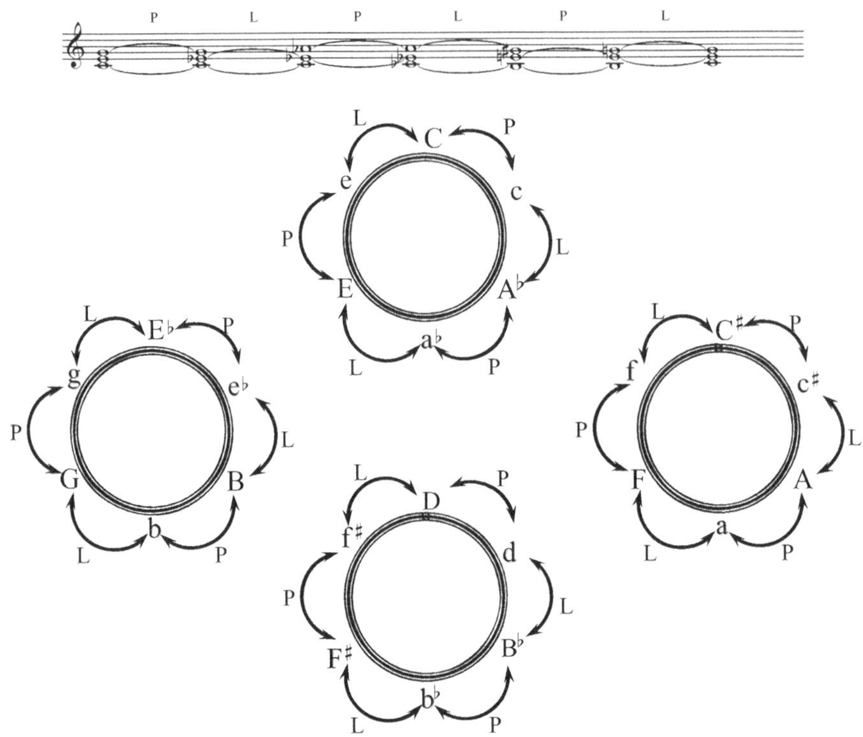

Figure 2.1. The Four Hexatonic Systems (C = C major; c = C minor, etc.) [Cohn, 1996].

that support the circles in question. The objective is to extend neo-Riemannian theory to address a broad range of circular musical objects.

In what follows, both the usual T/I and the S/W groups will be discussed. Right-to-left orthography is the traditional convention for the T/I group, while left-to-right orthography is often adopted for the *Schritt/Wechsel* group. To use both conventions simultaneously could lead to confusion, so in this paper, a right-to-left orthography convention will be adopted for all group actions.

2.2 Enumeration of UFFCs

I refer to circles such as those in figures 2.1–2.3 as *Uniform Flip-Flop Circles* (UFFCs) and notate them in text as wrap-around sequences of uppercase (for major triads) and lowercase (for minor) letters enclosed in parentheses. Thus, the top circle of figure 2.1, reproduced in figure 2.4, is notated: (C–c–A♭–a♭–E–e). Circles and

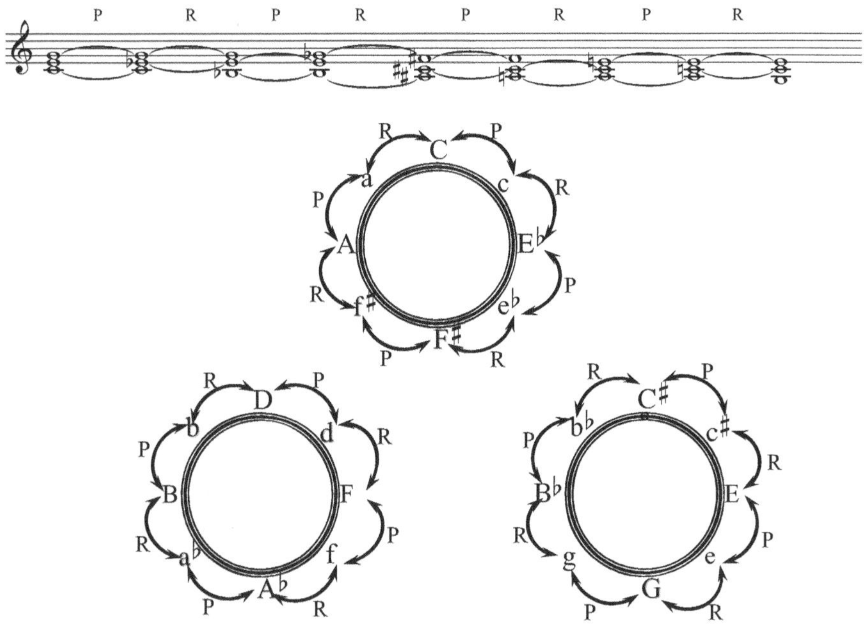

Figure 2.2. The Three Octatonic Systems [Cohn, 1997].

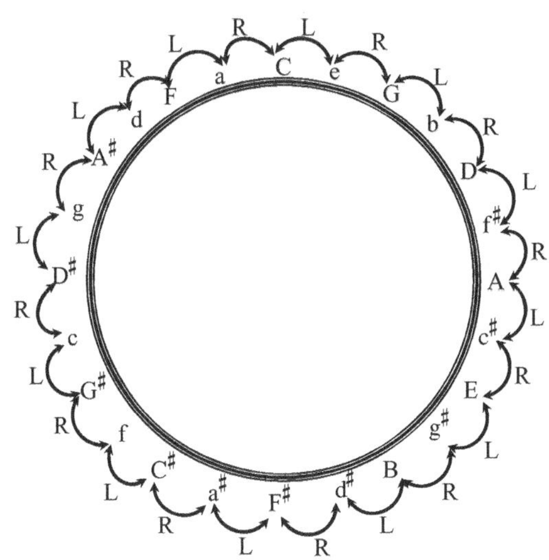

Figure 2.3. The L-R Loop.

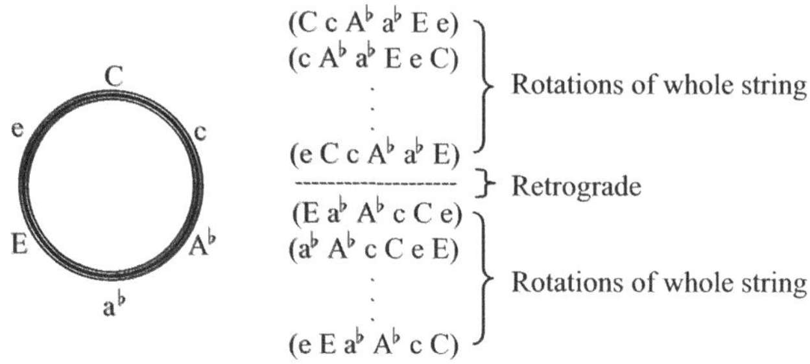

Figure 2.4. Circular Rotations and Retrogrades of the circle (C c A♭ a♭ E e).

their notations are indifferent to rotation and retrograde, so there are 12 different notations, as shown in figure 2.4, that all represent the circle at the left of the example (one could establish a "prime form" for these notations, but it seems not worth the trouble in the present context). However, the string (C–c–E–e–A♭–a♭), for example, is distinct from the strings in figure 2.4 in that it cannot be rotated and/or retrograded to produce the original string.

To get a sense of how the enumeration proceeds, reconsider figure 2.1. Each of its four circles can be formed by superimposing two circles—one of three major triads and the other of three minor triads. The roots of the major triads form an augmented triad, and so do those of the minor triads. Moreover the triads are oriented in the same direction; moving clockwise, both are based on pitch-class transposition operator T_8. Here the two constituent circles of each UFFC are based on the same augmented triad, but that is not the case for all UFFC's. Each of the four distinct augmented triads, taken as a set of chord roots, supports a circle of major triads, and likewise a circle of minor triads. So there are 4 × 4 = 16 distinct pairs each consisting of a "trio" of major triads and a "trio" of minor triads. Further, each pair supports three different UFFCs when its two constituent circles, oriented in the same sense tranpositionally, are rotated with respect to one another. Consider the major triads with roots (C E A♭), paired with the minor triads with roots (d♭ f a), as shown figure 2.5. Rotation produces three circles for this pair, as it does for any of the 16 pairs; thus there are in all 16 × 3 = 48 UFFCs based on T_4.

In table 2.1, the shaded row reflects the count we have just completed. The first five cells of that row give, respectively, the value of the transposition operator for each component circle ($t = 4$); the number of component circles of each mode (4, corresponding to the 4 augmented triads); the number of elements in each component circle ($|C| = 3$), which in this case also gives the number of distinct rotations of the two component circles; the number of pairs of component circles (#C prs = 16); and the number of UFFCs (#UFFC = 48).

(C E A♭)
(d♭ f a) } (C d♭ E f A♭ a)

(C E A♭)
(a d♭ f) } (C a E d♭ A♭ f)

(C E A♭)
(f a d♭) } (C f E a A♭ d♭)

Figure 2.5. Circular Rotations and Retrogrades derived from (C G A♭) and (d♭ f a).

Table 2.1 UFFCs in the 12-pc universe

t	#C	|C|	#Cprs	#UFFC	#P	#UTTs
1	1	12	1	12	12	24
2	2	6	4	24	12	24
3	3	4	9	36	12	24
4	4	3	16	48	12	24
5	1	12	1	12	12	24
6	6	2	36	36	6	12
TOTAL				168	66	
0	12	1	144	144	12	12
GD TOT				312	78	144

key
t [cyclic interval of component circles]
#C [no. of component circles, each mode] = GCD (*t*, 12)
|C| [no. of elements, each component circle] = 12/#C
#C prs [no. of circle pairs, one of each mode] = $(\#C)^2$
#UFFC [no. of UFFCs] = |C| • #C prs = 12 • #C, for $t \neq 6$ (see note 1); = #C • #P, for all *t*
#P [no. of patterns] = |C| • #C, for $t \neq 6$ (see note 1)
#UTT [no. of UTTs] (see note 2)

1 From the equalities given for "#UFFC" (see above), it follows that # P = |C| • # C, for $t \neq 6$, as shown here. The case $t = 6$ is exceptional: because of the tritone, |C| = 2 does not give the number of distinct rotations of component circles with respect to one another (there is but one such rotation in this case); therefore each C pr corresponds to a unique UFFC; hence #UFFC = (|C| • # C prs)/2 = 36, and # P = (|C| • # C)/2 = 6.

2 For values of $1 \leq t \leq 6$, one-way (non-involutional) UTTs are counted. For $t = 0$, two-way UTTs (involutions) are counted.

To complete the enumeration of UFFCs in the 12-pc universe, we need to consider the remaining operators T_t, $1 \leq t \leq 6$. The value $t = 0$ produces "degenerate" UFFCs that alternate between one particular major and one particular minor triad, by way of a single *Wechsel* (shown in a separate row of table 2.1), and values of *t* greater than 6 merely replicate UFFCs with smaller values of *t*. Formulas for computation of values in the first five columns of table 2.1 are given in the key beneath the table. The column headed #UFFC shows the total number of UFFCs (168) and their distribution with respect to the six non-zero values of *t*. The remaining data in table 2.1 will be discussed later in the paper.

In the course of the enumeration sketched out above, we have lost touch with the idea of UFFCs based on the alternation of two distinct *Wechsel*, focusing instead on the UFFC as a composite of two constituent circles based on the same T_t. However, any such composite may be generated by the alternation of two distinct *Wechsel*. Conversely, the alternation of any two distinct *Wechsel* will produce a circle that can be decomposed into two constituent circles of the type treated above. Thus we may regard a UFFC from either perspective. Having dwelt on the notion of constituent circles, we now work our way back to the idea of alternating *Wechsel*, by way of the group-theoretic aspects of UFFCs.

2.3 UFFCs and the Schritt/Wechsel Group

The 12 mode-reversing *Wechsel*, when composed (combined) with one another, imply 12 distinct transformations between two triads of the same mode. Riemann defined the full set of 24 mode-reversing and mode-preserving transformations, first described in plainly group theoretical terms by Klumpenhouwer and hereafter referred to as the *Schritt/Wechsel* (S/W) group.

Each *Schritt* has the effect of transposing major triads by a particular pitch-class interval, and minor triads by the inverse of that interval. I represent the 12 *Schritte* by the capital letter S with subscript indicating the level of transposition; thus, as shown in the top two-thirds of figure 2.6, S_0 is the identity element that leaves its argument unchanged, S_1 is the *Schritt* that transposes a major triad up by one semitone, or a minor triad down by one semitone, S_2 is the *Schritt* that transposes a major triad up two semitones, or a minor triad down two semitones, etc.

The 12 *Wechsel* may in turn be represented as composites of *Schritte* with some arbitrarily chosen *Wechsel;* it is natural to choose P for this purpose. Thus, PS_n is the *Wechsel* that first transposes its argument by n semitones up or down, as stated above, and then converts it to a triad of opposite mode on the same root. To simplify the notation for the present purpose, I will write W_0 for PS_0, W_1 for PS_1, etc. This is comparable to writing I_n for pitch class inversion operators instead of T_nI; however, by convention, the inversion is performed first in that context. Thus, as shown in the bottom part of figure 2.6, W_0 is the *Wechsel* that first transposes its

S_0	S_1		S_{11}
C → C	C → C♯	....	C → B
C♯ → C♯	C♯ → D	....	C♯ → C
⋮	⋮		⋮
B → B	B → C	....	B → A♯
c → c	c → b	....	c → c♯
c♯ → c♯	c♯ → c	....	c♯ → d
⋮	⋮		⋮
b → b	b → a♯	....	b → c
W_0 (= PS_0)	W_1 (= PS_1)		W_{11} (= PS_{11})
C ↔ c	C ↔ c♯	....	C ↔ b
C♯ ↔ c♯	C♯ ↔ d	....	C♯ ↔ c
⋮	⋮		⋮
B ↔ b	B ↔ c	....	B ↔ a♯

$W_0 = P$ $\qquad W_4 = L$ $\qquad W_9 = R$

Figure 2.6. Action of the S/W group on major and minor triads.

For any triad X,

$$\begin{aligned} S_m S_n(X) &= S_{m+n}(X) \\ S_m W_n(X) &= W_{n-m}(X) \\ W_m S_n(X) &= W_{m+n}(X) \\ W_m W_n(X) &= S_{n-m}(X) \end{aligned}$$

(sums and differences mod 12)

Figure 2.7. The S/W group, composition of S_n and W_n.

argument by zero semitones and then converts it to a triad of opposite mode on the same root; W_1 is the *Wechsel* that transposes its argument by one semitone up or down and then converts it to a triad of opposite mode on the same root; etc. The Riemannian P, L, and R are given in this notation by W_0, W_4, and W_9, respectively.[6]

Formulas for composites in S/W are given in figure 2.7. The structure of the S/W group may be represented by a set of defining relations such as that given in figure 2.8. For the mathematically inclined, S/W is isomorphic to the dihedral group known as D_{12}, that is, the group of 24 congruence motions of the dodecagon.

$$(S_1)^{12} = (W_0)^2 = (W_1)^2 = S_0$$

Figure 2.8. S/W group in terms of generators and relations.

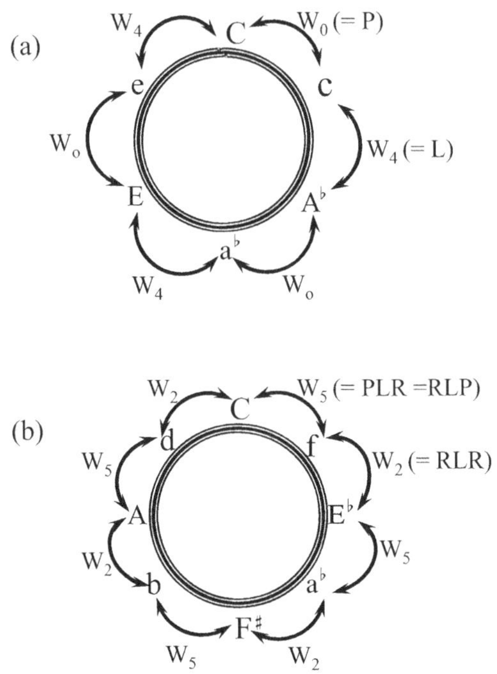

Figure 2.9. Two *Wechsel* Cycles.

With this notation, it is easy to represent the pair of *Wechsel* that supports any given UFFC. Consider once again the circle reproduced in figure 2.9a. Here, the alternating *Wechsel* P and L are re-labeled W_0 and W_4, respectively. In the UFFC of figure 9b, the two *Wechsel*, notated there as W_5 and W_2, are, respectively, in neo-Riemannian notation, PLR or RLP, and RLR. As Cohn and others have shown, it is worthwhile in many analytical contexts to conceive the *Wechsel* and *Schritte* as strings of P, L, and R, but our objectives here are better served by the more neutral notations W_n and S_n.

Now that we are focused once again on the concept of UFFCs formed from alternating *Wechsel*, it is appropriate to give this a localized technical term. I will call an alternating pair of *Wechsel* a *pattern*. To gauge the theoretical space of UFFCs, it is perhaps more helpful to enumerate patterns than UFFCs; this is

Table 2.2 Instantiation of UFFCs in extant music

t	patterns			
1	(W_4, W_5)	(W_9, W_{10})		
2	(W_3, W_5)	(W_6, W_8)	(W_8, W_{10})	
3	(W_1, W_4)	(W_9, W_0)		
4	(W_0, W_4)	(W_1, W_5)	(W_4, W_8)	(W_5, W_9)
5	(W_4, W_9)	(W_5, W_{10})		
6	(W_2, W_8)			

done in the next-to-last column of table 2.1. For non-zero values of t, the total number of patterns is 66 (i.e., 12 choose 2). For $t = 0$, each *Wechsel* is paired with itself; hence there are 12 patterns. For any value of t, the number of corresponding patterns is the number of pairs of subscripts for W_n that differ by t or $-t$, mod 12, or, what amounts to the same thing, the number of pitch-class pairs that form interval class t. This number is 12 for all values of t except $t = 6$, where we have a kind of "tritone exception" familiar to students of atonal set theory (a somewhat different approach to the enumeration of patterns is given as footnote 1 to table 2.1).

To what extent is the theoretical space of alternating *Wechsel* patterns used in actual music? Table 2.2 shows the results of a survey of neo-Riemannian literature, revealing that only 13 of 66 patterns (roughly 20%) are in examples, ranging from the eighteenth century through the early twentieth century, treated in that literature.[7] This tabulation appears to suggest a high degree of selectivity, presumably on the basis of voice-leading and other tonal criteria. However, many of the patterns appear only once in the theoretical literature searched. Additional instances of patterns will no doubt come to light. So the degree of compositional selectivity, its range, and plausible reasons underlying it, await further investigation.

Patterns account for chord-to-chord transitions as we make our way around a UFFC, but what of other pairwise relationships in a UFFC? For any UFFC, there is a unique smallest subgroup of S/W that supports all pairwise relationships among members of the circle. Such subgroups are isomorphic to dihedral groups—groups of congruence motions of regular polygons. Figure 2.10 reproduces the two UFFCs of figure 2.9, and gives, for each of them, a table showing pairwise connections by means of the appropriate subgroup of S/W. The table of figure 2.10a and the subgroup listed there support not only the pictured UFFC but all three UFFCs produced by rotating the two component circles of the pictured UFFC; a similar statement applies to figure 2.10b, where the subgroup supports four UFFCs.

It is easy to enumerate the dihedral subgroups of S/W in relation to UFFCs, using table 2.1. For any particular value of t, the second column of the table, headed #C, gives the number of smallest subgroups of S/W that support UFFCs

	C	E	A♭	c	e	a♭
C	S_0	S_4	S_8	W_0	W_4	W_8
E	S_8	S_0	S_4	W_8	W_0	W_4
A♭	S_4	S_8	S_0	W_4	W_8	W_0
c	W_0	W_8	W_4	S_0	S_8	S_4
e	W_4	W_0	W_8	S_4	S_0	S_8
a♭	W_8	W_4	W_0	S_8	S_4	S_0

Subgroup: { S_0, S_4, S_8, W_0, W_4, W_8 }

	C	E♭	F♯	A	d	f	a♭	b
C	S_0	S_3	S_6	S_9	W_2	W_5	W_8	W_{11}
E♭	S_9	S_0	S_3	S_6	W_{11}	W_2	W_5	W_8
F♯	S_6	S_9	S_0	S_3	W_8	W_{11}	W_2	W_5
A	S_3	S_6	S_9	S_0	W_5	W_8	W_{11}	W_2
d	W_2	W_{11}	W_8	W_5	S_0	S_9	S_6	S_3
f	W_5	W_2	W_{11}	W_8	S_3	S_0	S_9	S_6
a♭	W_8	W_5	W_2	W_{11}	S_6	S_3	S_0	S_9
b	W_{11}	W_8	W_5	W_2	S_9	S_6	S_3	S_0

Subgroup: { S_0, S_3, S_6, S_9, W_2, W_5, W_8, W_{11} }

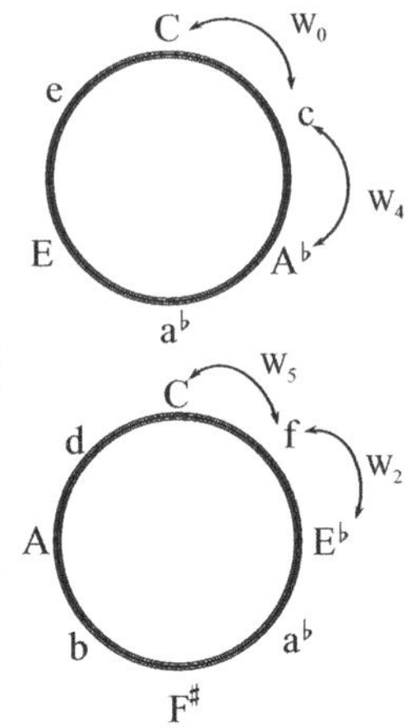

Figure 2.10. Actions of S/W subgroups. Each multiplication table shows the action of a subgroup of the S/W group on the set of triads in the adjacent UFFC. For two triads *X* and *Y*, the transformation that takes *X* to *Y* is found at the intersection of the row labeled *X* and the column labeled *Y*.

generated by that value of *t* (all such groups are dihedral, though the smaller ones are "degenerate" dihedral groups, as will be explained below). For any value of *t*, the number of elements in a UFFC is 2|C|; this quantity is also the order of the smallest subgroup that supports all pairwise connections among members of the UFFC. For *t* = 0, 2, 3, 4, 6, each corresponding dihedral subgroup supports UFFCs generated by a unique value of *t*. However, for *t* = 1 and *t* = 5, whose UFFCs embrace all 24 major and minor triads, the same subgroup is required, namely the full S/W group. Thus, the second column of table 2.1 accounts for the 28 dihedral subgroups of order 2 through 24, of the S/W group.[8] The S/W subgroups of figure 2.10 are *simply transitive* over the subsets of triads in their corresponding UFFCs; that is, for any pair of triads *X* and *Y* in the circle of figure

	C	E	A♭	c	e	a♭
C	T_0	T_4	T_8	T_7I	$T_{11}I$	T_3I
E	T_8	T_0	T_4	$T_{11}I$	T_3I	T_7I
A♭	T_4	T_8	T_0	T_3I	T_7I	$T_{11}I$
c	T_7I	$T_{11}I$	T_3I	T_0	T_4	T_8
e	$T_{11}I$	T_3I	T_7I	T_8	T_0	T_4
a♭	T_3I	T_7I	$T_{11}I$	T_4	T_8	T_0

Figure 2.11. Action of the T/I Subgroup {T_0, T_4, T_8, T_3I, T_7I, $T_{11}I$} on the Set {C, E, A♭, c, e, a♭}.

2.10a, there is one and only one transformation that takes *X* to *Y*, and the same is true for figure 2.10b. Indeed for any UFFC, the smallest subgroup that supports connections among all members of the circle is simply transitive over the set of elements in the UFFC. This fact renders any UFFC suitable for treatment as a Generalized Interval System.[9] I will comment on the analytical relevance of such subgroups in the final section of the paper.

2.4 UFFCs and the T/I Group

As noted above, the S/W group is isomorphic to the dihedral group of order 24, or D_{12} in mathematical notation. The set of 24 transpositions and inversions of atonal set theory (the T/I group) is also isomorphic to D_{12}, and hence to the S/W group.

Consider figure 2.11. Like figure 2.10a, figure 2.11 shows how a subgroup may account for transformations among all the triads of any of the three UFFCs containing the major and minor triads on roots C, E, and A♭. The subgroup of figure 2.10a and that of figure 2.11 have the same structure: they are isomorphic to D_3, the dihedral group of order 6. However, while figure 2.10a shows a subgroup of the S/W group, figure 2.11 shows a subgroup of the T/I group. For any UFFC, there are subgroups of identical structure—one a subgroup of the S/W group and the other of the T/I group—that account for transformations that take us from any triad of the UFFC to any other triad. So why do we need the S/W group when we could employ the more familiar T/I group to construct the UFFCs?

The answer lies with the notion of pattern. If we conceive of UFFCs as arising from the alternation of a pair of inversions, then in general the S/W groups and its subgroups will serve. Figure 2.12 shows a UFFC that we looked at above (C c A♭ a♭ E e). Marked outside the circle is the pattern of two *Wechsel*, W_0 and W_4, that alternate to construct the UFFC. What if we employ the subgroup of figure 2.10a to move around this same UFFC? As shown inside the circle, three trans-

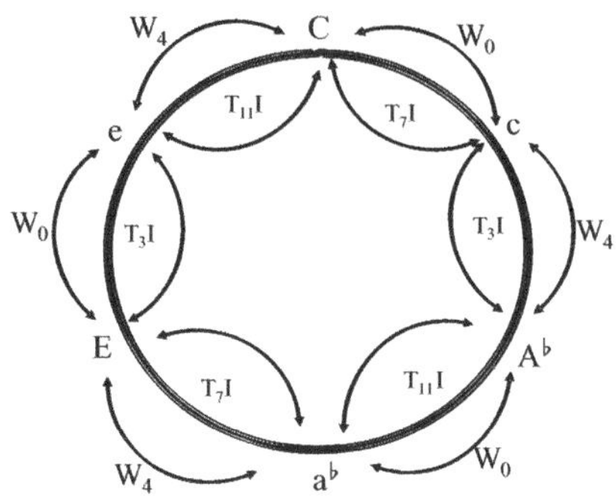

Figure 2.12. Groups Acting on the (C c A♭ a♭ E e) Circle.

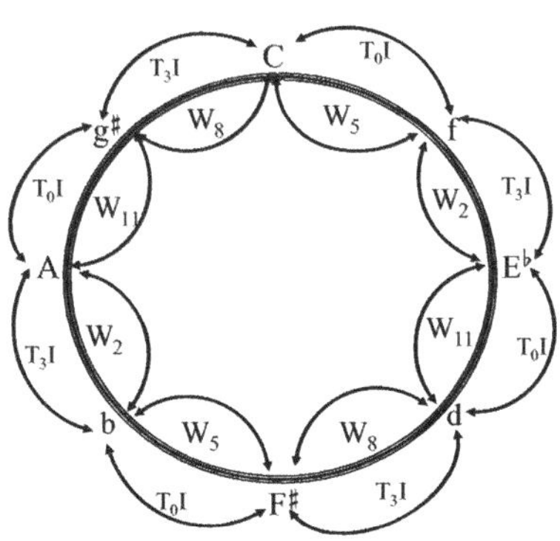

Figure 2.13. Groups Acting on the (C, f, E♭, d, F♯, b, A, g♯) Circle.

formations, T_{11}, T_3, and T_7, are required to generate it. Similarly for any UFFC, save those where $t = 6$: if we are limited to the T/I group, then more than two inversional transformations are required to generate the chords of the circle.

The question naturally arises: what if we export to the T/I group our notion of pattern developed for the S/W group? Suppose we select two inversions, say I_0 and I_3, and an arbitrary starting triad, say C major. The result of alternating these two inversions is shown in figure 2.13, where I_0 and I_3 are marked outside the circle. Note that each of the constituent circles, one of major triads and the other of minor, is regular, but the circles are "reversed" with respect to one another: as we move clockwise from one major triad to the next, the triads are transposed up by three semitones (i.e., $t = 3$), but when we do the same for minor triads, they are transposed down by three semitones (i.e., $t = 9$). It is as though two com-

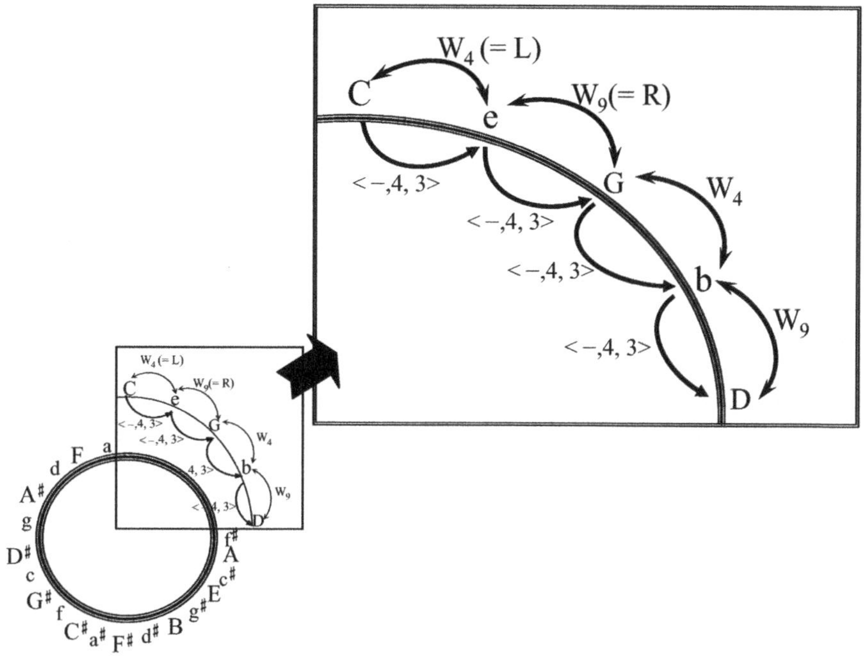

Figure 2.14. Reduction of figure 2.3 with Additional Notation.

ponent circles were superimposed so as to make a UFFC, but one was flipped over. In this case, we have a pattern of two inversions drawn from the T/I group. Now, suppose we approach this circle with the S/W group instead of the T/I group; as shown inside the circle, four different *Wechsel* are required to generate the adjacencies.[10]

2.5 UFFCs and Hook's UTTs

It is plain that any UFFC is indeed uniform in the following sense: as we proceed around the circle in either direction, the roots of major triads are all transposed by the same interval to produce the roots of the next minor triads, and the minor triads' roots are similarly transposed to produce the major triads' roots. Consider figure 2.14, which reproduces figure 2.3 with added notation. As shown outside the circle, the alternating *Wechsel* are W_4 and W_9, equated with L and R, respectively. Can we conceive such a circle as being generated by a single transformation? Jonathan Kochavi proposed that we can; following Kochavi, all we need do in the case of figure 2.14 is to specify that, proceeding clockwise,

$S_0: < +, 0, 0 >$	$W_0: < -, 0, 0 >$
$S_1: < +, 1, 11 >$	$W_1: < -, 1, 11 >$
$S_2: < +, 2, 10 >$	$W_2: < -, 2, 10 >$
.	.
.	.
.	.
$S_{11}: < +, 11, 1 >$	$W_{11}: < -, 11, 1 >$

Figure 2.15. *Schritte* and *Wechsel* represented as UTTs.

major triads are transformed by L (or, more fussily, by an L-like transformation, since L gives up its involutional status in this application) and minor triads by R; or counter-clockwise, by R and L, respectively.[11] In the notation of Julian Hook we write < –, 4, 3 >, as shown on figure 2.14, where the minus sign denotes change of mode and the numbers 4 and 3 represent the intervals of transposition applied, respectively, to major triad roots and minor triad roots when moving to the next triad clockwise.[12] Or we could write < –, 9, 8 >, the inverse of < –, 4, 3 >, to represent uniform counterclockwise motion about the same circle. These two transformations are, by the way, identical to Lewin's MED and SUBM transformations, conceived in a rather different setting.[13]

Hook's notation describes a set of Uniform Triadic Transformations (UTTs), as he calls them. *Schritte* and *Wechsel* may be represented as UTTs as shown in figure 2.15: My S_1, for example, is equivalent to Hook's < +, 1, 11 >, where the plus sign signifies mode preservation; my W_1 is equivalent to Hook's < –, 1, 11 >. I will represent elements of the S/W group in terms of the subscripted S's and W's as set forth above; however, Hook's notation will be useful in contrasting two perspectives on UFFCs.

Returning to the circle of figure 2.14, let us compare two transformational schemes, one based on W_4 and W_9, the other based on Hook's < –, 4, 3 >. Since *Wechsel* are involutions, W_4 and W_9, as they appear on figure 2.14, work in both directions. On the other hand, Hook's UTTs are not, in general, involutions; the transformation < –, 4, 3 > does not reverse itself when applied twice. As a consequence, the groups that these two schemes produce are different. The pair of transformations W_4 and W_9 induces the dihedral group of order 24. By contrast, < –, 4, 3 > induces the cyclic group of order 24; any move from, say, the C major triad to any triad in the UFFC of figure 2.14 may be defined in terms of < –, 4, 3 > repeated an appropriate number of times, and the same is true for its inverse, < –, 9, 8 >.

There is a tension between these two readings of a UFFC, one as a chain of paired involutions and the other as a chain of repeated one-way transformations. Which approach is preferable? The answer, I think, depends upon one's objectives, and one's perceptions in a particular musical context.[14] I will return to this question in the final section of the paper.

Hook's notational scheme allows for 144 mode-reversing UTTs (each of the two numerical entries may range from 0 to 11). How do these correspond to UFFCs? We saw that the UFFC of figure 2.14 could be approached by way of UTTs: repeated application of < –, 4, 3 > or < –, 9, 8 > will generate the circle. The same is true for any UFFC based on the pair W_4 and W_9; indeed a similar statement applies to any pattern that supports a non-degenerate UFFC: the pattern corresponds to a unique inverse-related pair of mode-reversing non-involutional UTTs. Referring again to table 2.1, this one-to-two correspondence is evident in the last two columns of the table, where, for each value of t, $1 \leq t \leq 5$, 12 patterns correspond to 24 one-way UTTs, and, for $t = 6$, 6 patterns correspond to 12 such UTTs. For degenerate UFFCs (those with $t = 0$), the situation is different. The only mode-reversing UTTs that apply are identical to *Wechsel;* each of the 12 mode-reversing UTTs that are involutions corresponds to just one pattern. Altogether, these correspondences account for the 144 mode-reversing UTTs in relation to the 78 (66 + 12) patterns. However, since each pattern comprises two *Wechsel,* except for the degenerate cases with $t = 0$ which involve a single *Wechsel,* we can see an overall consistency in the correspondence between *Wechsel* and mode-reversing UTTs. It is pair-to-pair (two-to-two) for non-degenerate UFFCs and uni-directional UTTs, and one-to-one for degenerate UFFCs and involutional UTTs.

It is clear that Hook's notation allows also for 144 mode-preserving UTTs. Along with the 144 mode-reversing UTTs, these have been discussed in detail by Hook. Suppose we conceive of UFFCs in terms of cyclic groups supported by Hook's transformations, or more specifically, in terms of simply transitive cyclic subgroups of Hook's group of 288. Which mode-preserving UTTs must these subgroups include? In this conception, the supporting UTT subgroup must include, as one of its subgroups, the group generated by < +, *t*, *t* >. And these are precisely the mode-preserving UTTs required. For example, the UFFC of figure 10b corresponds to the full subgroup generated by < –, 5, 10 > (or its inverse < –, 2, 7 >). That is the group { < +, 0, 0 >, < –, 5, 10 >, < +, 3, 3 >, < –, 8, 1 >, < +, 6, 6 >, < –, 11, 4 >, < +, 9, 9 >, < –, 2, 7 > }, which in turn includes the cyclic subgroup of mode-preserving UTTs { < +, 0, 0 >, < +, 3, 3 >, < +, 6, 6 >, < +, 9, 9 > }.

2.6 UFFCs in Extra-Triadic Spaces

Exploring in equal-tempered universes with other than 12 pitch classes, Cohn identified classes of trichords amenable to connection through extended versions of P, L, and R.[15] Subsequently, several theorists, including Clifton Callendar, Jack Douthett and Peter Steinbach, and Edward Gollin have investigated extensions of contextual transposition and inversion to set classes other than the usual triads, especially the seventh chords of tonal harmony.[16] In the 12-pc universe, the extension of the S/W group-theoretic apparatus to scs with 24 pcsets is straightforward, though interesting questions of voice-leading, group structure, and notation arise; the same can be said for set classes of chords that lack non-trivial symmetry in universes with any number of pitch classes.

Julian Hook has suggested more general extensions, which are perhaps best understood through his notation for UTTs.[17] As we saw above, that notation comprises an ordered triple: a plus or minus sign indicating mode preservation or reversal, followed by two numerical values giving pc intervals of transposition that apply according to the mode of the argument-triad. As described above the plus/minus notation represents binary opposition between major and minor, but in Hook's more general conception, the plus/minus notation can represent any dichotomy we wish, for example, dominant and half-diminished seventh chords, as in the work cited above, or 5-3 and 6-3 chords, which will be discussed below. Further, the dichotomy of types need not be chords from the same sc. Indeed, the opposing classes need not involve chords at all.

Moreover, in Hook's general scheme there may be as many different types as we wish (with each type cycling over the same range, e.g., the 12 pcs), with the plus/minus sign replaced by a permutation of types. Instead of enlarging here upon the musical and mathematical ramifications of this, I will limit the final section of this paper to an extension of the UFFC concept to the diatonic world—an extension that flows naturally from UFFCs in 12-space, as explored above, and is at the same time in the spirit of Hook's proposed expansion.

Consider figure 2.16a, a familiar sequence from the first movement of Mozart's Piano Sonata in C Major, K. 545. Beginning with a tonic 6-3 chord, the sequence alternates 6-3 and 5-3 chords in a circle of 5ths, regaining the tonic, now in 5-3 position, on the second chord of measure 4. This is modeled in figure 2.16b, on the lower-left circle, where, reading clockwise from the top numeral, we see I, IV′, VII, III′, VI, II′, V, and then I′ at the bottom of the example. Instead of using "prime" in alternation with "no prime," we could write 6-3 and 5-3 alternately, reflecting the progression of figure 2.16a in more detail, but the model of figure 2.16b is designed to accord with any diatonic circle-of-fifths progression which alternates between two types susceptible to description as binary states, in terms of figured bass, differing registral presentations, or non-pitch parameters. We will return presently to the other symbols on the example.

In figure 2.17 we have a harmonic progression based on Pachelbel's much-abused Canon, which contains a segment of a complete circle of diatonic chords. Here, since all chords are in 5-3 position, figured bass does not capture the alternate states of the progression. To do so, we can look at outer-voice intervals (10–5–10–5, etc.), or, adapting the notation of Chapman to the diatonic context, we can represent the vertical structures as shown beneath the music.[18]

In essence, what we see in figures 2.16 and 2.17 are UFFCs in diatonic space; a logical next step is to enumerate the UFFCs in 7-space. This is easily done since circles of seven diatonic pitch classes, or scale steps, take one of just three essentially different forms—circles of steps (i.e., scales), circles of thirds, and

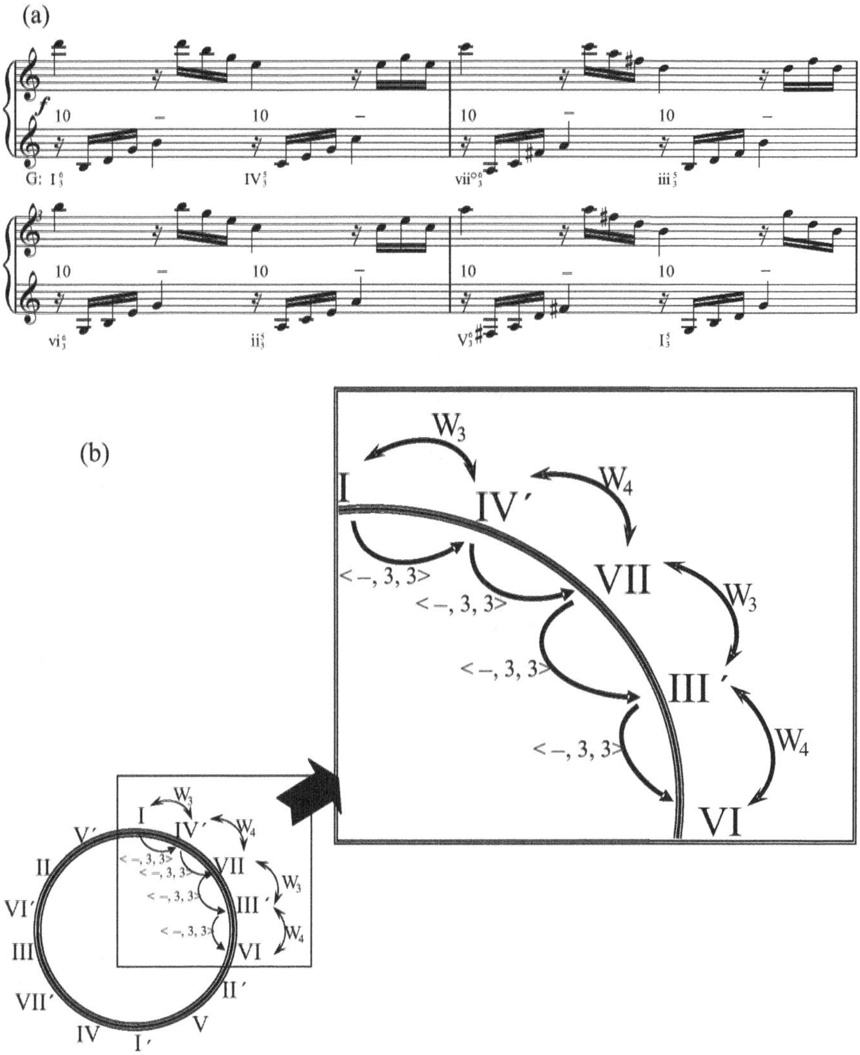

Figure 2.16. First Movement of Mozart's Piano Sonata in C Major, K. 545.

circles of fifths, running in either direction, as shown in the upper half of table 2.3, where the seven pitch classes are numbered 0 through 6. Since 7 is prime, each of these circles contains all 7 diatonic pitch classes (this must be one of the more underappreciated facts of music theory). Recall that a UFFC has two component circles based on the same transposition operator. Pairs of the three

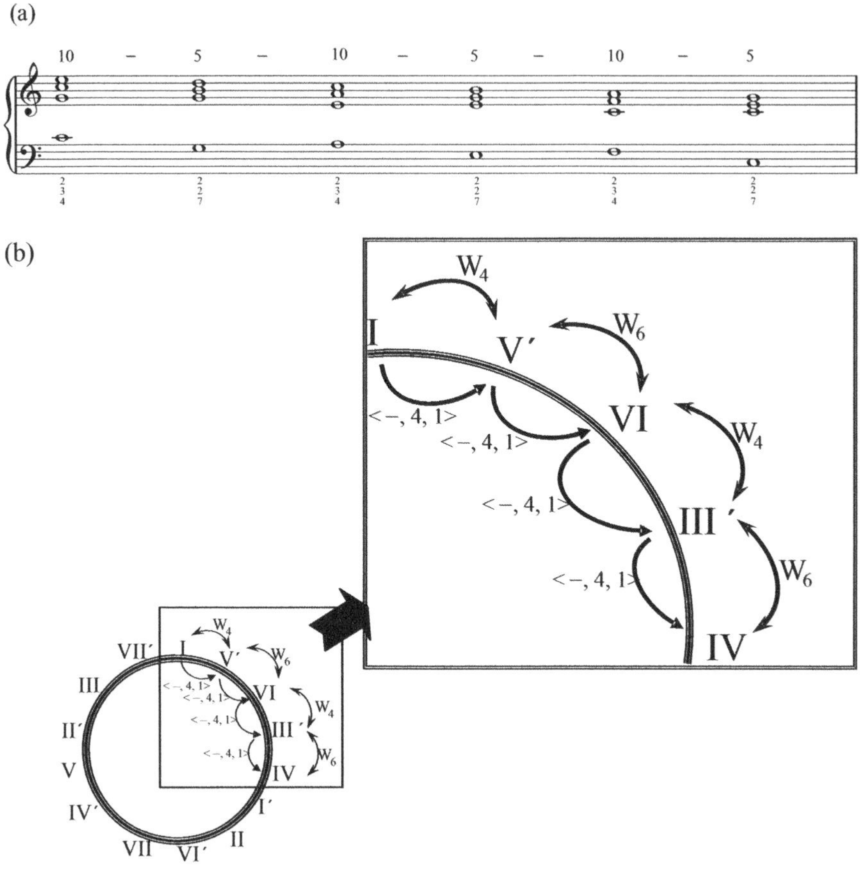

Figure 2.17. Harmonic Progression Based on Pachelbel's Canon.

diatonic circles provide the component circles for any diatonic UFFC, as shown in the lower half of table 2.3. For each of the three pairs of component circles, we have seven possible rotations of the two component circles, yielding in all 21 diatonic UFFCs, each consisting of 14 elements—the seven diatonic pcs and their "double." Perhaps we should call them 21 "typeless" UFFCs, in the absence of any specified dichotomy of types.

In replacing the Roman numerals of figures 2.16b and 17b with the Arabic numerals of table 2.3, we make a transition from numerals that symbolize scale steps of a fundamental bass to numerals that may symbolize only themselves in a melodic line, as in the ensuing analysis of Brahms, or that may symbolize arbitrary arrays of diatonic pcsets, as for example the 14 specific members of the

Table 2.3 UFFCs in 7-space

component circles														
$t = 1$ (2nds/7ths)	(0	1	2	3	4	5	6)							
$t = 2$ (3rds/6ths)	(0	2	4	6	1	3	5)							
$t = 3$ (4ths/5ths)	(0	3	6	2	5	1	4)							
sample UFFCs														
$t = 1$ (2nds/7ths)	(0	4'	1	5'	2	6'	3	0'	4	1'	5	2	6	3')
$t = 2$ (3rds/6ths)	(0	6'	2	1'	4	3'	6	5'	1	0'	3	2'	5	4')
$t = 3$ (4ths/5ths)	(0	5'	3	1'	6	4'	2	0'	5	3'	1	6'	4	2')

diatonic sc of prime form (0, 1, 3) in C major: {{C, D, F}, {C, E, F}, {D, E, G}, {D, F, G}, . . ., {B, C, E}, {B, D, E}}.[19]

Continuing in parallel to our exploration of UFFCs in 12-space, it is natural at this point to look at the group underlying circles such as those of figures 2.16 and 2.17, along the lines of the S/W group in 12-space. The singular S/W-type group supporting all such circles—I will call it S/W(7)—is isomorphic to D_7, the group of congruence motions of the regular 7-sided polygon, or the dihedral group of order 14.[20] Taking W to be the involution that maps scale step 0 to 0', or vice versa, scale step 1 to 1', or vice versa, etc., and loosely using Riemann's nomenclature, we can label the 14 elements of this group S_0, S_1, . . . S_6, and W_0, W_1, . . ., W_6, analogously to the S/W group in 12-space. Figure 2.18a shows the action of S/W(7), a simply transitive action, on the bi-lateral set of scale steps, symbolized here with Arabic numerals 0–6, instead of Roman numerals. Basic structural information on S/W(7) is contained in table 2.4.

Looking back now at figures 2.16b and 17b, we see that the diatonic UFFCs there are generated by patterns—W_3 with W_4 in figure 2.16b, and W_4 with W_6 in figure 2.17b—patterns that nicely capture the sense of alternating moves, a feature not captured, in the case of figure 2.16, by the description "circle of fifths." As shown inside the circles of figures 2.16b and 17b, these UFFCs may be generated, respectively, by the Hook-type transformations $< -, 3, 3 >$ and $< -, 4, 1 >$ in mod 7, revealing once again the tension between readings based on pairs of alternating *Wechsel* and those based on a single transformation.[21]

Consider the opening of Brahms' Fourth Symphony, given at the top of figure 2.19. The melody traces a circle of thirds in alternate time values, short–long–short–long. Figure 2.19a represents the melody as diatonic pitch classes 0–6, with the initial note mapped to 5 to match scale-step numbers (except for scale step 7 which maps to 0). I emphasize that the bilateral aspect of the transformational network is based on binary rhythmic states, not on a dichotomy of chord types.[22] The two component circles, or half-circles, of the UFFC are given as the upper and lower rows of boxed numerals, respectively. With reference to S/W(7), W_5 and W_2 alternate on the surface; W_1 captures the move from the initial 5 (pitch

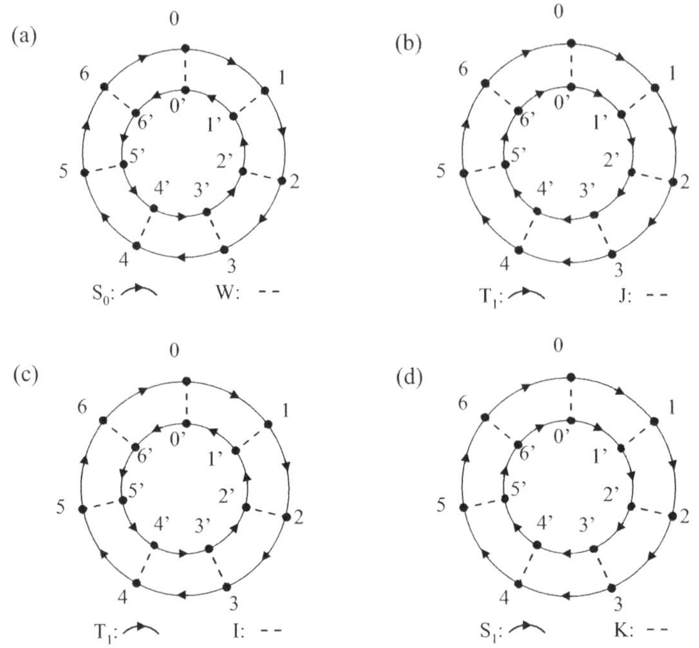

Figure 2.18. Groups of (un)usual Transpositions with (un)usual Inversions in 7-Space.

Table 2.4 Groups of (un)usual transpositions with (un)usual inversions in 7-space

Group	structure	transposition	inversion: labels*	
S/W(7)	D_7	Schritte	W:	$W_n = WS_n$
T/J(7)	$\mathbf{Z}_2 \times \mathbf{Z}_7 \cong \mathbf{Z}_{14}$	usual	J (=W):	$J_n = T_nJ = JT_n$
T/I(7)	D_7	usual	I:	$I_n = T_nI$
S/K(7)	$\mathbf{Z}_2 \times \mathbf{Z}_7 \cong \mathbf{Z}_{14}$	Schritte	K (=I):	$K_n = S_nK = KS_n$

* All inversion labels use right-to-left notation (e.g., T_nI means first invert, then transpose).

B_5) to the registrally connected 6' (C_6), likewise the move from 4 to 5' in the consequent subphrase, completing the double-neighbor motion; W_0 represents the regaining of the initial 5; S_3 represents the relationship between adjacent short values in the component circles, and S_4 the relationship between adjacent long values; and, finally, S_1 is the *Schritt* that takes a long note in the first half to a corresponding long note in the second half of the passage, while S_6 functions similarly for corresponding short notes from the first to the second half.

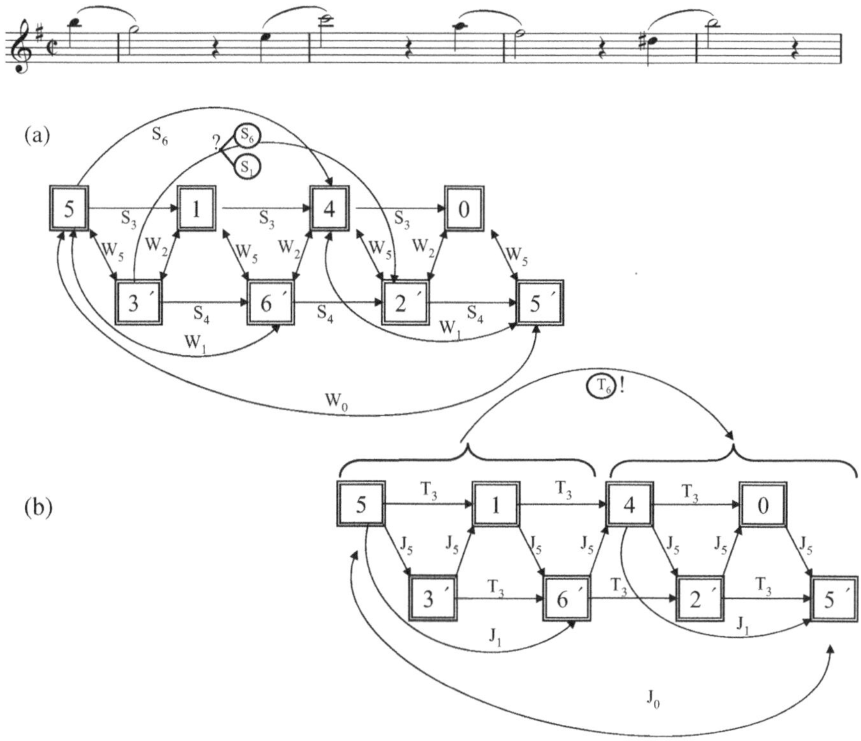

Figure 2.19. Melody from the opening of Brahms's Fourth Symphony.

All is well mathematically, but some things are amiss analytically. To begin with, the alternation of moves on the surface, while consistent with our analysis of circles thus far, is questionable; the surface is, after all, a circle of constantly descending thirds in 7-space. But a more serious problem with the analysis is this: the second half of the melody is obviously a transposition of the first half, down a step, within the harmonic minor scale, yet two different transformations from our group of 14 are required to express this transposition—S_1 for long notes and S_6 for short notes, as shown with a question mark at the top of figure 2.19a.

Fortunately, a remedy is available. Figure 2.18b shows the action of a different group on the system in question, and table 2.3 again provides technical details. $T_0, T_1, \ldots, T_6$ are the usual transpositions of the scale steps while $J_0, J_1, \ldots, J_6$ are composites of the transpositions with W, here relabeled J. They are exchange transformations that reverse the binary state of the scale-step argument and increase its numerical value by 0–6 (mod 7), respectively. Since J (= W) commutes with any T-transformation, we can think of a J-transformation as a reversal

(c)

(d)

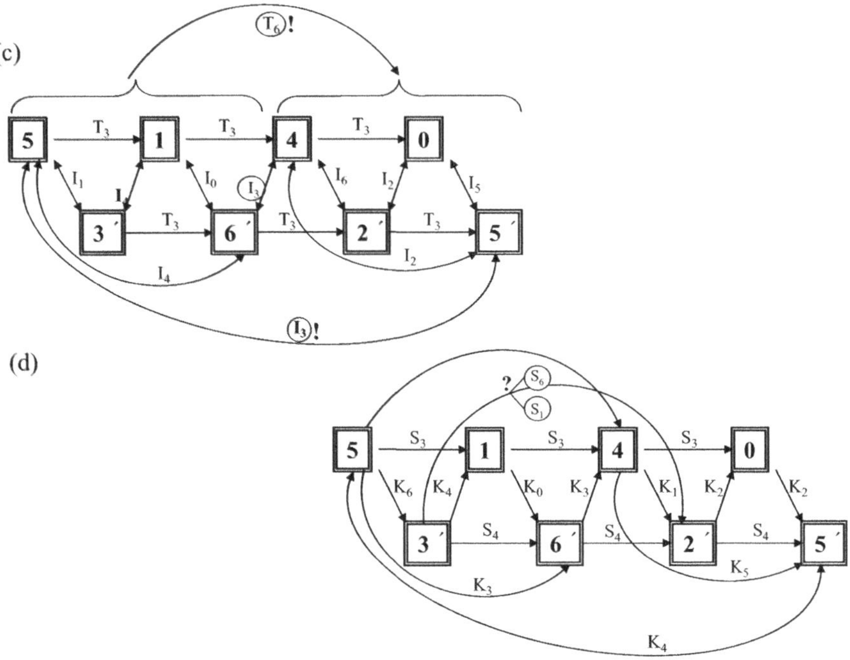

Figure 2.19. (*continued*).

of state and a transposition (in either order). Indeed the 14 T- and J-transformations all commute with one another; they form a commutative group, which I call T/J(7).

I have used the symbol "J" in honor of David Lewin who, I believe, first showed the musical relevance of such a commutative group.[23] Though I adopt Lewin's strategy in my use of the J-transform, there is an important structural difference between his group of 24 transformations and T/J(7). The former lacks an element that generates the entire group of 24, while the latter has several—in fact any of the J-transformations save J_0 will do so.[24]

In figure 2.19b, the transformational analysis based on S/W(7) is replaced by analysis based on T/J(7). The surface is now a continual application of J_5, and T_6 applies from any note in the first subphrase to its correspondent in the second subphrase or, in other words, takes the whole first subphrase to the second. As a less significant gain, the previously disparate moves to adjacent short and long notes, S_3 and S_4, are replaced by the consistent T_3. At the same time, the matching W_1's of figure 2.19a are replaced by matching J_1's, and W_0 is replaced by J_0. Thus, the two problems noted above disappear, while nothing is lost.

A reworking of the analyses of figures 2.16b and 2.17b with T/J(7) would produce comparable correctives. In figure 2.16b, the transformation J_3 would apply throughout from one chord to the next, replacing the alternating W_3's and W_4's, while T_6 would reflect the plainly audible downward step motion from one pair of chords to the next (the scheme of figure 2.16b would call for W_6 between adjacent 6-3 chords and W_1 between adjacent 5-3 chords). In figure 2.17b, the situation is a little different. From the standpoint of T/J(7), successive pairs of chords are related consistently by T_5, which accords with intuition, but there is no single J-transform that transports from one chord to the next throughout, for the reason that, on the surface, there is no circle of steps, thirds, or fifths, as in figures 2.16 and 2.19.[25]

Earlier, we touched briefly on the usual T/I group and its application to UFFCs, particularly the fact that, except where $t = 6$, more than two inversions are required for circumnavigation. Might there be some analytical advantage to such a multiplicity of inversions on the surface? To test this idea, we construct the group T/I(7), based on the usual notions of transposition and inversion, but in 7-space. Figure 2.18c shows the action of this group on the binary system of scale steps, and figure 2.19c shows the analysis of the Brahms melody with T/I(7). Compared to figure 2.19b, the reading of figure 2.19c preserves the advantages of usual transposition but gives up the constant inversional transformation on the surface, substituting for it the cyclic sequence of inversions I_1, I_4, I_0, etc. If the subscripts for these transformations are converted to centers of inversion, they produce the sequence (4, .5), (2, 5.5), (0, 3.5), (5, 1.5), (3, 6.5), (1, 4.5), (6, 1.5).[26] This sequence reflects the constant descent by thirds in the first members of each pair: 4, 2, 0, 5, 3, 1, 6. Also, the descent by step from the antecedent to consequent subphrases is evident in the initial and final triples of the foregoing succession (4, 2, 0), (3, 1, 6). A further advantage to the reading with T/I(7) is that the two I_3's (circled on figure 2.19c) expose the double-neighbor motion around the initial and final pc 5.

We have considered groups in 7-space entailing usual transpositions and inversions, and unusual transpositions and inversions, *Schritte* and *Wechsel* (reconceived in 7-space). We have combined the usual transpositions with both kinds of inversions to form T/I(7) and T/J(7), but we have combined the *Schritte* only with *Wechsel*. The remaining combination of transpositions and inversions, combining *Schritte* with a usual inversion, produces the group labeled here S/K(7), whose action on the dual scale steps is given in figure 2.18d, and whose application to the Brahms melody is given in figure 2.19d[27] (technical details for T/I(7) and S/K(7) are given in table 2.4). If the reading based on S/K(7) contributes anything new of interest, it would seem to lie with the sequence of inversions on the surface, K_6, K_4, K_0, etc. As in the case of figure 2.19c, alternate members of the sequence may be interpreted as representing the stepwise transposition between subphrases, but this time as a series of one-way transformations, conveying, perhaps, a sense of directedness not evident in figure 2.19c.

We have used four groups in the above analysis: S/W(7), T/J(7), T/I(7), and S/K(7). In order to gain insight into the relationships among these groups, it may be helpful to compare smaller versions of groups of the same types. Figures 2.1, 2.17, and 2.19 nicely illustrate a general point about UFFCs: it is not only the alternation of states on the surface that is amenable to study by means of transformational group theory; various musical factors tend to expose other relationships among elements in the circle—as we have seen in figure 2.19, with pitch proximity, rhythmically reinforced adjacencies in the component circles, and pitch transposition.

As the four groups constructed above are used with a single melody, they expose different features, in some cases highlighting successive repetitions of single transformations, in other cases repetitions of pairs of transformations, and in others cycles of seven transformations. I would subscribe to the view that the various groups bring to light diverse musical perspectives, and we can appreciate, even savor, the tensions among these, feeling no need to select any particular approach as absolutely preferred.

I hope that these explorations of circles in a neo-Riemannian setting and in the context of diatonic sequences suggest that the uniform alternation of states in various musical spaces is a significant, perhaps elemental, phenomenon deserving of sharper theoretical focus and broader analytical application.

Notes

I am pleased to acknowledge the capable help of my research assistant, So-Yung Ahn, who prepared graphics for many of the examples, interpreted an earlier version of this paper when it was presented in Korea, November 1999, translated that version for publication in Korean, and responded insightfully as ideas for the paper developed. Additional information regarding this topic can be found in Ahn 2003.

I am grateful as well to David Clampitt, Jack Douthett, Julian Hook, and Jonathan Kochavi, for helpful advice on the mathematics of this paper, and for thoughtful comments in general. Discussions with Kochavi regarding diatonic sequences, addressed from a different perspective in his ongoing work, were especially valuable.

1. Figure 2.1 is a reproduction of Cohn 1996 with notations adjusted for the present purpose.
2. Riemann 1880.
3. Cohn 1997.
4. Lester 1992.
5. Klumpenhouwer 1994.
6. This notation is essentially that proposed by Gollin 1998, 203–4; however, in the spirit of Riemann, Gollin defines W_0 to be the inversion that maps the root of a major triad to the fifth of the corresponding minor triad. Thus Gollin's W_n equals my W_{n-7}; as pointed out by Gollin, any two notations comparable in this way must differ by some constant.
7. The data for table 2.2 were initially compiled by Ahn 2003.
8. For any given value of *t*, all corresponding subgroups not only have the same order, they are also isomorphic to one another. This follows immediately from the

fact that for any positive integer *n*, there is a unique abstract dihedral group D_n. Sternberg 1994, 29 refers to D_1 and D_2 as *degenerate dihedral groups*: D_1 is isomorphic to C_2 (the cyclic group of order 2) and D_2 is isomorphic to the Klein four-group—the smallest non-cyclic group, familiar to students of atonal set theory as the group consisting of the four generalized operations on a 12-tone row, that is {T, I, R, RI}. The 28 subgroups counted in the second column of table 2.1 include the 12 degenerate subgroups corresponding to $t = 0$ (isomorphic to D_1, containing a single *Wechsel* in addition to the identity) and the 6 degenerate subgroups corresponding to $t = 6$ (isomorphic to D_2, containing a pair of *Wechsel* along with the S_6 and the identity).

There is a slight discrepancy in the usage of "degenerate" as applied above to D_1 and D_2 and my usage of the same term as applied to UFFCs supported by subgroups of S/W that are isomorphic to D_1 and D_2. It seems reasonable to regard a UFFC with just two elements as degenerate, but UFFCs with four elements seem legitimately circular and are therefore included in the main body of table 2.1 as non-degenerate.

9. David Lewin 1987.

10. The matter of different analytical perspectives gained from different group-theoretic approaches is treated in detail by Clampitt 1997 and 1998, on whose work the present discussion relies heavily. A pertinent group-theoretic concept here, called *anti-isomorphism* by David Lewin, is discussed in Lewin 1987, Clampitt 1997, and Clough 1998.

11. Kochavi 1997.

12. Hook 1999.

13. Lewin 1982.

14. Cohn 1996, Clampitt 1997 and 1998, Hook 2002, Hyer 1995, and Kochavi 1998.

15. Cohn 1997.

16. Callendar 1998, Douthett and Steinbach 1998, and Gollin 1998.

17. Hook 2002.

18. Chapman 1981.

19. The appendix gives a formal definition of UFFC, recognizing the generalization outlined here. While the definition is not applied directly in this paper, it may provide insight to the analysis contained herein, and may serve as a basis for further work.

20. As noted above, because 7 is prime, there is a paucity of component circles that form UFFCs. For the same reason, the group D_7 is impoverished in its progeny of subgroups; its only non-trivial ones are a single subgroup isomorphic to $\mathbf{Z}_7$ and seven subgroups of order 2, each containing a single *"Wechsel"* along with the identity element. These seven subgroups support degenerate circles of two elements, as do the comparable groups in 12-space. Though they spawn degenerate circles, I do not mean to say that these binary subgroups are without musical relevance. For example, the crashing alternations of tonic and dominant, each with its own voicing, that conclude countless Classical movements, are modeled by such a subgroup.

21. The diagrams of figure 2.18 are "structure diagrams" akin to those of Kochavi 1998, with sets of points laid out concentric circles instead of parallel rows.

22. Kochavi, in unpublished work, proposes for such contexts the more general binary opposition: first position/second position. David Clampitt, in private correspondence, suggested that the binary opposition of metric locations, in juxtaposition with pitch-class space, might be advantageously treated with Lewin's cross-product GIS.

23. Lewin 1993.

24. In mathematical parlance, Lewin's group has the structure $\mathbf{Z}_2 \times \mathbf{Z}_{12}$, while T/J(7), the cyclic group of 14 transformations constructed here, has the structure $\mathbf{Z}_2 \times \mathbf{Z}_7 \cong \mathbf{Z}_{14}$. Mathematical aspects of Lewin's group and other comparable groups are treated in Clough 1998 and Kochavi 1998.

25. Hook's < –, 4, 1 > (or its inverse < –, 6, 3 > serves to navigate the surface of figure 2.17b. In mod 7, the group induced by < –, 4, 1 > is isomorphic to $\mathbf{Z}_{14}$, hence to T/J(7). However, the two groups act differently on the object-set of 14 elements.

26. In this sequence the symbol (a, b) indicates a dual center of inversion, comparable to those in the 12-pc universe. In mod 12 the two members of a dual center are always separated by $6 = 12/2$. Here, in mod 7, $a - b = b - a = 3.5 = 7/2$.

27. See Clough 1998 for further information, in particular the discussion of Lewin's 1987 "anti-isomorphism."

Appendix 2.1

Definition. Given a set $S = \{a_0, a_1, a_2, \ldots, a_n, b_0, b_1, b_2, \ldots, b_n\}$, a *uniform flip-flop circle* (UFFC) in S is a closed chain of the form

$(a_i, b_j, a_{i+t}, b_{j+t}, a_{i+2t}, b_{j+2t}, \ldots, a_{i-t}, b_{j-t})$, where i, j, t are integers,

$0 \leq i, j, t \leq n - 1$, and subscripts are reduced mod n.

$(a_i, a_{i+t}, a_{i+2t}, \ldots, a_{i-t})$ and $(b_j, b_{j+t}, b_{j+2t}, \ldots, b_{j-t})$ are the *component circles* of the UFFC.

t is the *transposition index.*

$j - i$ is the *offset.*

It is convenient to study UFFCs in terms of their subscripts. Thus a dualistic chain of subscripts of the form $(i, j', i + t, (j + t)', i + 2t, (j + 2t)', \ldots, i - t, (j - t)')$ serves to represent a UFFC.

Chapter Three

Pitch-Time Analogies and Transformations in Bartók's Sonata for Two Pianos and Percussion

Richard Cohn

This paper describes an unusually strong relationship between pitch and rhythm in the first movement of Béla Bartók's Sonata for Two Pianos and Percussion, composed in 1937. The movement contains four distinct themes, three of which are used in dialogue with the classical sonata tradition. The fourth theme is a nine-note motto from the movement's *Lento* opening which reappears as an up-tempo ostinato in the development section. Following in the Beethoven tradition,[1] the four themes are strongly individuated in both their tonal characteristics and their rhythmic profiles. Were we to represent the themes on a harmonic map, and, independently, on a rhythmic map, the two maps would be similar enough to be viewed as realizations of a single underlying design.

The idea of intimate pitch-time affinities in Bartók's music is likely to strike many readers as improbable. Such affinities are characteristic of a self-conscious, precompositional approach that flowered only after Bartók's death, an approach associated with the mechanical application of a Platonist imagination to music, rather than a musical imagination per se.[2] This approach is difficult to reconcile with what we know of Bartók's attitudes and methods. His claims that "My entire music . . . is determined by instinct and sensibility," and "I have never created new theories in advance, I have always hated such ideas" are well supported by anecdotal evidence, and by the lack of significant documentary evidence to the contrary.[3] Yet Bartók scholars have long intuited correspondences between his treatment of the two domains.[4] Ernö Lendvai suggests that Fibonacci ratios governed Bartók's harmonic as well as durational structures,[5] and Janos Kárpáti finds mistuned structures in rhythm as well as pitch. Elliott Antokoletz has noted that expansion of both pitch and durational intervals is a feature of Bartók's developmental technique, and suggests that "organic" pitch transformations in the Sonata for Two Pianos and Percussion have counterparts in the realm of durations. And, in a recent study of rhythmic conflicts in the Sonata's first movement, Daphne Leong makes several specific suggestions concerning where such parallels might be sought.[6]

Any attempt to explore such intuitions formally must confront the partial incommensurability of the spaces occupied by pitch and time. In the tonal/metric "common practice," both pitch and duration are organized into cycles (octaves, measures), oriented to particular points (tonics, downbeats), with the remaining elements (diatonic or chromatic pitch classes, time points or beat classes) dispersed equally around the cycle. The problem arises in the size of the cycles in the respective domains. Pitch cycles are fixed at 7 (for diatonic music) or 12 (for chromatic). Metric cycles vary in size, ranging, at a minimum, across the smaller products of the powers of two and three. When the metric cycle has 12 elements, it is possible to create a one-to-one relationship between the two domains.[7] In all other cases, any attempt to map pitch and time structures onto each other must be captured instead via fuzzy correspondences, fluid analogies, and many-to-one mappings: transformations rather than operations.[8]

3.1

I begin by reviewing a familiar pitch-class distinction, and adapting it to the metric domain. I rely on a conceptual framework advanced by Jeff Pressing,[9] although some of the terminology developed here is new. The six classes of pitch intervals partition into two types on the basis of their generative behavior. "Generation" in this context refers to the recursive application of an interval; this is the sense in which the perfect fourth generates the diatonic collection. *Open* interval classes (1, 5) are prime with respect to the group order, and hence generate the entire pitch-class universe. A brake on the generative process can be applied approximately halfway through the cycle, producing significant collections of intermediate cardinality, including the maximally even pentatonic and diatonic sets. The process by which each of those collections is generated is open, in the sense that its cyclic journey is left incomplete. To the extent that such collections are conceived as cyclically closed, it is by substituting a perturbed version of the generating interval (e.g., the diminished fifth in the case of the diatonic collection) as a "short cut" back to the point of origin.[10] *Closed* interval classes (2, 3, 4, and 6) are divisors of 12. They fulfill their cyclic generative potential without generating the entire pitch-class universe, whose remaining elements are generated by distinct co-cycles. With the exception of the ic2-generated whole-tone scale, cycles generated by closed interval classes are too small to be of collectional interest. Rather, interesting collections are produced by combining co-cycles: any two ic3 cycles form an octatonic collection, and two adjacent ic4 cycles form a hexatonic collection.

The open/closed duality also applies to generated rhythmic ("beat-class") patterns in a metric environment. Given a metric universe of c beat classes, there are $c/2$ distinct durational interval classes where c is even, $(c-1)/2$ where c is odd. Each durational interval may serve as a generator g. An *open* durational interval g is co-prime to c. Following the practice introduced into music theory by John Clough, we indicate this with the functional expression gcd $(c, g) = 1$, indicating that 1 is the greatest common divisor of c and g. An *open* interval g generates the beat-class aggregate if left unchecked. Since that aggregate lacks an individual rhythmic personality, a brake is applied part-way, so that maximally even collections of intermediate cardinality emerge. For example, if $c = 8$ and $g = 3$ (unit = ♪) and a brake is applied after two ♩. spans are generated, then the ♩. ♩. ♩ pattern results. The final ♩ plays a role analogous to the diminished fifth in the diatonic collection: it "artificially" and prematurely completes a cyclic process. A *closed* durational interval g shares a common divisor with c (that is, gcd $(c, g) > 1$), and thus generates a complete cycle without completing the beat-class aggregate. If $c = 8$ and $g = 4$, then the 𝅗𝅥 𝅗𝅥 pattern results. Since this cycle may not be sufficiently distinct to hold compositional interest, it may be combined with an adjacent co-cycle, resulting in, for example, an iambic ♩. ♪♩. ♪pattern.

Daphne Leong has studied the relationship between closed and open durational intervals in the first movement of Bartók's Sonata for Two Pianos and Percussion, which has a time signature of 9/8. If $c = 9$, then 2 is an open interval and 3 is closed. Leong demonstrates that, in the course of the movement, rhythmic patterns generated by open intervals become displaced by, or progress to, those generated by closed patterns.[11] These progressions occur on multiple structural levels but undergo a reversal in the final measures of the coda. Her study demonstrates, among other things, the malleability of the sonata-form framework. Historically used to show contrast, progression, transformation, and modulation in the tonal and thematic domain, it can show similar processes in the metric domain as well.

Figure 3.1 presents the movement's principal generated beat-class sets, together with their variants. The 2-generated, open rhythmic mode comes in two complementary forms, shown at figure 3.1a. Bartók frequently combines them into a two-voice hocket that completes the beat-class aggregate. Adapting from Leong, E and D refer respectively to the five- and four-note forms, and P to their combination (as a mnemonic aid, note that E and D are the fifth and fourth letters of the alphabet). Each pattern has a subdivided variant, shown at figure 3.1b. Such variants are characteristic of Balkan dance patterns, where a step is replaced by a "hop-lift" combination.[12] The 3-generated, closed rhythmic mode is presented in two complementary forms, shown at figure 3.1c: a three-attack series and its iambic six-attack complement. Here the complementary relationship is abstract, as Bartók does not make significant use of their hocketing potential. The four generated patterns presented in figure 3.1a and 3.1c are united by the property of maximal evenness; indeed, together they constitute the mod-9 ME sets of intermediate cardinality.[13]

Figure 3.1a. Open (2-generated) maximally even sets, mod 9.

Figure 3.1b. Variant versions of the two open patterns.

Figure 3.1c. Closed (3-generated) maximally even sets, mod 9.

The duality between the open and closed modes has a potential for extra-musical interpretation. The closed mode is inherent in the conception of 9/8 meter as practiced in Western Europe, the region in whose music Bartók was trained, and where he sought his reputation. The open mode is characteristic of 9/8 meter as it is practiced by musicians and dancers in Eastern Europe and Northern Africa, the regions where Bartók, through his extensive fieldwork, sought his inspiration. (For this reason, I shall sometimes substitute *Western* and *Balkan*, respectively, for *closed* and *open*.)[14] The open/closed binary on maximally even rhythmic sets constitutes a "violent hierarchy," in the Derridean sense; one of the terms dominates the other. The relative values of the hierarchical components, however, depend on which direction one is travelling along the Danube. Upstream from Vienna, the closed mode is the natural division of 9/8, while the open mode is exotic, primitive, and difficult. Downstream from Bratislava, the open mode is natural, the closed mode stodgy and imperial. The Sonata's first movement can be read as a map of Bartók's music-cultural psyche, or as an allegory of the cultural position of Hungary in the first years of the twentieth century.[15]

Figure 3.2. Measures 420–32. © COPYRIGHT 1942 BY HAWKES & SON (LONDON) LTD. Reproduced by kind permission of Boosey & Hawkes Music Publishers Ltd. Continued on next page.

3.2

Figure 3.2 is a passage from near the end of the coda. A study of this passage will introduce us to the interaction between the two rhythmic modes, and their potential to parallel events in the domain of pitch. Leong observes a shift from a closed to open mode of organization through the last 24 measures of the movement. [16] Our concern is with the mechanism of that shift. As we pick up the music at measure 420, a ♩. tactus strongly articulates a symmetric 3 × 3 partitioning of each $\frac{9}{8}$ measure. The tactus is anacrustically anticipated,

Figure 3.2. (*continued*).

creating an iambic composite rhythm. This iambic rhythm was initiated in measure 326, when the third theme of the recapitulation was first prepared, and has continued through a fugato (starting in measure 332), a dissolution into a contrapuntal wash (m. 360), the re-gathering of fugal fragments over an ostinato pattern (m. 383), an apotheosis of fugal fragments in parallel sixths (m. 405) leading to an impasse (m. 413), and a second re-gathering that leads to the passage at hand. Throughout these 94 measures, there has occurred not a single attack on beats 2, 5, or 8, i.e. on the interior time-point of each triple group.[17]

In this context, the attacks that occur on the fifth beat of measure 422 (circled in figure 3.2) are marked for attention. They sound where Bartók has conditioned us to expect silence, and trigger the rhythmic course of the final measures. At first only Piano I responds. While Piano II continues to sail with the iambic winds, Piano I articulates a series of marcato E♭s that reinforce the dotted-quarter pulse but withhold the iambic anacrustic attacks. Beginning at measure 424, the E♭s spawn a fragmentary tune, which is echoed at a tritone transposition. Beginning at measure 428, the two transpositions overlap canonically. In each of its entries, the tune runs aground on an after-beat, an interior second beat of a group of three (each of which is circled in figure 3.2).

The accumulation of these after-beats, in response to the D natural at measure 422, ultimately knocks the wind out of the iambic sails. The moment of final exhaustion occurs when Piano II curbs its iambic inertia on the eighth beat of measure 432, sounding an {A♭, B} after-beat. In the final measures (not shown in the figure), the open Balkan rhythms seize the rudder, and guide the movement quickly into its final port. Measure 433 initiates a series of melodic fragments that melodically expand and durationally compress the tune that was introduced in 424. Most of these fragments articulate rhythmic cell <u>D′</u> ; several of them, including the final one, articulate cell <u>E</u> instead. The ♩. pulse is partially restored, however, with the entrance of the timpani at 437. At the final cadence, the timpani presents an iambic rhythm, while the Balkan rhythms persist in the piano; the music ends with its two rhythmic modes equally represented.

The rhythmic modulation coincides with one in the pitch domain. This is a modulation not between tonal centers, in the classical sense, but rather between modes of pitch organization, what Fred Lerdahl has referred to as a space-shifting hypermodulation.[18] The music from measure 417 is octatonic. As I have shown elsewhere, its appearance has rhetorical power as the realization of a long chain of implications.[19] The octatonic collection begins to break apart at the approach to the *poco allargando* at measure 432, and is displaced at Tempo I by chromatic fragments that mark out a fifth-based, quasi-diatonic orientation (the *martellato* chords at measures 434 and 436 are octatonicism's "last gasp" in this movement).

The pitch modulation is initiated by the same event that triggered the rhythmic modulation. The D natural on the fifth beat of measure 422 is foreign to the octatonic collection that has been accumulating since measure 413 and sounding since 417.[20] The prevailing octatonic collection continues to sound in Piano II through measure 429 (with the exception of two transient {F/A♭} dyads), and is reinforced by the series of tune-segments in Piano I. With a single exception, each after-beat note that terminates a tune-segment is from outside the collection.[21] The cumulative impact of these non-octatonic pitches transforms the collection back to the fifth-based quasi-diatonicism that began at measure 433 and concludes the movement, just as the after-beats that contain these pitches provoke the Balkan rhythms that are restored at the same moment.

In both the pitch and rhythmic domains, a pattern is established: a single outside event intrudes; the intrusion is echoed, the echoes become increasingly pervasive, and eventually the pattern is dissolved and replaced by a different organizational mode. As I shall now argue, the analogy is not limited to these broad outlines; it extends to the detailed physiognomy of the patterns. The octatonic collection is closed in the same sense as the Western rhythmic mode, and the diatonic collection is open in the same sense as the ♩-generated Balkan rhythms.

To render this argument precisely, and to optimize its potential for generalization, it will be useful to adopt a more formal approach. which will help compare mod-12 pitch-class space with the less familiar mod-9 beat-class space. The rhythmic pattern at the beginning of figure 3.2 is characterized by a mod-9 beat-class set, IAMB = {0, 2, 3, 5, 6, 8}. Here are three distinct but co-extensive characterizations of IAMB:

1) **Step-Interval Structure.** IAMB's step-interval series is [2,1,2,1,2,1], reflecting the alternation of values consisting of the unit (♪) and its double (♩).
2) **Cyclic Union.** IAMB unites two cycles generated by beat-interval 3: the attacks that articulate the ♩. pulse {0, 3, 6}, and their anacrustic preparations {2, 5, 8}.
3) **Complement.** IAMB's complement, {1, 4, 7}, is a complete interval-3 generated cycle.

Each of these characterizations applies to the mod-12 pitch-class set that dominates from measure 417 through measure 432, OCT = {0, 1, 3, 4, 6, 7, 9, 10}:

1) **Scale.** OCT's step-interval series is [2,1,2,1,2,1,2,1], reflecting the alternation of values consisting of the unit (semitone) and its double (whole step).
2) **Cyclic union.** OCT unites two adjacent cycles generated by pitch-interval 3: {0, 3, 6, 9} and {1, 4, 7, 10}
3) **Complement.** OCT's complement, {2, 5, 8, 11}, is a complete interval-3 generated cycle.

These identical characterizations suggest an intimate relationship between the rhythmic and pitch materials of the music of figure 3.2. Yet the distinctive properties of 9- and 12-element cyclic space preclude any possibility that this relationship is one of isomorphism. Not every structural property of IAMB, mod 9, is also a property of OCT, mod 12. Consider the following, where, as before, *c* represents the size of the cycle:

1) **Step-Interval Structure.** IAMB's step-interval series is $\left[\frac{c}{3}-1, 1, \frac{c}{3}-1, 1, \frac{c}{3}-1, 1\right]$ reflecting the alternation of the unit (♪) with the value that is one unit smaller than the cycle-trisector (♩ + ♪ = ♩.)

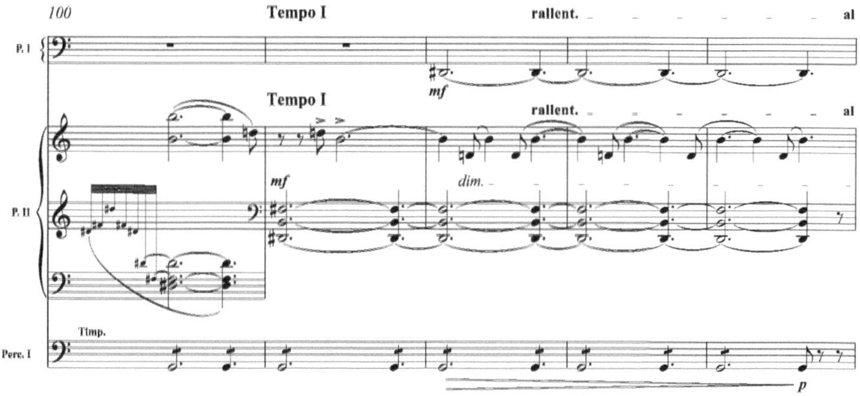

Figure 3.3a. Measures 100–104.

Figure 3.3b. Measures 123–27.

2) **Cyclic union.** IAMB unites two adjacent cycles generated by interval $\frac{c}{3}$.

When applied to a pitch-class universe mod 12, these properties apply not to the octatonic collection, but to the hexatonic collection-class, HEX = prime form {0, 1, 4, 5, 8, 9}. HEX's step-interval structure is [3, 1, 3, 1, 3, 1], where 3, the minor third, is one unit smaller than the cycle-trisecting major third. HEX also unites the major-third-generated cycles {0, 4, 8} and {1, 5, 9}.

Figure 3.3c. Measures 381–85. © COPYRIGHT 1942 BY HAWKES & SON (LONDON) LTD. Reproduced by kind permission of Boosey & Hawkes Music Publishers Ltd.

The iambic rhythm is thus allied equally with hexatonic and octatonic collections. We can profit from this observation by studying the iambic rhythms as they first emerge at measure 100, when the third-theme area begins to coalesce (figure 3.3a). The {B, D, D♯, F♯} tetrachord in Piano II is neutral with respect to the hexatonic and octatonic collections, representing the largest subset-class common to both. But the G in the timpani implies a hexatonic collection, which requires only Bb for its completion. This completion occurs at measure 123, where a hexatonic maelstrom aborts an incipient fugato (figure 3.3b). A new hexatonic collection, at a different transposition, is presented at the opening of the second, more fully realized fugato, mm. 133–36. At the analogous passage in the recapitulation, the same transposition of the hexatonic collection recurs as an ostinato pattern, in alternation with its hexatonic complement (mm. 381 et seq., figure 3.3c).

Thus Bartók aligns the iambic rhythm with both hexatonic and octatonic pitch collections in the movement, but moves from the former to the latter. This transformation reflects a broader shift in the movement from a dyad-class 4 to a dyad-class 3 orientation.[22] To frame this shift in a way that suits the present analysis: at the beginning, the focus is on the interval that divides the octave into three equal parts; by the end, the focus is on the interval that divides it into four. In a system of twelve elements, these two dyad-classes are in a relationship of inherent tension between "size of part" and "quantity of parts": interval 4 divides the aggregate into three parts; interval 3 divides it into four.[23] When there are nine elements, this tension is dissolved: as a square number, nine is the product of two identical factors. The iambic rhythm, mod 9, can be interpreted as resolving the hexatonic/octatonic opposition, by injecting it into a domain where their features lose their mutual distinctiveness. In the rhythmic domain, the octatonic/hexatonic difference becomes subsumed into a higher-level equivalence.

This has a direct musical consequence. The octatonic and hexatonic collections share with the iambic beat-class set the property of alternating two units, one (L) of which is larger or longer, the other (S) of which is smaller or shorter, where the specific proportion of L to S remains undetermined. A musical prototype for this conception lies close at hand: many of the dance musics of southeastern Europe use cycles that are defined by a pattern of long and short durational units whose proportional relationship is subject to variation, even within a single performance.[24] Jeff Pressing, citing Alice Singer's work on Macedonian dance, proposes "an *analogue* transformation, . . . a rule of correspondence between patterns of unequal [length, which] may be represented by coding the patterns in terms of elements of long (L) and short (S) duration."[25] Hexatonic and octatonic collections can be interpreted as equivalent, then, by translating these rules of correspondence from dance steps to pitch-class steps.

The same relationship can be expressed in terms of a homomorphism **Q** that maps the three smallest pitch step-intervals (expressed below as the first three integers) onto the two smallest durational intervals:

$3 \xrightarrow{Q}$ ♩

$2 \xrightarrow{Q}$ ♩

$1 \xrightarrow{Q}$ ♪

Section 3.4 of this study will modify these rules of inter-domain correspondence, expanding their application to aspects of thematic pitch-time affiliations in Bartók's Sonata.

3·3

If Bartók wished his Sonata to cultivate in the beat-class realm an oppositional tension analogous to the one that it develops in the domain of pitch, then its analogical source could not be located in the relation of hexatonic and octatonic collections, whose distinctness disappears in the mod-9 rhythmic domain. Instead, the 3-generated iambic rhythms that close the exposition and recapitulation find their foil in the 2-generated Balkan rhythmic mode that opens and closes the *Allegro.* This mode emerges in the *Lento assai* introduction, as the music gathers energy for the *Allegro molto.* The $\underline{\text{D}}$ pattern is first explicitly presented in measures 21–23 (figure 3.4a).[26] Measures 26–31 contain an ostinato pattern in three voices, two of which present the $\underline{\text{E}}$ pattern, the third of which presents the $\underline{\text{D}'}$ pattern. At the opening of the *Allegro,* $\underline{\text{D}}$ and $\underline{\text{E}}$ are combined into the $\underline{\text{P}}$ hocket pattern, but only in the melodically active, odd-numbered measures. The pattern does not immediately recur: the piano extends the $\underline{\text{E}}$ pattern to the following downbeat, forming an $\underline{\text{E}'}$ pattern, and then is tacet for the remainder of the measure while the timpani sustains the ♩ pulse, only re-engaging P at the downbeat of the subsequent odd-numbered measure (figure 3.4b).[27]

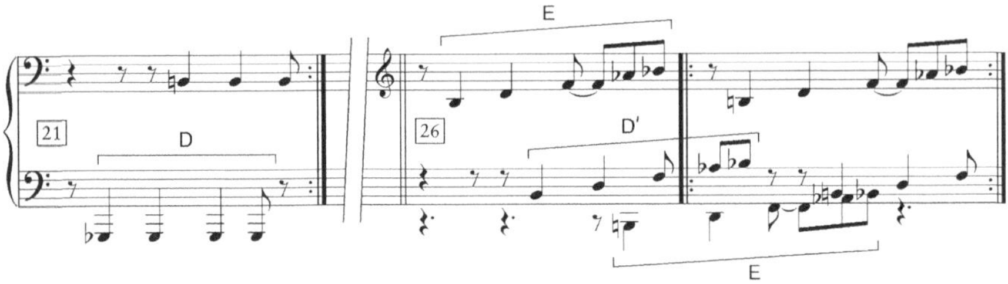

Figure 3.4a. Measures 21–31.

Figure 3.4b. Measures 32–35.

Figure 3.4c. Measure 63.

The first cyclic treatment of the Balkan rhythms occurs in measures 65–68, at the counter-statement of the first theme, where Piano I consistently sounds the E cell (figure 3.4c). A transition, consisting of an imitative sequence, follows immediately. The first four imitative segments (mm. 69–70, figure 3.4d) form the Balkan D′ pattern, [01246]. The following four segments, beginning at measure 71, contain a new hexarhythmic beat-class set that we will refer to as F, of prime form [012457]. Beginning at measure 72, the segments are inverted and

Figure 3.4d. Measures 69–74. © COPYRIGHT 1942 BY HAWKES & SON (LONDON) LTD. Reproduced by kind permission of Boosey & Hawkes Music Publishers Ltd.

Figure 3.4e. Measures 80–82. © COPYRIGHT 1942 BY HAWKES & SON (LONDON) LTD. Reproduced by kind permission of Boosey & Hawkes Music Publishers Ltd.

extended, sounding a heptarhythmic pattern that we will refer to as G, of prime form [0124568]. This pattern gives way at measure 80, where a clear and unconflicted ♩. pulse is sounded for the first time in the movement, in a harmonically static passage that prepares for the second theme (figure 3.4e).

Figure 3.5 presents a "space" that will help to capture the logic of the beat-class progression from measure 65 to measure 80, and to position it within the context of a broader rhythmic modulation across the exposition. All solid edges in the graph connect cardinality-distinct set classes, as in the classic network representations introduced into music theory by Allen Forte.[28] Dotted edges connect cardinality-equivalent set classes related by a minimal beat-class displacement. These constitute the rhythmic analog to the "P-relations" of neo-Riemannian harmonic theory.[29] The graph partitions into two distinct sub-graphs. The upper sub-graph, running along the top margin, represents the 3-generated universe through a Kh-inclusion network connecting [036] to its iambic complement,

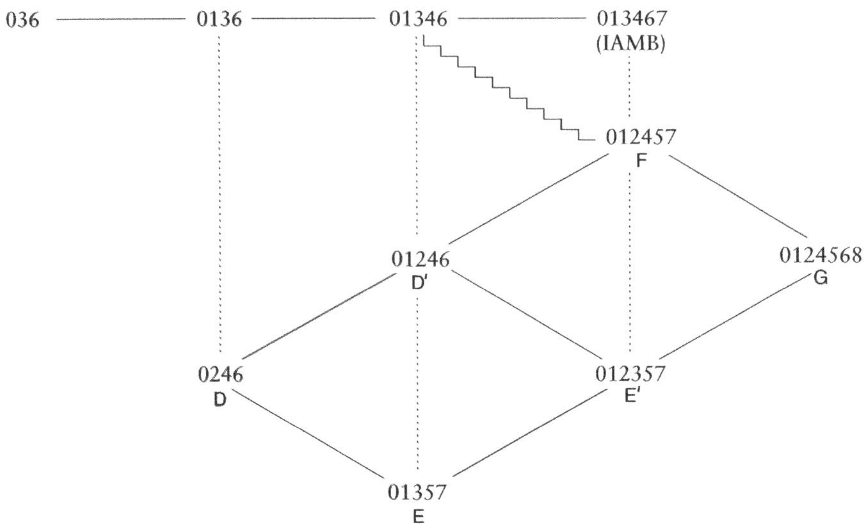

Figure 3.5. Inclusion/Displacement Graph of Beat-Class-Set Classes.

[013467]. The lower sub-graph combines two overlapping diamond-shaped graphs; the left one contains the four Balkan rhythms of figures 3.1a and 3.1b, and the right one contains the F and G rhythms of the transition. The two subgraphs share a single solid edge which connects [01346] to its [012457] superset. The stepped shape of this edge is intended solely to clarify the sub-graph partition discussed above.

The first theme group is positioned to the lower left, and culminates with the cycling of E at measure 65. The graph indicates that E is remote from the 3-generated universe, relating to it by neither inclusion nor minimal displacement. The substitution of D′ for E at measure 69 brings us to within a single displacement of an iambic pentarhythm. F for D′ at measure 71 brings us one step closer yet: this new hexarhythm includes both the Balkan D' and the iambic pentarhythm (represented by figure 3.5's crooked edge); moreover, it is a single displacement from the iambic hexarhythm that represents the 3-universe in its most concentrated form. G for F at measure 72 then deflects this trajectory: its heptarhythm abstractly includes all of the previous Balkan rhythms, while including the iambic pentarhythm that represents the next section.

At measure 80, we are poised for a new theme in a closed, 3-generated rhythmic mode. But the fulfillment of this expectation is deferred until the iambic fugato begins at measure 100. This music is preceded by a sixteen-measure parallel period that combines aspects of both rhythmic modes. We shall explore the rhythmic properties of this "second" theme in some detail in part 4 of the paper; for the moment it is sufficient to note that the entire antecedent phrase (mm. 84–90) is accompanied by an ostinato figure in Piano II consisting of the four-attack Balkan D pattern. Pattern D is again cycled when the second theme sounds at the close of the exposition (mm. 161–70), after the sounding of the iambic third-theme material; here it is the product of the composite rhythm of the canonic strands. Similar open-mode cycles emerge in the latter part of the development, beginning at measure 235, and in the recapitulation: at the end of the theme 1 material (mm. 286–89), in the composite rhythm of theme 2 (mm. 301–24), and in the final measures of the movement, beginning at measure 437.

Just as the 3-generated rhythmic mode is abstractly aligned with octatonic and hexatonic collections generated by closed pitch intervals 3 and 4, so the Balkan 2-generated rhythmic mode is abstractly aligned with structures generated by the open pitch intervals 1 and 5. Structures generated by those open intervals abound in this movement, both separately and together. The opening measures of the *Lento assai* are dominated by a motto that marks out and then fills interval-class-1-generated bands of pitch-space, creating chromatic sets of cardinality 7 and 9. Interval-class 5 is first featured in the opening of the *Allegro*; the first theme (mm. 32–41) features a number of quartal ([05], [027] and [0257]-type) harmonies, particularly at phrase boundaries, and several of its melodic phrases have a pentatonic cast.[30] During the development section, the two open interval classes become increasingly affiliated. The beginning of the development (mm. 175–94) features quartal chords combined with a melodically prominent semitonal neighbor figure. Similarly, when the chromatic motto of the opening measures is brought back as a propulsive ostinato, it planes through [027] harmonies, expressed first as stacked fourths (mm. 208 et seq.), then fifths (mm. 239 et seq.). The combination of ic1 and ic5 reaches a peak in the final measures of the development (mm. 264–73), where a five-octave band of pitch-space is divided into twelve perfect fourths and chromatically saturated, in a stricter version of the technique familiar from the opening movement of the *Music for Strings, Percussion, and Celeste,* composed during the previous year.[31]

The logic of the oppositional structure that we have been developing implies that open-generated rhythms and chromatic or diatonic scales be superposed, in the same way as the closed rhythms and pitch collections of the third theme were. Such a superposition is discernible, not as clearly as in the former case, but increasingly so as the movement progresses. At their first sounding (26–31), the Balkan rhythms are affiliated with a pentachordal octatonic subset (see figure 3.4a). The first theme (32–41) is harmonically hybrid: closed octatonic and hexatonic collections are represented by the prominent [037] and [014] trichords, while an open ic5 orientation is suggested by the pentatonic melodies of the third and fourth phrases, and by the [027] and [0257] chords featured at phrase boundaries. Likewise hybrid is the counter-statement (mm. 65–68), where the Balkan rhythms are first cycled: it features [037] minor triads, this time in (closed) octatonic juxtapositions, against a background of (open) running parallel fifths. The

Figure 3.6a. Chromatic (open) pitch combined with 3-generated (closed) rhythm.

Figure 3.6b. Octatonic (closed) pitch combined with 2-generated (open) rhythm.

second theme, which also cycles Balkan rhythms, features parallel [037] major triads, whose juxtaposition creates no consistent scalar affiliations.

In the development, Bartók avoids the simultaneous sounding of open modes in both domains by exploiting a pitch-class/order-number disjunction that is latent in the introductory motto of the movement. Although the two extended ostinato cyclings of that chromatic motto saturate the beat-class aggregate, Bartók individuates them in both pitch and rhythm through registral partitionings and selected bass-octave doublings. In the first cycling (figure 3.6a), a closed rhythmic mode is combined with an open pitch one. The ♩. pulse is initially projected by the doubling of the metrically accented notes of the ostinato. Yet these same doubled notes project the ic1-generated chromatic cell <E, G, F♯, F>. In the second cycling (figure 3.6b), the situation is exactly reversed. Through most of this section, the ostinato is rhythmically articulated as an open-mode Balkan P partition, hocketing the <u>E</u> cell (right hand) with the <u>D</u> cell (left).[32] Yet this registral bisection of the motto results in a partitioning of the pitch classes into subsets of the closed-mode octatonic collection: [01347] is represented in the right hand, [0236] in the left hand.

Open (1- and 5-generated) pitch structures become affiliated with open (2-generated) beat-class structures for the first time in the final ten measures of the development, where the perfect-fourth partitioning and chromatic saturation of pitch space is carried out through a set of overlapping <u>E</u> and <u>D′</u> motives, culminating in the P-partition hocket at measure 273 (figure 3.7a). Rhetorically, however, this passage is no culmination; it is a structural upbeat for the

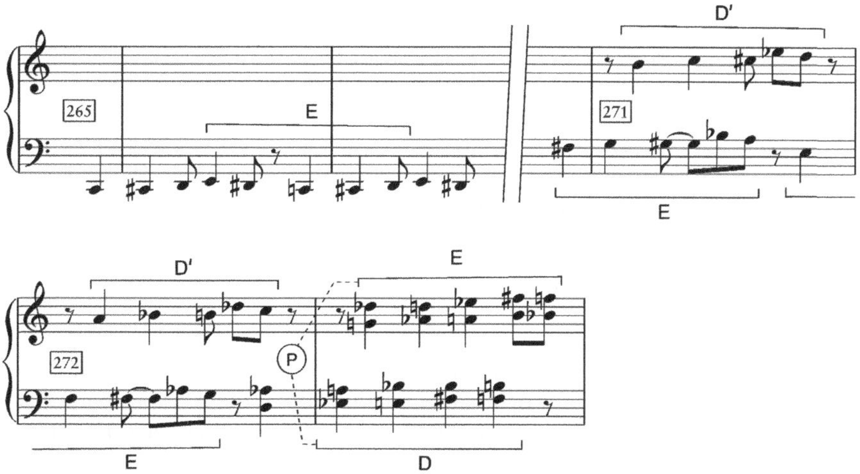

Figure 3.7a. Balkan cells in measures 265–73. © COPYRIGHT 1942 BY HAWKES & SON (LONDON) LTD. Reproduced by kind permission of Boosey & Hawkes Music Publishers Ltd.

Figure 3.7b. Balkan cells in measures 286–87, © COPYRIGHT 1942 BY HAWKES & SON (LONDON) LTD. Reproduced by kind permission of Boosey & Hawkes Music Publishers Ltd.

recapitulation that begins at measure 274. Here the harmonic structure of the first theme is "purified": the diverse harmonies of the exposition are transformed into parallel fifths. But, as in the treatment of the motto in the development section, Bartók takes away with one hand what he gives with the other: the rhythm is resolutely Western, lacking any discernible Balkan properties. Open structures in both domains are again conjoined at measures 283–88 (figure 3.7b), where the Balkan D′ pattern is restored, in a passage that features pentatonically embedded fifth-based harmonies, replacing the octatonically embedded minor triads appearing at the equivalent passage of the exposition (mm. 65–68). The most extended cross-domain combination of prime-generated materials is reserved for the final measures of the movement, already examined in part 2 of this paper

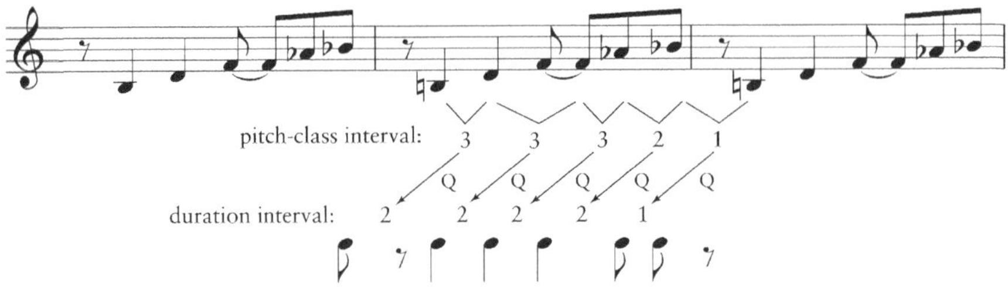

Figure 3.8. Q transformation applied to measures 26–31.

(figure 3.2). While pattern D′ cycles, the melody fills a band of chromatic space, and interval-class 5 controls the imitative relationship of the two voices as well as the harmonies at the approach to the final cadence. These final measures culminate a process that plays out across much of the movement.

3·4

What then of the pitch organization of those Balkan-rhythm passages that occur at the beginning of the movement (as in figure 3.4)? Section 3 characterized their pitch structure in negative terms, noting only what they failed to be. This section develops a more productive characterization, by extending the Q homomorphism introduced at the end of part 2.

We can begin by observing some affinities between pitch and rhythm in the final measures of the *Lento assai* (mm. 26–31), in connection with figure 3.8. Adjacent pitches are pitch-separated by three semitones, from B up to A♭, followed by a compressed interval of two semitones, A♭ to B♭. Similarly, adjacent attacks are time-separated by two beats, from B up to A♭, followed by a compressed interval of one beat. The pitch-time affinity suggested by this characterization comes into focus by comparing the step-class series of the respective pitch-class and beat-class sets, in their respective modular universes. Mod-12 set class 5-31, prime form [0, 1, 3, 6, 9], has an ordered step-interval series of [1, 2, 3, 3, 3]; mod-9 beat-class set class E, prime form [0, 1, 3, 5, 7], has an ordered durational series of [1, 2, 2, 2, 2]. The mapping that takes 5-31 to durational series E is familiar: it is the Q function introduced in section 2.

5-31 is not the only set class that is mapped on to E by Q. There is one other pentachord that permutes its step intervals: 5-32, prime form [0, 1, 4, 6, 9], whose step-interval series is [1, 3, 2, 3, 3]. Paul Wilson has shown that 5-32 figures prominently at the middleground of the Balkan-affiliated first theme (mm. 32–40). The melodic pitches that begin and end its four phrases constitute ordered set <C, E♭, A♭, B, F♯, G>, whose first five pcs form a version of 5-32. This pentachord surfaces in measures 41–56, where it serves as an ostinato accompaniment to the

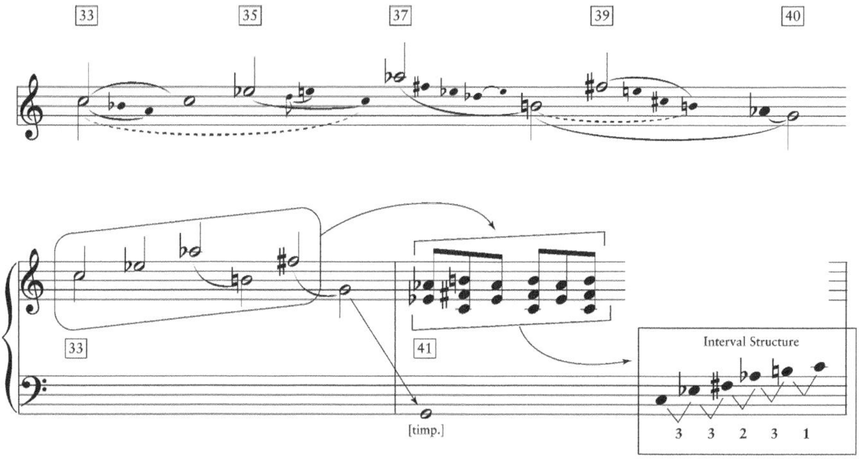

Figure 3.9. Composing out and compression of {C, E♭, F♯, A♭, B} in measures 31 ff. (After Paul Wilson).

E′ pattern sporadically appearing in the snare drum. Figure 3.9 summarizes this connection.[33]

Suitably adapted, the Q function can also shed light on the second theme, which is first heard at measure 84, and throughout the movement consists of parallel major triads. The theme, in its initial form, consists of four segments of two measures each. The first three of these segments start with a 𝅗𝅥 ♩ ♩. measure, beat-class set {0, 4, 6} mod 9, which we shall refer to by the symbol C. In Bartók's initial conception of this theme, this rhythm repeated through all eight measures of the theme. At some point before composing the recapitulation, however, he evidently revised these measures, creating a variety of rhythmic configurations, some of which constitute rotations of the head measure's rhythm.[34]

Rhythmic pattern C, [046] mod 9, shares several abstract features with the major triad, [047] mod 12, that it accompanies through the second theme. Although neither set is generated, both combine the intervals that generate the principal structures in their respective domains. The three intervals of the major triad generate octatonic, hexatonic, and diatonic collections respectively, each of which embeds multiple major triads. The major triad thus occupies a neutral position with respect to the principal pitch polarities of the movement. Rhythmic pattern C manifests a similar neutrality in the rhythmic domain: it contains the intervals that generate both of the movement's rhythmic modes. C is abstractly included in D, and indeed Bartók exploits this inclusion literally at the canonic passage beginning at measure 161. Yet, as figure 3.10a shows, C is also included in IAMB. IAMB and C also share a more abstract affiliation: both are comprised of 2:1 proportions. In the former case, these are concatenated (♩ + ♪); in the

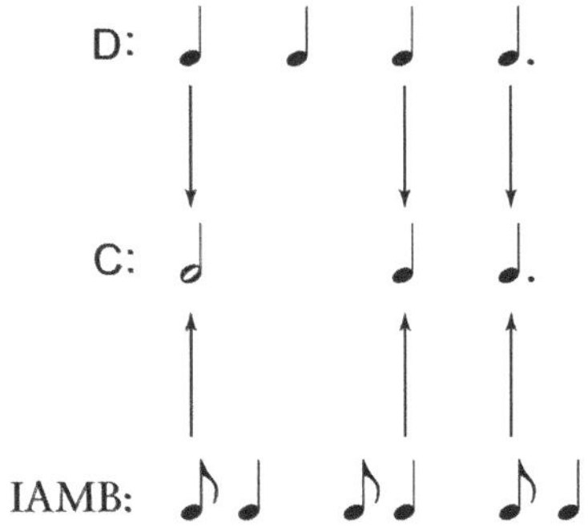

Figure 3.10a. C as subset of D and IAMB.

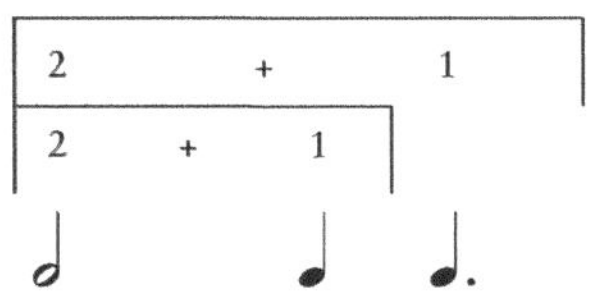

Figure 3.10b. Nested 2:1 proportions in C.

latter case they are nested (figure 3.10b).[35] The neutrality of both entities—the major triad in the pitch domain, the C cell in the rhythmic—is logical: the music of the second theme stands literally midway between the Balkan/fifth-based music of the first theme and the iambic/third-based music of the fugal third theme.

Further affinities between the major triad and rhythmic cell C become apparent through their step-interval structures: both are of the form $\left[\frac{c}{3}-1, \frac{c}{3}, \frac{c}{3}+1\right]$ for their respective modular universes. We have already posited a contextual affinity in the Sonata between the smallest of these intervals: Q maps pitch-class interval 3 to duration interval 2. To account for the remaining intervals, we need only propose a more general version of Q, as follows:

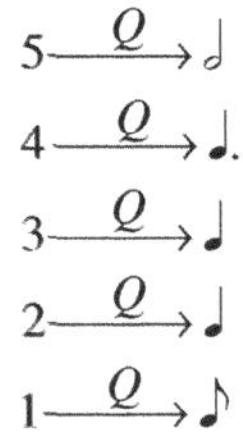

Q is an idiosyncratic transformation of limited compositional and analytical value; it maps only a small minority of pitch-class sets into beat-class sets of cardinality 9. Q's only virtue is that it maps the most prominent pitch-class sets of the first movement of Bartók's Sonata for Two Pianos and Percussion into the beat-class sets with which they are compositionally paired. In this sense it functions like the similarly idiosyncratic "wedging to E" that David Lewin has proposed for the Schoenberg song "Angst und Hoffen."[36] The chromatic opening of the *Lento*, the [01369] approach to the *Allegro*, the [01469] pentachord articulated by the first theme in the exposition, the major triads featured in the second theme of the exposition and recapitulation, the hexatonic collection featured when the third-theme music is introduced in the exposition, and the octatonic collection featured when the third-theme music bids farewell in the coda are all accompanied by cycled beat-class patterns that are their respective Q images.

The Q transformation, and the cross-domain affiliation of generated open and closed sets, constitute two distinct strategies for relating pitch-class and beat-class structures. In neither case is the mapping from pitch to time always a clean one. Moreover, the strategies sometimes clash, yielding divergent results. But to some degree, inconsistencies are to be expected when cycles of twelve pitch classes are affiliated with cycles of nine beat classes. To a greater degree, they result from the fluidity of creative thinking, particularly of a musical thinker who shunned systems or suppressed their traces. This study echoes Leong's in suggesting that, even without a usefully refined descriptive language or a mode of systematic inquiry, Bartók explored the properties of duration and meter as intensively as he did those of harmony, melody, and mode. Moreover, it provides tangible evidence that Bartók integrated his procedural knowledge of the two domains, allowing them to interact and cross-fertilize.

Notes

1. Aspects of this movement are particularly reminiscent of the Waldstein first movement, which Bartók performed on his first piano recital at age 11 (Stevens 1993). The *Allegro* begins with a propulsive C-major theme, and moves to a more lyrical, reverse-arch E-major theme (m. 84), which is transposed in the recapitulation to A major (m. 292) and C (m. 309).

2. Stempel 1979, 357, exemplifies this commonly held view: "The isomorphic premise . . . is something which fares better in the silent contemplation of its numbers than in the hubbub of its musical experience."

3. Somfai 1996, 10–11.

4. One possible opening is created by Bartók's professed admiration for Henry Cowell, the one contemporaneous composer who systematically explored pitch-time parallels. Bartók and Cowell became acquainted in 1923 (Stevens 1993, 67), and Cowell's correspondence indicates that they spoke at length about various aspects of composition (Suchoff 2001, 101). Four years earlier Cowell had drafted a treatise (Cowell 1930) that explored pitch-time analogies to an unprecedented extent, although it was evidently not until the 1930's that he implemented these ideas in his

compositions (Nicholls 1990, 140 ff.). Although worth pursuing, at present this is a slender thread on which to hang an argument persuasive on musicological grounds alone.

5. Lendvai 1971 does not mention specific pieces, and it is difficult to take his claim seriously, given the tendentious way he prosecutes it; nonetheless, the intuition behind the claim bears some scrutiny.

6. Kárpáti 1995, Antokoletz 1984 and 2000, Leong 1999.

7. An example from an equally surprising source is the opening of Ravel's "Une barque sur l'océan" (*Miroirs*, 1904–5), whose 12-beat ostinato <G♯, F♯, E, C♯> is articulated by a 2 + 2 + 3 + 5 division of the beat-class cycle, as if the pianist were playing a descending chromatic scale on a piano most of whose hammers had been disabled.

8. On the distinction between operations and transformations see Lewin 1987, 3.

9. Pressing 1983.

10. The artificial/natural distinction is from Schenker 1954 [1906], 42–43 who generates the diatonic collection in this way.

11. Leong 1999 uses the terms "mod 2" and "mod 3" where I use "open" and "closed."

12. Singer 1974, 393, 396.

13. Clough and Douthett 1991. Rhythmic interpretations of maximal evenness are explored briefly in Lerdahl 2001, 286–87, and more comprehensively in London 2004.

14. The Balkan qualities of pattern D are cited in Bartók's essay on "The So-Called Bulgarian Rhythm" (Suchoff 1992). Moreover, the complementary P hocket occurs at the opening of one of the "Dances in Bulgarian Rhythm" (Mikrokosmos 152). Whereas some of Bartók's other Bulgarian rhythms represent extrapolations that do not actually occur in music of the region, those of Mikrokosmos 152 are authentic (Rice 2000). Since the rhythmic patterns that Bartók calls "Bulgarian" are common to a wider geographic region, Rice suggests "Balkan" or "Balkan/Turkish" rhythm as the preferred term. This class of durational patterns is also frequently referred to in the ethnomusicological literature by the term *aksak*, Turkish for "limping."

15. For some compelling speculations on the meanings encoded into Balkan rhythms for Hungarian composers, see Willson 2002. For an account of the youthful Bartók poised between two worlds, and the political symbolism of rhythm in early twentieth-century Hungary, see Hooker 2001.

16. Leong 1999, 175.

17. Bartók comes close at two anomalous points, measures 349 and 357, where he partitions the dotted quarter into quadruplet melismas; this durational value occurs nowhere else in the movement.

18. Lerdahl 2001, 280.

19. The build-up to this octatonic moment is detailed in Cohn 1991. For a related but distinct approach to the movement's octatonicism, see Wilson 1992.

20. The dually anomalous status of the D natural at measure 422 is reflected on page 19 of Bartók's initial draft of the movement, a copy of which was made available to me when it was still in the collection of Peter Bartók, where it was catalogued as #75FSS1; the manuscript is now held by the Sacher Foundation. To the left of the notehead is a natural sign which is gratuitous, as there are no proximate D♭s to be cancelled. This suggests that Bartók initially wrote a D♭ at the fifth beat of measure 422. Indeed, the accidental resembles a flat to which a downward hook has been added (Bartók frequently wrote his natural signs in this way, so this alone is not

conclusive). At the same time, there are signs that Bartók struggled with the rhythm in this measure. The D and the previous C, to which it is beamed, flank a symbol that may represent a partially over-written eighth rest, suggesting an initial placement of the D on beat 6. The evidence of Bartók's initial intentions, however, is inconclusive. I cite with gratitude the advice of Klára Móricz and László Somfai in interpreting this measure of the draft.

21. The exception is the B♭ in Piano I at measure 432, which participates as an after-beat along with the four pitch classes that lie outside the prevalent octatonic. The set of after-beats {B, D, F, A♭, B♭} subtly refers to the pitch-class collection of measures 26–31, the music that directly precedes the opening of the Allegro, and the moment when the Balkan rhythms initially coalesce and gain traction. See figure 3.4a.

22. Cohn 1991.

23. One aspect of this tension is that the chromatic scale can be expressed as a cross-product of the diminished-seventh chords and the augmented triads (Weitzmann 1853).

24. Singer 1974, 386. Other pertinent constructs are the model of rhythmic contour proposed by Marvin 1991, and Kárpáti's mistuning model (1975 [1967]).

25. Pressing 1983, 43.

26. D̲ is, however, already implicit in the location of the semitonal dyads in the opening motto. On this topic, see Petersen 1994. Leong 1999 suggests that a 3-generated interpretation of the motto is equally plausible. Both interpretations are made explicit when the motto returns in the development; see figure 3.6.

27. Leong 1999 provides a more comprehensive analysis, with examples.

28. Forte 1973.

29. Figure 3.5 thus is a rhythmic analog to the scale-class graphs in Callender 1998.

30. This is discussed in some detail by Kárpáti 1994, 413–14.

31. This retransitional passage was not part of Bartók's initial conception; he inserted it in 1940, after the initial performances but before the publication. See Somfai 1996, 196–98. The initial draft features pentatonic melodic fragments, stacked-fifth harmonies, and a straightforward cycling of Balkan rhythmic cells, over a dominant pedal in the timpani. Except for the dominant pedal, each of these features recurs at the music of measure 286, which in the initial draft occurs only eight measures into the recapitulation. Bartók's dissatisfaction with his initial conception may have stemmed from this resemblance, which would have weakened the demarcation between the development and the recapitulation.

32. The 3-generated partitioning also appears in this section, although not as prominently. See Petersen 1994, 40–41.

33. Wilson 1992, 145.

34. See page 6, system 2 (antecedent phrase) and system 3 (consequent phrase) of the draft cited above in note 20.

35. Lendvai 1971 has emphasized that embedded proportional reflections of this sort are common in Bartók's music. So too are motivic augmentations and diminutions: Leong 1999, 65, notes that the opening of the iambic fugal theme at measure 105 embeds a three-fold augmentation of P̲.

36. Lewin 1987, 124.

Chapter Four

Filtered Point-Symmetry and Dynamical Voice-Leading

Jack Douthett

4.1 Introduction

Until now, diatonic systems and neo-Riemannian transformations have generally been considered separately, and group and graph theoretic approaches have dominated the neo-Riemannian and transformational theory literature.[1] In this paper, an alternative approach will be explored; techniques similar to those used in the study of *dynamical systems* in science will be adopted to study neo-Riemannian theory and its connection to diatonic theory.

Dynamical systems are probably best known today for the fractals they sometimes generate (e.g., Koch's snowflake, the Dragon curve, Mandelbrot's set, the Julia set, etc.), but fractals are only part of this field of study. As Strogatz puts it in his text on nonlinear dynamics and chaos, dynamics ". . . is the subject that deals with change, with systems that evolve in time. Whether the system in question settles down to equilibrium, keeps repeating in cycles, or does something more complicated, it is dynamics we use to analyze the behavior."[2]

Dynamical systems related to the topics that will be discussed here will utilize *point-symmetric concentric circles* that rotate through time and *stroboscopic portraits*, which record events at particular time intervals. These dynamical systems can be thought of as sequence generators which, with the appropriate choice of *control parameters*, produce *periodic orbits* (cycles) of scales and chords well known in both diatonic and neo-Riemannian theory.

Douthett and Steinbach's *Relation Definition* will also be used:[3]

Let X and Y be pcsets. Then X and Y are $P_{m,n}$-related if there exists a set $\{x_k\}_{k=0}^{m+n-1}$ and a bijection $\tau : X \to Y$ such that $X \setminus Y$ (the set of pcs in X that are not in Y) $= \{x_k\}_{k=0}^{m+n-1}$, $\tau(x) = x$ if $x \in X \cap Y$, and

$$\tau\left(x_k\right)=\begin{cases} x_k \pm 1 \pmod{12} & \text{if } 0 \le k \le m-1 \\ x_k \pm 2 \pmod{12} & \text{if } m \le k \le n-m-1. \end{cases}$$

The requirement that τ be a bijection implies that X and Y have the same cardinality. In addition, since $\tau(x)=x$ when $x \in X \cap Y$ and $X \setminus Y = \{x_k\}_{k=0}^{m+n-1}$, $Y \setminus X = \{\tau(x_k)\}_{k=0}^{m+n-1}$. Musically, this implies that two pcsets are $P_{m,n}$-related if one can be obtained from the other by leaving all common pcs fixed, moving m pcs by ic1 (a half step), and moving the remaining n pcs by ic2 (a whole step). For example, G and e are $P_{0,1}$-related, since one can be obtained from the other by moving a single note by a whole step. On the other hand, E and a are $P_{2,0}$-related, since one can be obtained from the other by moving two notes by half steps.

4.2 Triadic Cycles

Before exploring dynamical systems and their connection to musical structure, it is necessary to formally define two types of triadic cycles: diatonic triadic cycles and parsimonious cycles.

Diatonic Triadic Cycles: In his 1991 paper on cyclically generated chords, Eytan Agmon discusses universes with odd cardinalities and generated chords within these universes whose cardinalities are just over or just under half the size of the universe. Agmon points out that for the family of chords with these cardinalities and generated by the appropriate interval, any member of this family can move to any another member of the family by step motion. This can be observed with the triads (generated by the third) embedded in any given diatonic set. In a diatonic context, this collection of triads can be thought of as a set class modulo 7. If the roots of two triads embedded in a diatonic set differ by a third or sixth, then one triad can be obtained from the other by moving a single note by a diatonic step (a half step or a whole step); if the roots of the triads differ by a fifth or a fourth, then two notes must move by diatonic steps, and if the roots differ by a seventh or a second, all three notes must move by diatonic steps. In general, if triads embedded in the same diatonic set are $P_{m,n}$ -related, then one can be obtained from the other by moving $m+n$ notes by diatonic steps. These relationships can be seen in what will be called "diatonic triadic cycles."[4]

Cycles of triads embedded in a diatonic set in which the interval between the roots of adjacent triads is a third will be called *mediant-submediant (diatonic triadic) cycles.* In these cycles, adjacent triads have precisely 2 notes in common and are either $P_{1,0}$-related (e.g., C and e) or $P_{0,1}$-related (C and a); any triad in these cycles can be obtained from an adjacent triad by moving a single note by a diatonic step. The mediant-submediant cycle associated with the C major diatonic set is given in figure 4.1. There are 12 such cycles, one for each diatonic set.

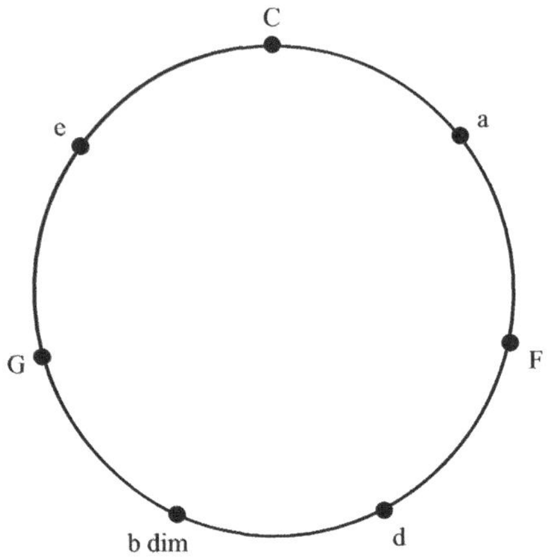

Figure 4.1. The C-Major Mediant-Submediant Cycle.

The *dominant-subdominant (diatonic triadic) cycles* differ from the above cycles in that the roots of adjacent triads are a fourth apart. Adjacent triads have precisely 1 note in common and are $P_{1,1}$- or $P_{0,2}$-related; two notes are required to move by diatonic steps to get from a given triad to an adjacent triad. Figure 4.2 shows the dominant-subdominant cycle associated with the C major diatonic set.

The last diatonic cycles to be discussed here will be called *supertonic-leading tone cycles*. In these cycles, the roots of adjacent triads are a second apart; the triads have no pcs in common. The supertonic-leading tone cycle associated with the C major diatonic set is shown in figure 4.3. Every pair of adjacent triads in these cycles are $P_{2,1}$-, $P_{1,2}$-, or $P_{0,3}$-related, implying that all three notes must move by diatonic steps to get from a given triad to an adjacent triad.

Parsimonious Triadic Cycles: With tuning systems that were equal-tempered came a new triadic freedom; triads no longer had to be associated with a small collection of diatonic scales (i.e., closely related keys). This allowed composers such as Beethoven, Brahms, Schubert, Wagner, and many others to write triadic sequences essentially independent of diatonic influences. Passages might lose their diatonic sense in sequences of chords emphasizing chromaticism and maximum common tone content. Such passages are often difficult to analyze with traditional functional harmony, as Cohn illustrates in his paper on maximally smooth cycles and hexatonic systems.[5]

In his 1997 article, Cohn introduces the term *parsimony* as related to harmonic triads (triads from Forte's sc 3-11). Two harmonic triads of opposite modality are *parsimonious* if they have maximum pc commonality (two pcs in common). Transformations that relate parsimonious triads can be found in what has become known as the *Riemann group*, initially explored by Klumpenhouwer.[6]

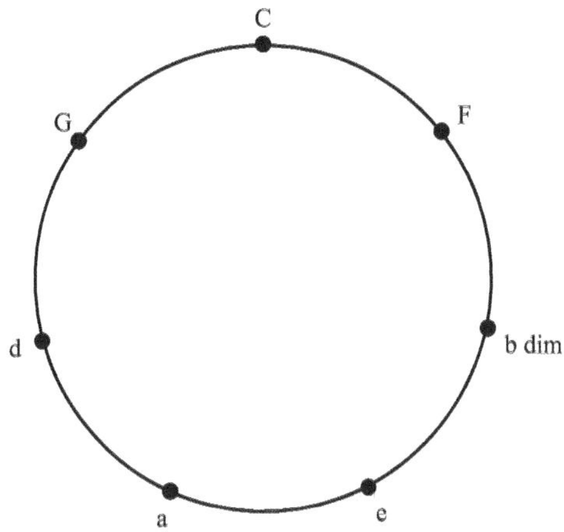

Figure 4.2. The C-Major Dominant-Subdominant Cycle.

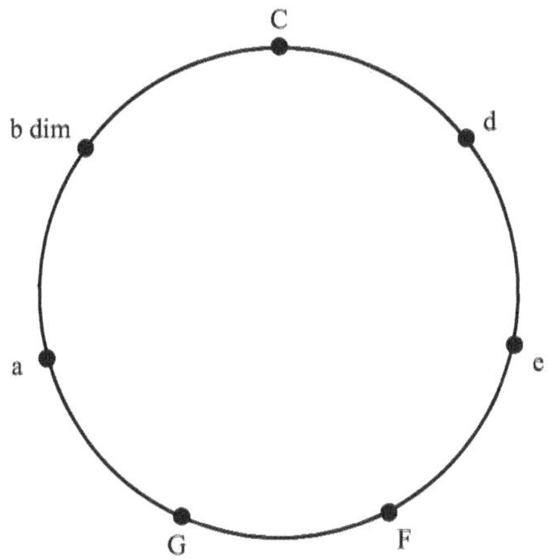

Figure 4.3. The C-Major Supertonic-Leading Tone Cycle.

These transformations are called *Parallel* (*P*), *Leading Tone* (*L*), and *Relative* (*R*). The P transformation exchanges triads that differ in modality but have the same root; whence the pcs common to both triads are ic5-related. The L transformation exchanges triads of opposite modality with common pcs related by ic3, and R exchanges triads of opposite modality that have common pcs related by ic4. These transformations are shown in figure 4.4, which is a section of the toroidal form of the *Oettingen/Riemann Tonnetz.* The pcs are represented by vertices and the triads are represented by the triangles. These transformations, as well as

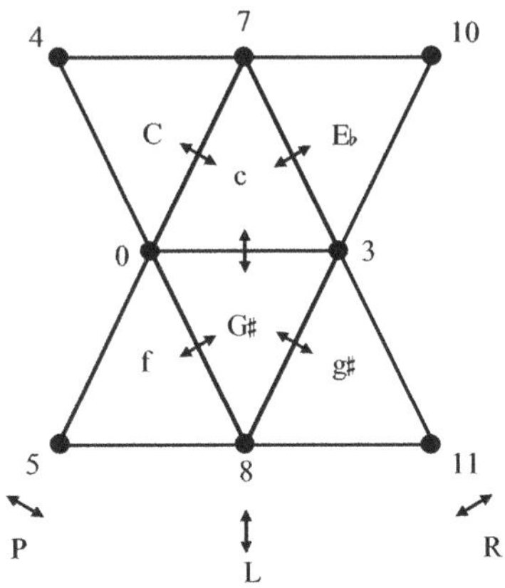

Figure 4.4. The Neo-Riemannian Transformations P, L, and R.

other Riemann transformations, turn out to be useful in the analysis of passages that have "lost their diatonic way."

The *parsimonious cycles* that will be discussed here are generated by an alternating pair of parsimonious transformations. *LP-cycles* are cycles in which adjacent triads are either L-related or P-related. This restriction results in four cycles of six triads each, which are shown in figure 4.5. For these cycles, all adjacent triads are $P_{1,0}$-related. Cohn refers to these cycles as *hexatonic sub-systems*, since the union of the triads in any given cycle is an all-combinatorial hexachord from Forte's sc 6-20.[7] These hexachords are listed in figure 4.5 below their corresponding LP-cycle.

Non-trivial cycles (cycles of length 3 or more) of pcsets from the same sc in which adjacent pcsets are $P_{1,0}$-related are called *Cohn cycles.* The LP-cycles are Cohn cycles. There are two fundamental types of Cohn cycles, *unidirectional* Cohn cycles and *toggling* Cohn cycles. The LP-cycles are an example of the latter. The origin of the term "toggling" can be seen in the voice-leading motion among the triads in the LP-cycles. Note in Example 5a that the note E♭ in the c triad moves up to the note E, resulting in the C triad. This note remains fixed through the next 2 triads and then returns to E♭ (= D♯) in the g♯ triad. This "toggling" voice-leading behavior can be observed in the other voices of the triads around this cycle; observe the toggling of the notes B and C and the notes G and G♯. It makes sense, then, to call such cycles *toggling* Cohn cycles. *Unidirectional* Cohn cycles will be discussed in section 4.4.

The *PR-cycles* are cycles in which adjacent triads are either P-related or R-related. There are three such cycles, each with eight triads. These cycles are shown in figure 4.6, and adjacent triad relations alternate between $P_{1,0}$ and $P_{0,1}$. Because the union of the triads in any given PR-cycle yields an all-combinatorial

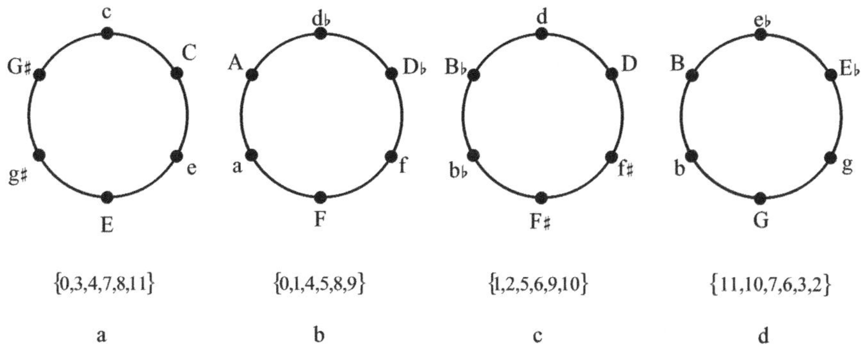

Figure 4.5. The LP-Cycles.

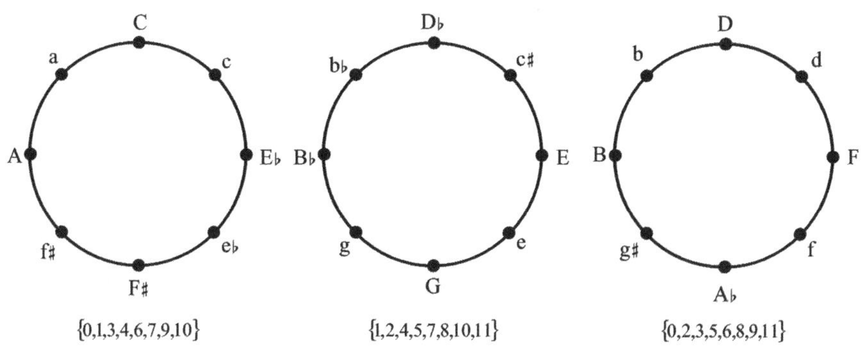

Figure 4.6. The PR-Cycles.

octachord from sc 8-28, Cohn refers to these cycles as *octatonic sub-systems.* These octachords are listed below their corresponding PR-cycle in figure 4.6.

The largest parsimonious cycle—which includes all 24 harmonic triads—is the *LR-cycle,* shown in figure 4.7. As in the PR-cycles, adjacent triad relations alternate between $P_{1,0}$ and $P_{0,1}$.

4.3 Filters and Beacons

A circle, together with a collection of points, is *point-symmetric* if the points are distributed evenly about the circumference of the circle. To begin construction of a dynamical system, consider two concentric circles of different radii. The outside circle has 12 holes equally spaced about its circumference and

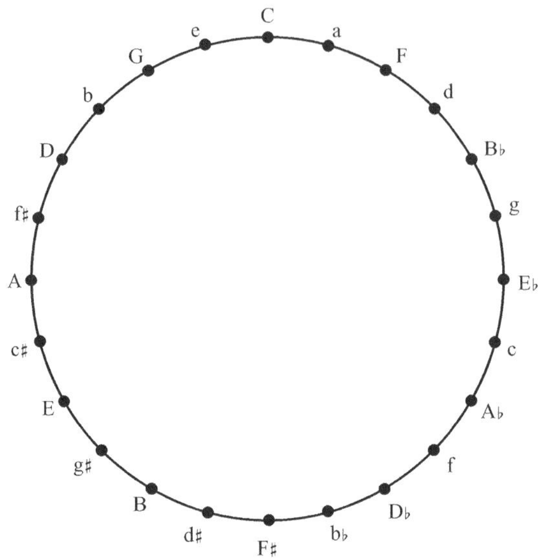

Figure 4.7. The LR-Cycle.

numbered 0 through 11, and the inside circle, called the *beacon,* has 7 *lamps,* equally spaced about its circumference and numbered 0 through 6. Each lamp transmits a beam *normal* (perpendicular) to the circumference of the beacon. There are two simple rules that apply to a beam when it hits the outside circle:

1. If the beam hits a hole on the circumference of the outside circle, the beam travels through the hole.
2. If the beam hits the inside wall of the outside circle, the beam moves counterclockwise on the circumference of the outside circle and travels through the first hole it encounters.

In this way, the outside circle acts as a type of *filter,* modifying slightly the paths of the beams. For the configuration in figure 4.8a, called the *7 through 12 dynamical system,* the set of *beam numbers* (numbers corresponding to the holes that the beams pass through on the outside circle) is $\{0, 1, 3, 5, 6, 8, 10\}$, the D♭ major (diatonic) set.

In this case, beam 0 passes through hole 0, and the other beams pass through the first holes counterclockwise to their collisions with the circumference of the outside circle. As the beacon is rotated clockwise, the beam numbers stay the same until the beacon has passed through an angle of $4\frac{2}{7}^\circ$ or $\frac{1}{7\cdot 12} = \frac{1}{84}$th of a revolution. At this point, beam 4 hits hole 7, changing the set of beam numbers to $\{0, 1, 3, 5, 7, 8, 10\}$, which is the A♭ major set, as shown in figure 8b. After

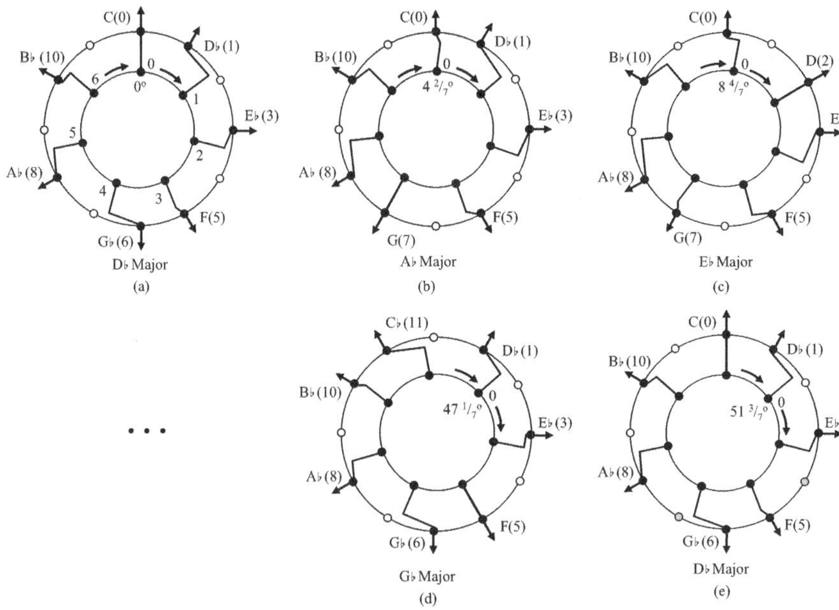

Figure 4.8. The 7 through 12 Dynamical Configuration.

rotating another $4\,{}^{2}/_{7}{}^{\circ}$, the beam set changes to $\{0, 2, 3, 5, 7, 8, 10\}$, the E♭ major set; see figure 8c. Continued rotation generates the diatonic sets in the cycle of fifths, and the cycle is completed when beam 6 lines up with hole 0, as in figure 4.8e. At this point, lamp 0 has passed through an angle of $51\,{}^{3}/_{7}{}^{\circ}$, or ${}^{1}/_{7}{}^{\text{th}}$ of a revolution. One can see that the cycle is complete, since the pattern of unlabeled (without lamp and hole labels) concentric circles and beams is exactly the same as the initial pattern; compare figures 4.8a and 4.8e. When the beacon completes one revolution (360°), the cycle of fifths has been repeated 7 times.[8]

Next, consider a beacon rotation that yields the C major set in figure 4.9a. Change the lamps to holes on the beacon, and add a new beacon with 3 lamps inside the old beacon as in figure 4.9b. This *3-through-7-through-12 dynamical configuration* has two filters, and the set of beam numbers on the outside circle is $\{0, 4, 7\}$, the triad C. When the beacon is rotated ${}^{1}/_{3\cdot 7} = {}^{1}/_{21}{}^{\text{th}}$ of a revolution clockwise, the beam set changes to the minor triad a in figure 4.9c. Continued rotation produces the triads F, d, b dim, G, and e, and the cycle begins again. This is the mediant-submediant cycle shown in figure 4.1. If the beacon with three lamps is replaced by a beacon with four lamps, clockwise beacon rotation produces a cycle of the seventh chords embedded in C major: Dm7, FM7, Am7, CM7, Em7, G7, and Bø7. As with the cycle of triads, adjacent seventh chords are either $P_{1,0}$-related or $P_{0,1}$-related.

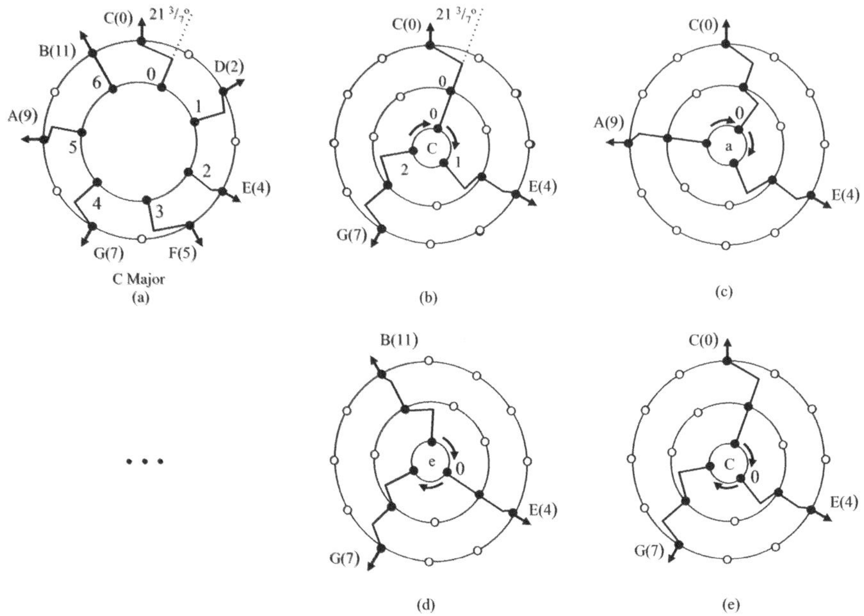

Figure 4.9. Embedded Triads in the C Major Scale.

For the last example in this section, we fix the beacon and rotate the 7-hole filter in the 3-through-7-through-12 dynamical configuration, as shown in figure 4.10. The cycle of triads generated in this way is C, e, E, g♯, G♯, c, and back to C. This is one of the LP-cycles shown in figure 4.5. This suggests an intriguing relationship between diatonic triadic cycles and parsimonious triadic cycles. As will be seen in what follows, diatonic and parsimonious triadic cycles differ only by the *control parameters* (to be defined later) of the 3-through-7-through-12 dynamical configuration. Moreover, the control parameters for parsimonious triadic cycles suggest what will be referred to later as a *stroboscopic diatonicism* within these cycles.

4.4 Maximally Even Sets and the Dynamics of Diatonic Sets

As it turns out, there is a strong connection between filtered point-symmetry and pcsets known as *maximally even sets*. The initial work on the theory of maximally even sets was done by Clough and Douthett.[9] In their formalism they adopt the *floor function*, also known as the *greatest integer function*:

$$\lfloor x \rfloor = \text{the greatest integer less than or equal to } x.$$

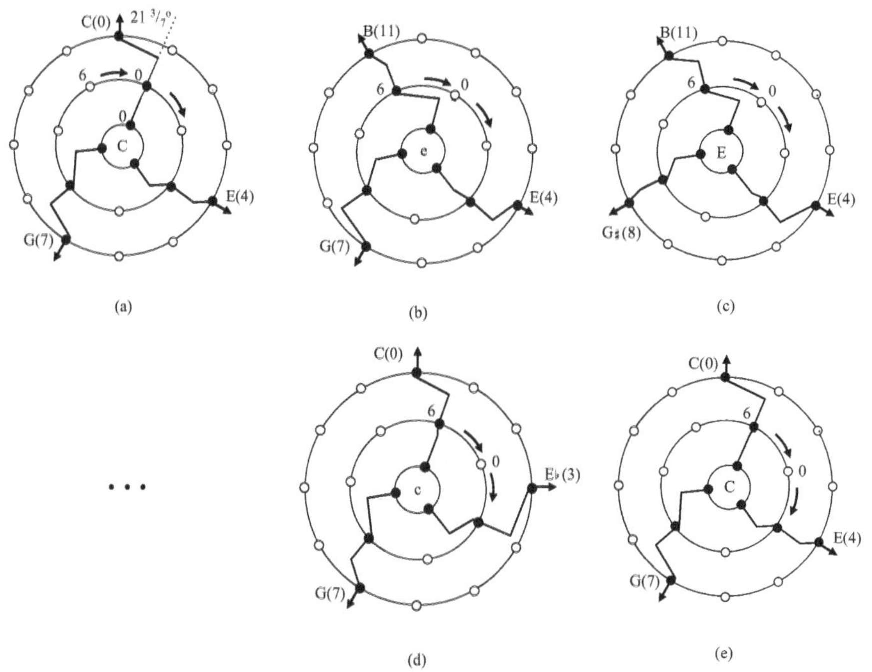

Figure 4.10. A 3-through-7-through-12 Hexatonic System (LP-Cycle).

For example, $\lfloor 3.9 \rfloor = \lfloor 3.5 \rfloor = \lfloor 3.1 \rfloor = \lfloor 3.0 \rfloor = 3$. They let c represent the *chromatic cardinality* (number of divisions to the octave), d represent the *diatonic cardinality* (number of pcs in the scale), and m—which they call the *mode index*—be a non-negative integer less than or equal to $c-1$. Then their *J-function* is defined as follows:

$$J_{c,d}^{m}(k) = \left\lfloor \frac{ck+m}{d} \right\rfloor$$

where k is an integer between 0 and $d-1$, inclusive. The *maximally even set* with given parameters c, d, and m is the set

$$J_{c,d}^{m} = \left\{ J_{c,d}^{m}(k) \right\}_{k=0}^{d-1} = \left\{ J_{c,d}^{m}(0), J_{c,d}^{m}(1), J_{c,d}^{m}(2), \cdots, J_{c,d}^{m}(d-1) \right\}.$$

This algorithm is known as the *maximally even set algorithm*, and the symbol $J_{c,d}^{m}$ is called the *J-representation* of the corresponding maximally even set. For fixed c and d (chromatic and diatonic cardinalities), there is a unique sc whose members are maximally even sets. To be precise, for a given c and d, the sc of maximally even sets is

$$\left\{ J^{0}_{c,d}, J^{(c,d)}_{c,d}, J^{2(c,d)}_{c,d}, \ldots, J^{c-(c,d)}_{c,d} \right\},$$

where (c,d) is the greatest common divisor of c and d. In the modulo 12 universe, well-known maximally even sets include augmented triads, diminished seventh chords, "black key" pentatonic scales, diatonic scales, and the all-combinatorial octatonic and enneatonic sets from scs 8-28 and 9-12, respectively.

As mentioned in section 4.2, a *Cohn cycle* is a cycle of 3 or more pcsets from the same sc in which adjacent pcsets are $P_{1,0}$-related. *Toggling Cohn cycles* were also discussed in that section. A *unidirectional Cohn cycle* is a Cohn cycle in which every pcset in the cycle can be determined from the counterclockwise adjacent pcset by moving a pc in the same direction by a half step (hence the term *unidirectional*). Lewin showed that a Cohn cycle is unidirectional if and only if the cycle includes every pcset in the sc.[10] The best known musical cycle is the cycle of fifths, which is an example of a unidirectional Cohn cycle: G major can be formed from C major by moving the note F *up* to F♯; D major can be formed from G major by moving the note C *up* to C♯; etc. This generates a cycle in which every diatonic set is included.

David Clampitt has shown that the members of a sc can form a unidirectional Cohn cycle if and only if the sets are maximally even and c and d are coprime.[11] In the case of the diatonic sc, the sets $J^{m}_{12,7}$ where m ranges from 0 to 11, inclusive, constitute the 12 diatonic sets: the set $J^{0}_{12,7}$ is the D♭ diatonic set; $J^{1}_{12,7}$ is the A♭ diatonic set; $J^{2}_{12,7}$ is the E♭ diatonic set; etc. Two diatonic sets are $P_{1,0}$-related (closely related keys) if and only if the mode indices of their *J*-representations differ by 1 (mod 12). Unidirectional Cohn cycles can be thought of as a type of "generalized cycle of fifths." Clampitt's observation of the connection between unidirectional Cohn cycles and maximally even sets implies that the maximally even algorithm is a convenient algorithm for generating these cycles.

The connection between maximally even sets and filtered point-symmetry can be seen in figure 4.11, which is a blowup of the top part of figure 4.8a. Suppose the length of the circumference of the filter is 12. If this distance is measured clockwise starting at hole 0, then the hole numbers are the same as the hole distances from hole 0. Beam 0 hits hole 0 and travels through. Beam 1 hits the filter at a distance of $12/7 = 1\,5/7$ from hole 0 and moves counterclockwise until it encounters a hole, which will have hole number $\lfloor 1\,5/7 \rfloor = 1$. Beam 6 hits the circumference of the filter at a distance of $12 \cdot 6/7 = 72/7 = 10\,2/7$ and passes through the first hole counterclockwise to its collision, which is hole $\lfloor 10\,2/7 \rfloor = 10$. In general, beam k will hit the circumference of the filter at a distance of $12k/7$ and travels counterclockwise until it encounters a hole, which will have hole number $\lfloor 12k/7 \rfloor$. It follows that the beam set is

$$\left\{ \left\lfloor \frac{12 \cdot 0}{7} \right\rfloor, \left\lfloor \frac{12 \cdot 1}{7} \right\rfloor, \left\lfloor \frac{12 \cdot 2}{7} \right\rfloor, \ldots, \left\lfloor \frac{12 \cdot 6}{7} \right\rfloor \right\} = J^{0}_{12,7}.$$

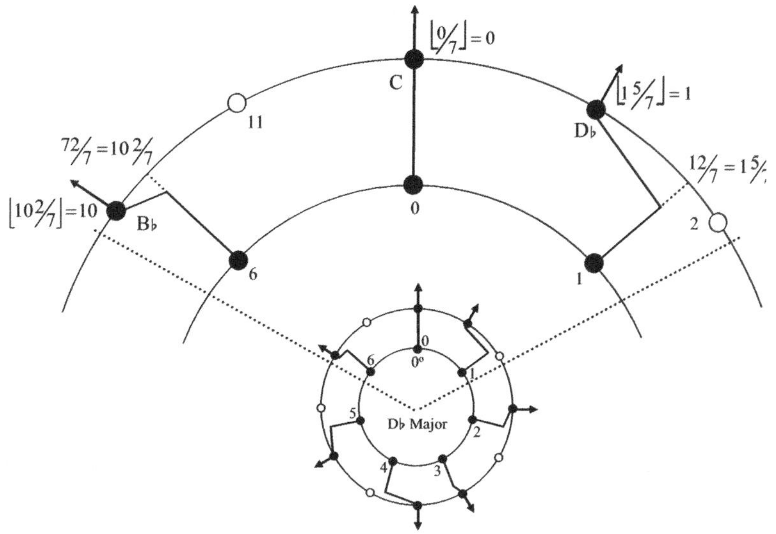

Figure 4.11. A Blowup of the Top Part of the D♭ Major Scale in figure 4.8a.

The other sets in figure 4.8 can be traced out in the same way and yield the sets $J^1_{12,7}$, $J^2_{12,7}$, $J^3_{12,7}$, etc.

Although Clough and Douthett always assume the mode index is an integer, this is not a necessary constraint. If the beacon in figure 4.8a is rotated clockwise by $\frac{1}{168}$ th of a revolution ($2\frac{1}{7}$°), then all the beams collide with the filter's circumference, hitting the circumference at distances of $\frac{12\cdot 0}{7}+\frac{1}{14}$, $\frac{12\cdot 1}{7}+\frac{1}{14}$, ..., $\frac{12\cdot 6}{7}+\frac{1}{14}$. The set of hole numbers immediately counterclockwise of these collisions is

$$\left\{\left\lfloor\frac{12\cdot 0+0.5}{7}\right\rfloor,\left\lfloor\frac{12\cdot 1+0.5}{7}\right\rfloor,\left\lfloor\frac{12\cdot 2+0.5}{7}\right\rfloor,\ldots,\left\lfloor\frac{12\cdot 6+0.5}{7}\right\rfloor\right\}.$$

If non-integers are allowed as mode indices, then the *J*-representation of the above set is $J^{0.5}_{12,7}$. When the calculations in the above set are executed, the resultant set is the same as $J^0_{12,7}$; that is, $J^{0.5}_{12,7}=J^0_{12,7}$. As the mode index *approaches* 1, the beam set remains the same. It is not until the mode index *reaches* 1 that the beacon rotates far enough ($4\frac{2}{7}$°) to change to the beam set to the configuration in figure 4.8b. Thus, for x in the interval $[0,1)$ (i.e., $0\le x<1$), $J^0_{12,7}=J^x_{12,7}$. Similarly, if $x\in[1,2)$ then $J^1_{12,7}=J^x_{12,7}$; if $x\in[2,3)$ then $J^2_{12,7}=J^x_{12,7}$; etc.

Before applying all this to dynamical systems, there are a few small adjustments that need to be made in the maximally even sets algorithm. Clough and Douthett assumed that $0\le m\le c-1$ and $0\le k\le d-1$. In doing so, the

J-functions, which represent pcs, are guaranteed to be within the pitch-class space. This restriction will be relaxed, and, in order to guarantee that the J-functions remain within pitch-class space, the J-functions must be reduced modulo c. Now replace m with t in the maximally even algorithm, and think of t as representing time in seconds. Then for each t, the diatonic set $J_{12,7}^{t+1}$ occurs one second after the diatonic set $J_{12,7}^{t}$, and these two diatonic sets will be $P_{1,0}$-related (closely related keys). This constitutes a dynamical system that evolves over time in which the cycle of fifths is a *periodic orbit* (cycle) *of period 12*. Whence, every 12 seconds the cycle of fifths begins again. We use the notation $\left\langle J_{12,7}^{t} \right\rangle_{12}$ to represent this system and will call this representation a *dynamical representation of the cycle*. In this dynamical representation, the subscript 12 outside the angled brackets is the time duration needed to complete one cycle. Since the beacon rotates $\frac{1}{84}$ th of a revolution per second ($4\frac{2}{7}$° per sec.), the *beacon's frequency* is $f = \frac{1}{84}$ cycles per second (cps).

It should be noted that the cycle of fifths (or any cycle that has a dynamical representation) does not have a unique dynamical representation. For example, $\left\langle J_{12,7}^{t+C} \right\rangle_{12}$, where C is any real constant, is a dynamical representation for the cycle of fifths. The choice of C determines the diatonic set on which the cycle begins (the *initial* diatonic set). Moreover, $\left\langle J_{12,7}^{-t} \right\rangle_{12}$ is also a dynamical representation of the cycle of fifths. In this case, the cycle is generated counterclockwise. This flexibility can be of use, as will be illustrated later when triadic sequences in the music of Brahms and Beethoven are discussed.

Next, we impose what is known as a *stroboscopic portrait* (or *stroboscopic record*) on the system. A "snapshot" is taken every second, and the beam sets are recorded. The record of beam sets over time is called the *stroboscopic portrait*. The stroboscopic portrait of diatonic scales for the dynamical representation $\left\langle J_{12,7}^{t} \right\rangle_{12}$ is the sequence of scales in the third row of table 4.1. The second row gives the stroboscopic portrait of their J-representations. Other pertinent information such as the time the snapshot is taken (starting at $t = 0$), clockwise angle of lamp 0 from north, and the beacon frequency is also given in table 4.1.

Now suppose this frequency is doubled, and $f = \frac{2}{84} = \frac{1}{42}$ cps. Snapshots taken every second will skip every other diatonic set in the cycle of fifths, resulting in a cycle of length 6 in figure 4.12. The time duration needed to complete this cycle is 6 seconds, and a dynamical representation for this cycle is $\left\langle J_{12,7}^{2t} \right\rangle_{6}$. This is one of two cycles of diatonic sets in which adjacent sets are $P_{2,0}$-related, and its stroboscopic portrait is given in table 4.2. Note that when the frequency is doubled, so is the coefficient of t in the dynamical representation (compare $\left\langle J_{12,7}^{t} \right\rangle_{12}$, which has frequency $\frac{1}{84}$ cps with $\left\langle J_{12,7}^{2t} \right\rangle_{6}$, which has frequency $\frac{1}{42}$ cps). In general, this coefficient, called the *frequency number* and notated $\hat{f}$, varies directly as f. More specifically, the relationship between f and $\hat{f}$ is given by the equation

$$f = \frac{\hat{f}}{cd},$$

Table 4.1 Stroboscopic Portrait of the Cycle $\left\langle J^{t}_{12,7} \right\rangle_{12}$

t	0	1	2	. . .	10	11	12
$\left\langle J^{t}_{12,7} \right\rangle_{12}$	$J^{0}_{12,7}$	$J^{1}_{12,7}$	$J^{2}_{12,7}$	. . .	$J^{10}_{12,7}$	$J^{11}_{12,7}$	$J^{12}_{12,7}$
Diatonic Scales	C♯ Major	A♭ Major	E♭ Major	. . .	B Major	F♯ Major	C♯ Major
Lamp 0 Angle from North	0°	$4\tfrac{2}{7}^{\circ}$	$8\tfrac{4}{7}^{\circ}$	. . .	$42\tfrac{6}{7}^{\circ}$	$47\tfrac{1}{7}^{\circ}$	$51\tfrac{3}{7}^{\circ}$

$f = \tfrac{1}{84}\ \text{cps}\left(4\tfrac{2}{7}^{\circ}\ \text{per sec.}\right)$

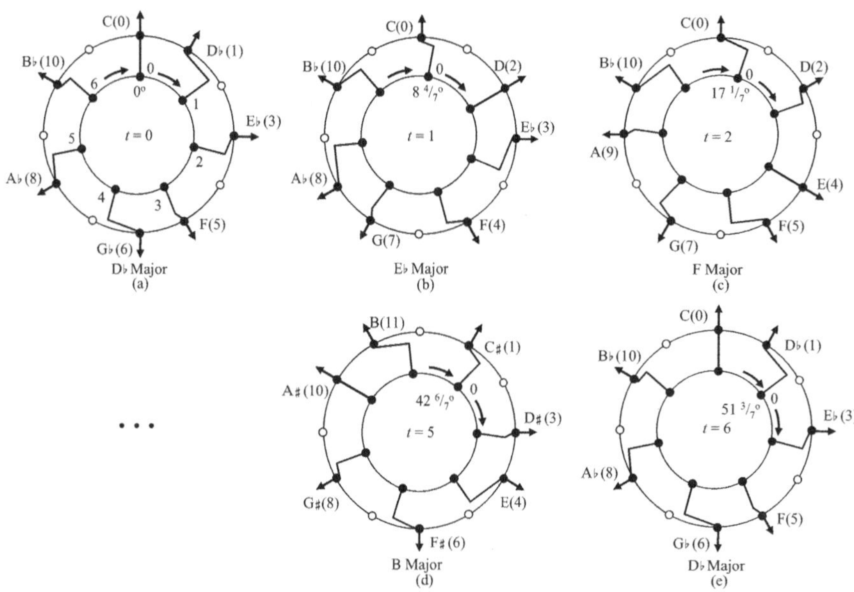

Figure 4.12. $P_{2,0}$-Related Diatonic Sets with $f = \tfrac{1}{42}$ cps ($8\tfrac{4}{7}^{\circ}$ per sec).

where c is the number of holes in the filter and d is the number of lamps on the beacon. Thus, for the cycle of fifths, the frequency number is $\hat{f} = 1$, and for the cycle of $P_{2,0}$-related diatonic sets, the frequency number is $\hat{f} = 2$. The frequencies are called *control parameters.* With only a beacon and one filter and a fixed frequency f, the *period* (cycle length) is the smallest positive integer T satisfying the equation

$$T\hat{f} \equiv 0 \pmod{c}.$$

Table 4.2 Stroboscopic Portrait of the Cycle $\left\langle J_{12,7}^{2t} \right\rangle_{12}$

t	0	1	2	3	4	5	6
$\left\langle J_{12,7}^{2t} \right\rangle_{6}$	$J_{12,7}^{0}$	$J_{12,7}^{2}$	$J_{12,7}^{4}$	$J_{12,7}^{6}$	$J_{12,7}^{8}$	$J_{12,7}^{10}$	$J_{12,7}^{12}$
Diatonic Scales	C♯ Major	E♭ Major	F Major	G Major	A Major	B Major	C♯ Major
Lamp 0 Angle from North	0°	$8\,\frac{4}{7}$°	$17\,\frac{1}{7}$°	$25\,\frac{5}{7}$°	$34\,\frac{2}{7}$°	$42\,\frac{6}{7}$°	$51\,\frac{3}{7}$°

$f = \frac{1}{42}$ cps $\left(8\frac{4}{7}° \text{ per sec.}\right)$

For the cycle of fifths, $c = 12$, $d = 7$, and $f = \frac{1}{84}$ cps (i.e., $\hat{f} = 1$), and the smallest positive integer satisfying the above equation is $T = 12$. For the cycle of $P_{2,0}$-related diatonic sets, $f = \frac{1}{42}$ cps ($\hat{f} = 2$), and the smallest positive integer satisfying the above equation is $T = 6$. Note that if the frequency (and hence, the frequency number) is an irrational number (e.g., $\sqrt{2}$, π, ...) then there is no integer solution for T in the equation above. This results in a non-periodic sequence of diatonic sets. Whence, while $\hat{f} = 1$ yields a cycle of diatonic sets with period 12, $\hat{f} = \sqrt{1.0000000001}$ (an irrational number very close to 1) results in a sequence of diatonic sets that is not periodic.[12] In fact, if the frequency is changed by any small amount at all, the period of the orbit, if one exists, will change dramatically. Hence, small changes in the control parameters can make big differences in the result. This is common behavior in dynamical systems.[13]

As mentioned above, the cycle of $P_{2,0}$-related diatonic sets shown in figure 4.12 and table 4.2 is one of two such cycles. To get the other cycle of $P_{2,0}$-related diatonic sets, the frequency remains the same, but the *phase of the beacon* (clockwise angle of lamp 0 from north at $t = 0$) must be changed. If the beacon is rotated clockwise by $\phi = 4\,\frac{2}{7}°$ before the clock is started (the *phase* of the beacon), then at $t = 0$ the beam set is $J_{12,7}^{1}$, which represents the A♭ diatonic set in figure 4.13a. At $t = 1$ the beacon has rotated $\frac{1}{42}$ th of a revolution ($8\,\frac{4}{7}$°), and the beam set is $J_{12,7}^{3}$, as shown in figure 4.13b. In effect, the stroboscope has "filtered out" $J_{12,7}^{2}$. This continues until the cycle is completed at $t = 6$; see figure 4.13e. Without labels, the configuration in figure 4.13e is the same as that in figure 4.13a, signaling the completion of the cycle. A dynamical representation of this cycle is $\left\langle J_{12,7}^{2t+1} \right\rangle_{6}$, and its stroboscopic portrait is given in table 4.3. Note that the superscript inside the angled brackets, $2t+1$, is a linear function. The coefficient of t is the frequency number, and the constant term, called the *phase number*, is related to the phase of the beacon. The phase number in this dynamical representation is $\hat{\phi} = 1$. With this phase number, the initial diatonic set (the set at $t = 0$) is A♭ major. If a different odd integer is chosen as the phase number,

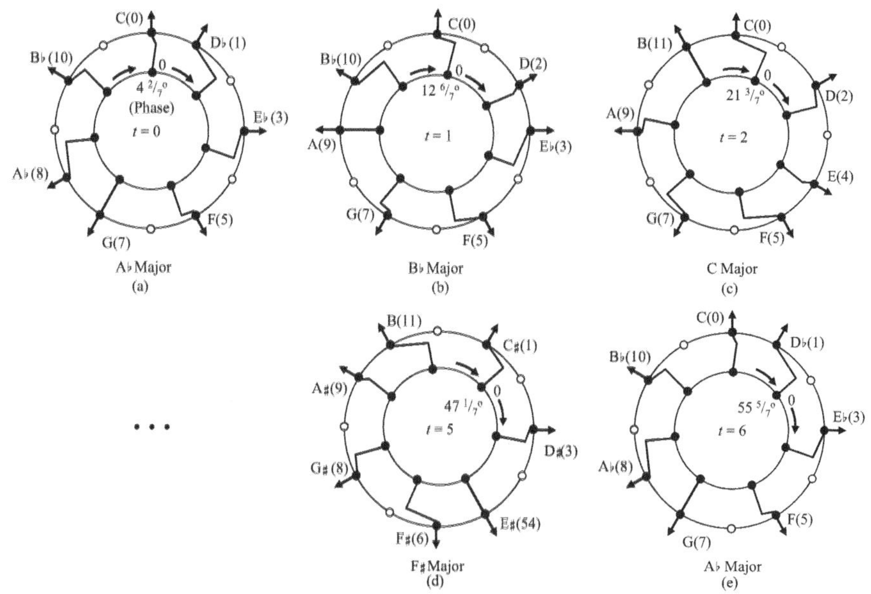

Figure 4.13. P_2-Related Diatonic Sets with $f = \frac{1}{42}$ cps and $\phi = 4\frac{2}{7}°$.

Table 4.3 Stroboscopic Portrait of the Cycle $\left\langle J_{12,7}^{2t+1} \right\rangle_{12}$

t	0	1	2	3	4	5	6
$\left\langle J_{12,7}^{2t+1} \right\rangle_6$	$J_{12,7}^{1}$	$J_{12,7}^{3}$	$J_{12,7}^{5}$	$J_{12,7}^{7}$	$J_{12,7}^{9}$	$J_{12,7}^{11}$	$J_{12,7}^{13}$
Diatonic Scales	A♭ Major	B♭ Major	C Major	D Major	E Major	F♯ Major	A♭ Major
Lamp 0 Angle from North	$4\frac{2}{7}°$	$12\frac{6}{7}°$	$21\frac{3}{7}°$	$30°$	$38\frac{4}{7}°$	$47\frac{1}{7}°$	$55\frac{5}{7}°$

$f = \frac{1}{42}$ cps $\left(8\frac{4}{7}° \text{ per sec.}\right)$ and $\phi = 4\frac{2}{7}°$

the same cycle will result, although the initial diatonic set might be different. The phase, ϕ, and the phase number, $\hat{\phi}$, relate as follows:

$$\phi = 360\frac{\hat{\phi}}{cd}$$

(the phases and the phase numbers in figures 10 and 12 and tables 4.1 and 4.2 are 0°). In addition to the frequency, the phase is also a control parameter. Small changes of the phase (or phase number) are not always as critical as small changes in the frequency. For example, $\left\langle J_{12,7}^{2t}\right\rangle_6 = \left\langle J_{12,7}^{2t+\hat{\phi}}\right\rangle_6$ for any $\hat{\phi} \in [0,1)$ (i.e., $0^\circ \leq \phi < 4\,\frac{2}{7}{}^\circ$).

Note that what was called the *mode index* (an integer between 0 and $c-1$, inclusive) has been replaced by a *mode function*. In this case, the mode function is a linear function of the form $m(t) = \hat{f}t + \hat{\phi}$. In general, for any fixed *c*, *d*, phase ϕ, and *rational* frequency *f*, a dynamical representation of the resultant cycle is $\left\langle J_{c,d}^{\hat{f}t+\hat{\phi}}\right\rangle_T$ where $\hat{f}$, $\hat{\phi}$, and *T* are calculated by the equations above.

4.5 Iterated Maximally Even Sets and Multiple Filters

With more than one filter, the dynamics gets a bit more complicated. Simple maximally even sets no longer do the job, and what are known as *n^{th}-order maximally even sets* must be introduced. In the early 1990s, these sets were referred to informally.[14] But it was not until 1997 that Clough, Cuciurean, and Douthett gave a formal definition.[15] Since there are multiple filters, it will be convenient to let d_0 play the role that *c* played previously and d_1 play the role of *d*. An *n^{th}-order J-function* is defined as follows:[16]

$$J_{d_0,d_1,d_2,\cdots,d_n}^{m_1,m_2,\cdots,m_n}(k) = J_{d_0,d_1}^{m_1}\left(J_{d_1,d_2}^{m_2}\left(J_{d_2,d_3}^{m_3}\left(\cdots J_{d_{n-1},d_n}^{m_n}(k)\right)\right)\right),$$

where d_j and m_j are integers, $d_0 > d_1 > d_2 > \cdots > d_n > 0$, $0 \leq m_j \leq d_{j-1} - 1$, and $0 \leq k \leq d_n - 1$. Then the *n^{th}-order maximally even set* with these parameters is given by

$$\begin{aligned} J_{d_0,d_1,d_2,\cdots,d_n}^{m_1,m_2,\cdots,m_n} &= \left\{ J_{d_0,d_1,d_2,\cdots,d_n}^{m_1,m_2,\cdots,m_n}(k)\right\}_{k=0}^{d_n-1} \\ &= \left\{ J_{d_0,d_1,d_2,\cdots,d_n}^{m_1,m_2,\cdots,m_n}(0), J_{d_0,d_1,d_2,\cdots,d_n}^{m_1,m_2,\cdots,m_n}(1), \cdots, J_{d_0,d_1,d_2,\cdots,d_n}^{m_1,m_2,\cdots,m_n}(d_n - 1)\right\}. \end{aligned}$$

These n^{th}-order maximally even sets are also called *iterated maximally even sets*.

Diatonic Triadic Cycles: Noting that $J_{12,7}^5$ is the *J*-representation of the C major scale, the second-order maximally even sets $J_{12,7,3}^{5,0}$, $J_{12,7,3}^{5,1}$, …, $J_{12,7,3}^{5,6}$ are the *J*-representations of the triads embedded in the C major scale: C, a, F, d, b dim, G, and e, respectively.[17] These are the same triads, in the same order, as those in the mediant-submediant cycle discussed in the second section. One can see how these triads are related to the dynamics of the concentric circles by comparing the following calculation with figure 4.14, which is figure 4.9b:

$$\begin{aligned} J_{12,7,3}^{5,0} &= \left\{ J_{12,7}^5\left(J_{7,3}^0(0)\right), J_{12,7}^5\left(J_{7,3}^0(1)\right), J_{12,7}^5\left(J_{7,3}^0(2)\right)\right\} \\ &= \left\{ J_{12,7}^5\left(\left\lfloor \tfrac{0}{3}\right\rfloor\right), J_{12,7}^5\left(\left\lfloor 2\tfrac{1}{3}\right\rfloor\right), J_{12,7}^5\left(\left\lfloor 4\tfrac{2}{3}\right\rfloor\right)\right\} \end{aligned}$$

$$= \{ J^5_{12,7}(0), J^5_{12,7}(2), J^5_{12,7}(4) \}$$

$$= \left\{ \left\lfloor \tfrac{5}{7} \right\rfloor, \left\lfloor 4\tfrac{1}{7} \right\rfloor, \left\lfloor 7\tfrac{4}{7} \right\rfloor \right\}$$

$$= \{0, 4, 7\}.$$

First, suppose the 7-hole filter in figure 4.14 has a circumference length of 7, and measure this distance clockwise, beginning at hole 0 in the 7-hole filter. Then beam 0 passes through hole 0, and beams 1 and 2 hit the circumference of the 7-hole filter at distances of $2\tfrac{1}{3}$ and $4\tfrac{2}{3}$, respectively. These are the numbers inside the floor functions in line two of the calculation. The floor functions bring the beams to the closest hole counterclockwise of their collisions with the circumference, yielding the arguments of the *J*-functions in line three of the equation. If the outside filter has a circumference of length 12 and the distance is measured clockwise from hole 0, then the beams hit the circumference of this filter at distances of $\tfrac{5}{7}$, $4\tfrac{1}{7}$, and $7\tfrac{4}{7}$, the numbers inside the floor functions in line four of the calculations above. Finally, the floor functions bring the beams to the closest holes counterclockwise of their collisions with the circumference, resulting in the beam set in line five of the calculation.

By allowing m_j to be replaced by a real valued function of time, $m_j(t)$, and requiring the value of the jth-order *J*-functions $J^{m_j(t)}_{d_{j-1},d_j}(k)$ to be reduced modulo d_{j-1}, the cycle of triads embedded in the C major scale shown in figure 4.1 has a dynamical representation of $\left\langle J^{5,t}_{12,7,3} \right\rangle_7$; its stroboscopic portrait is given in table 4.4.

In general, we will call the outside filter *circle 0*. This filter has d_0 holes and remains fixed with the 0 hole in the north position. So, the frequency and phase of circle 0 are always 0 cps and $0°$, respectively. The next filter (beacon) will be called *circle 1*, and the number of holes (lamps) in this circle is d_1. Its frequency, f_1, and phase, ϕ_1, can be determined by $\hat{f}_1$ and $\hat{\phi}_1$ in the mode function $m_1(t) = \hat{f}_1 t + \hat{\phi}_1$, as discussed in the previous section. The frequency and phase of the next circle, called *circle 2*, is a little more complicated to calculate. Assuming the mode function of circle 2 is $m_2(t) = \hat{f}_2 t + \hat{\phi}_2$, calculations of the frequency and phase as in the previous section will result in a frequency and phase of circle 2 relative to circle 1, not necessarily relative to north. For example, in figure 4.9b the phase number of the mode function $m_2(t) = t + 0$ is $\hat{\phi}_2 = 0$ (the constant term), implying the phase of circle 2 relative to circle 1 is $0°$. But since the phase of circle 1 relative to north is $\phi_1 = 21\tfrac{3}{7}°$, the phase of that beacon relative to north is the sum of these two phases: $\phi_2 = 21\tfrac{3}{7}° + 0° = 21\tfrac{3}{7}°$. In general, the phase of circle n (relative to north), is the sum of the phases of circles k relative to circles $k-1$, $1 \le k \le n$. This is also true of the frequencies. Whence, if f_k^{rel} and ϕ_k^{rel} are the frequency and phase of circle k relative to circle $k-1$, then

$$f_k^{\text{rel}} = \frac{\hat{f}_k}{d_k d_{k-1}} \quad \text{and} \quad \phi_k^{\text{rel}} = 360\frac{\hat{\phi}_k}{d_k d_{k-1}}.$$

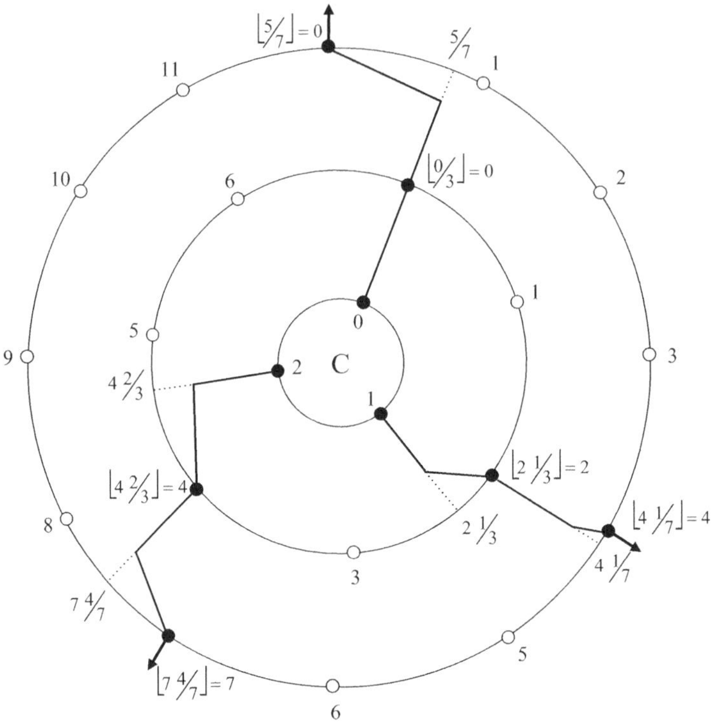

Figure 4.14. Blowup of figure 4.9b.

Table 4.4 Stroboscopic Portrait of the Cycle $\left\langle J^{5,t}_{12,7,3} \right\rangle_7$ (Mediant-Submediant Cycle)

t	0	1	2	3	4	5	6	7
$\left\langle J^{5,t}_{12,7,3} \right\rangle_7$	$J^{5,0}_{12,7,3}$	$J^{5,1}_{12,7,3}$	$J^{5,2}_{12,7,3}$	$J^{5,3}_{12,7,3}$	$J^{5,4}_{12,7,3}$	$J^{5,5}_{12,7,3}$	$J^{5,6}_{12,7,3}$	$J^{5,7}_{12,7,3}$
Diatonic Triads	C	a	F	d	b dim	G	e	C
Lamp 0 Angle from North	$21\,^3\!/_7{}^\circ$	$38\,^4\!/_7{}^\circ$	$55\,^5\!/_7{}^\circ$	$72\,^6\!/_7{}^\circ$	90°	$107\,^1\!/_7{}^\circ$	$124\,^2\!/_7{}^\circ$	$141\,^3\!/_7{}^\circ$

$f_1 = 0$ cps (0° per sec.) and $\phi_1 = 21\,^3\!/_7{}^\circ$

$f_2 = {}^1\!/_{21}$ cps ($17\,^1\!/_7{}^\circ$ per sec.) and $\phi_2 = 21\,^3\!/_7{}^\circ$

The frequency and phase of circle n (relative to north) are

$$f_n = \sum_{k=1}^{n} f_k^{\text{rel}} = \sum_{k=1}^{n} \frac{\hat{f}_k}{d_k d_{k-1}} \quad \text{and} \quad \phi_n = \sum_{k=1}^{n} \phi_k^{\text{rel}} = 360 \sum_{k=1}^{n} \frac{\hat{\phi}_k}{d_k d_{k-1}},$$

Table 4.5 Stroboscopic Portrait of the Cycle $\left\langle J^{5,2t}_{12,7,3} \right\rangle_7$ (Dominant-Subdominant Cycle)

t	0	1	2	3	4	5	6	7
$\left\langle J^{5,2t}_{12,7,3} \right\rangle_7$	$J^{5,0}_{12,7,3}$	$J^{5,2}_{12,7,3}$	$J^{5,4}_{12,7,3}$	$J^{5,6}_{12,7,3}$	$J^{5,8}_{12,7,3}$	$J^{5,10}_{12,7,3}$	$J^{5,12}_{12,7,3}$	$J^{5,14}_{12,7,3}$
Diatonic Triads	C	F	b dim	e	a	d	G	C
Lamp o Angle from North	$21\,3/7^\circ$	$55\,5/7^\circ$	90°	$124\,2/7^\circ$	$158\,4/7^\circ$	$192\,6/7^\circ$	$227\,1/7^\circ$	$261\,3/7^\circ$

$f_1 = 0$ cps (0° per sec.) and $\phi_1 = 21\,3/7^\circ$

$f_2 = 2/21$ cps $\left(34\,2/7^\circ \text{ per sec.}\right)$ and $\phi_2 = 21\,3/7^\circ$

Table 4.6 Stroboscopic Portrait of the Cycle $\left\langle J^{5,3t}_{12,7,3} \right\rangle_7$ (Supertonic-Leading Tone Cycle)

t	0	1	2	3	4	5	6	7
$\left\langle J^{5,3t}_{12,7,3} \right\rangle_7$	$J^{5,0}_{12,7,3}$	$J^{5,3}_{12,7,3}$	$J^{5,6}_{12,7,3}$	$J^{5,9}_{12,7,3}$	$J^{5,12}_{12,7,3}$	$J^{5,15}_{12,7,3}$	$J^{5,18}_{12,7,3}$	$J^{5,21}_{12,7,3}$
Diatonic Triads	C	d	e	F	G	a	b dim	C
Lamp o Angle from North	$21\,3/7^\circ$	$72\,6/7^\circ$	$124\,2/7^\circ$	$175\,5/7^\circ$	$227\,1/7^\circ$	$278\,4/7^\circ$	330°	$381\,3/7^\circ$

$f_1 = 0$ cps (0° per sec.) and $\phi_1 = 21\,3/7^\circ$

$f_2 = 1/7$ cps $\left(51\,3/7^\circ \text{ per sec.}\right)$ and $\phi_2 = 21\,3/7^\circ$

respectively (for circle 1, $f_1 = f_1^{\text{rel}}$ and $\phi_1 = \phi_1^{\text{rel}}$). These frequencies and phases are called the *control parameters* of the dynamical system. Thus, the control parameters of the configuration in figure 4.9b-e and table 4.4 are $f_1 = 0$, $f_2 = 1/21$, and $\phi_1 = \phi_2 = 21\,3/7^\circ$. The control parameters can be computed from their corresponding frequency numbers and phase numbers using the above equations. In table 4.4, the frequency numbers and phase numbers are $\hat{f}_1 = 0$, $\hat{\phi}_2 = 5$, $\hat{f}_2 = 1$, and $\hat{\phi}_2 = 0$. The frequency and phase of circle 1 are

$$f_1 = \sum_{k=1}^{1} \frac{\hat{f}_k}{d_k d_{k-1}} = \frac{0}{7 \cdot 12} = 0 \text{ cps and } \phi_1 = 360 \sum_{k=1}^{1} \frac{\hat{\phi}_k}{d_k d_{k-1}} = 360\left(\frac{5}{7 \cdot 12}\right) = 21\,\frac{3}{7}^\circ,$$

and for circle 2, the frequency and phase are

$$f_2 = \sum_{k=1}^{2} \frac{\hat{f}_k}{d_k d_{k-1}} = \frac{0}{7 \cdot 12} + \frac{1}{3 \cdot 7} = 1/21 \text{ cps} \quad \text{and}$$

$$\phi_2 = 360\sum_{k=1}^{2}\frac{\hat{\phi}_k}{d_k d_{k-1}} = 360\left(\frac{5}{7\cdot 12}+\frac{0}{3\cdot 7}\right) = 21\,{}^3\!/_7{}^\circ .$$

If the frequency for the beacon in the above is doubled, the resultant dynamical representation, $\left\langle J^{5,2t}_{12,7,3}\right\rangle_7$, represents the dominant-subdominant cycle in C major in figure 4.2, and if the frequency for the beacon is tripled, the dynamical representation, $\left\langle J^{5,3t}_{12,7,3}\right\rangle_7$, represents the supertonic-leading tone cycle in C major, as illustrated in figure 4.3. The stroboscopic portraits for these cycles are given in tables 4.5 and 4.6, respectively. If the signs of the frequencies (equivalently, frequency numbers) are reversed, the initial triads remain the same, but the cycles are generated in the opposite order. The initial triad in these cycles depends on the phase of the beacon. To get the cycles of triads embedded in the other diatonic scales, one need only change the phase of the 7-hole filter. For the embedded triads in D♭ major, choose $\phi_1 = 0^\circ$ ($\hat{\phi}_1 = 0$); for A♭ major, choose $\phi_1 = 4\,{}^2\!/_7{}^\circ$ ($\hat{\phi}_1 = 1$); for E♭ major, choose $\phi_1 = 8\,{}^4\!/_7{}^\circ$ ($\hat{\phi}_1 = 2$); etc.

Parsimonious Triadic Cycles: Now consider the LP-cycle in the last example discussed in section 4.3 (figure 4.10). In this case, the beacon is fixed, and circle 1 rotates clockwise. The phases of circle 1 and the beacon are the same as in the previous example, $\phi_1 = \phi_2 = 21\,{}^3\!/_7{}^\circ$ (equivalently, $\hat{\phi}_1 = 5$ and $\hat{\phi}_2 = 0$). A cycle of triads is completed when circle 1 rotates $51\,{}^3\!/_7{}^\circ$, which is where the unlabeled configurations of 7-hole filter and beacon are the same as the initial configuration; compare figure 4.10e with 4.10a. Since the cycle has length 6, the 7-hole filter must rotate ${}^1\!/_6{}^{\text{th}}$ of $51\,{}^3\!/_7{}^\circ$ ($=8\,{}^4\!/_7{}^\circ$) each second. Converting this into cps, the frequency of circle 1 is $f_1 = {}^1\!/_{42}$ cps. Thus, the frequency number of circle 1 is the solution to

$$ {}^1\!/_{42} = \frac{\hat{f}_1}{7\cdot 12}.$$

It follows that $\hat{f}_1 = 2$. Since the beacon is fixed, its frequency is $f_2 = 0$ cps. It follows that the frequency number of the beacon is the solution to

$$0 = f_2 = \sum_{k=1}^{2}\frac{\hat{f}_k}{d_k d_{k-1}} = \frac{2}{7\cdot 12}+\frac{\hat{f}_2}{7\cdot 3}.$$

Whence, $\hat{f}_2 = -0.5$ (although the beacon is fixed relative to north ($f_2 = 0$), the negative frequency number, $\hat{f}_2 = -0.5$, reflects the fact that the beacon rotates counterclockwise relative to circle 1). Putting the above together, a dynamical representation for this cycle is $\left\langle J^{2t+5,-0.5t}_{12,7,3}\right\rangle_6$, and its stroboscopic portrait is given in table 4.7.

As suggested in section 4.2, there are inherent difficulties in applying functional harmony to the "diatonically lost" LP-cycles. That said, there is a type of *stroboscopic diatonicism* suggested by the stroboscopic portrait in table 4.7. Note the first mode

Table 4.7 Stroboscopic Portrait of the Hexatonic System (LP-Cycle) $\left\langle J_{12,7,3}^{2t+5,-0.5t} \right\rangle_6$

t	0	1	2	3	4	5	6
$\left\langle J_{12,7,3}^{2t+5,-0.5t} \right\rangle_6$	$J_{12,7,3}^{5,0}$	$J_{12,7,3}^{7,-0.5}$	$J_{12,7,3}^{9,-1}$	$J_{12,7,3}^{11,-1.5}$	$J_{12,7,3}^{13,-2}$	$J_{12,7,3}^{15,-2.5}$	$J_{12,7,3}^{17,-3}$
Triads	C	e	E	g♯	A♭	c	C
Hole 0 Angle in Circle 1	$21\frac{3}{7}°$	$30°$	$38\frac{4}{7}°$	$47\frac{1}{7}°$	$55\frac{5}{7}°$	$64\frac{2}{7}°$	$72\frac{6}{7}°$
Triad Relation to Major Scale	I of C Major	ii of D Major	I of E Major	ii of F♯ Major	I of A♭ Major	ii of B♭ Major	I of C Major

$f_1 = \frac{1}{42}$ cps $\left(8\frac{4}{7}° \text{ per sec.}\right)$ and $\phi_1 = 21\frac{3}{7}°$

$f_2 = 0$ cps $(0° \text{ per sec.})$ and $\phi_2 = 21\frac{3}{7}°$

function in the dynamical representation, $m_1(t) = 2t + 5$. This function describes the relationship between circle 0 and circle 1. At $t = 0$, the clockwise angle of hole 0 from north in circle 1 is $21\frac{3}{7}°$, implying that the 7-hole filter is oriented in the C major scale position (temporarily, think of circle 1 as the beacon). Thus, the triad C can be considered as I of the C major scale, written *I of C major* (see row 5 in table 4.7). At $t = 1$, the 7-hole filter is oriented in the D major position, and the triad e can be interpreted as ii of D major. Similarly, the next triad, E, can be interpreted as I of E major, etc. Whence, the triads in this LP-cycle are, in a sense, "plucked" from the cycle of the $P_{2,0}$-related diatonic sets in figure 4.13.

If the strobe is turned off, a different *continuous diatonicism* becomes apparent in the LP-cycle. In this case, the 7-hole filter rotates continuously, traveling through all 12 diatonic sets before completing the LP-cycle. As illustrated in figure 4.15, hole 0 in the 7-hole filter begins at $21\frac{3}{7}°$ clockwise from north, with its orientation in C major. If the 7-hole filter rotates clockwise by any small amount, indicated by $21\frac{3}{7}+°$ in figure 4.15, the filter is still oriented in the C major position, but the triad has changed from C to e. At this point the triad e is iii of C major. When the filter rotates $4\frac{2}{7}°$ clockwise, the filter is oriented in the G major position, where e becomes vi of G major. After rotating another $4\frac{2}{7}°$, the triad e becomes ii of D major. In another $4\frac{2}{7}°$ the 7-hole filter is oriented in the A major position, where e changes to E, or V of A major. The 7-hole filter will *modulate* through the entire cycle of fifths before completing the LP-cycle, as illustrated in figure 4.15.

Larry Zbikowski suggests still another interpretation of this dynamical system.[18] Zbikowski observes that, without the 7-hole filter, the beam set is the C aug triad. Zbikowski suggests that the 7-hole filter *perturbs* the augmented triad. Depending on the orientation of the 7-hole filter, C aug is perturbed in one of 6 ways: three result in major triads that are $P_{1,0}$-related to C aug, and three result in minor

Clockwise Angle of Filter from North:	$21\frac{3}{7}°$	$21\frac{3}{7}+°$	$25\frac{5}{7}°$	$30°$	$34\frac{2}{7}°$	$38\frac{4}{7}°$	$38\frac{4}{7}+°$	$42\frac{6}{7}°$
Corresponding Key:	I of C Major	iii of C Major	vi of G Major	ii of D Major	V of A Major	I of E Major	iii of E Major	vi of B Major

Triad: C ⟷ e ⟷ E ⟷ g♯ ⟷

Clockwise Angle of Filter from North:	$47\frac{1}{7}°$	$51\frac{3}{7}°$	$55\frac{5}{7}°$	$55\frac{5}{7}+°$	$60°$	$64\frac{2}{7}°$	$68\frac{4}{7}°$	$72\frac{6}{7}°$
Corresponding Key:	ii of F♯ Major	V of D♭ Major	I of A♭ Major	iii of A♭ Major	vi of E♭ Major	ii of B♭ Major	V of F Major	I of C Major

Triad: (g♯) ⟷ A♭ ⟷ c ⟷ C ⟷

Figure 4.15. Continuous Relationship between the LP-Cycle and the Circle of Fifths.

triads that are $P_{2,0}$-related to C aug. This observation suggests what we will call the $\{0, 3, 4, 7, 8, 11\}$ *hexatonic Tonnetz* in figure 4.16a. The vertices in this *Tonnetz* represent the pcs in the hexachord, and each face represents the triad defined by the incident pcs (vertices). Each major triad in the *Tonnetz* has one edge in common with C aug, indicating that each of these triads shares two pcs with C aug. Moreover, the pcs not shared by the two triads are ic1-related. It follows that every major triad in the *Tonnetz* is $P_{1,0}$-related to C aug. Each minor triad in the *Tonnetz* shares precisely one pc with C aug, and the pcs not shared can be paired into two ic1-related dyads. Whence, each minor triad in the *Tonnetz* is $P_{2,0}$-related to C aug. The LP-cycle can be traced in figure 4.16a through adjacent faces in the "washer" around the C aug triad.

The *Tonnetz* in figure 4.16a is known as a *planar graph*, since no edges in the graph cross. When counting faces in this planar graph, not only are the faces inside the large circle counted, but the rest of the infinite plain—the area outside the large circle—is counted as a face as well. Noting that the vertices that define this face represent pcs 3, 7, and 11, this face represents E♭ aug.

The graph in figure 4.16a can be rearranged as the octahedron shown in figure 4.16b. As in figure 4.16a, the vertices in figure 4.16b represent pcs and the faces (triangles) of the octahedron represent triads. To avoid cluttering figure 4.16b, the names of the triads have been left out. The faces on the top of the octahedron represent the triads C aug, C, c, and A♭, while the faces on the bottom of the octahedron represent E, e, E♭ aug, g♯. This form of the $\{0, 3, 4, 7, 8, 11\}$ hexatonic *Tonnetz* may be a bit more appealing in that there is no "wandering

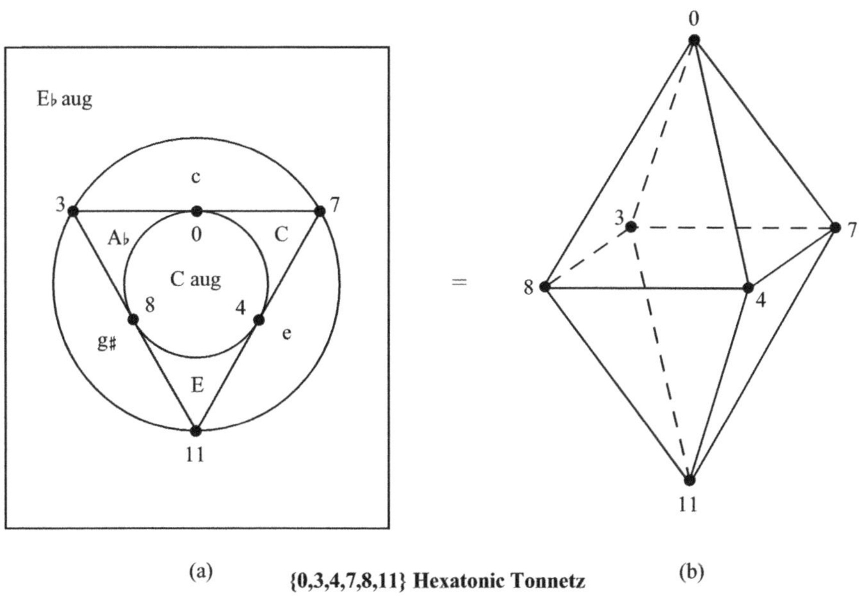

Figure 4.16. Equivalent Versions of the {0,3,4,7,8,11} Hexatonic *Tonnetz*.

triad" outside the octahedron; E♭ aug is now represented in the same way as the other triads, as a triangular face on the octahedron.

To the best of our knowledge, this octahedral representation of the hexatonic *Tonnetz* has not previously been used. However, the *dual* of this graph does appear in an article by Douthett and Steinbach.[19] Their graph can be obtained from the hexatonic *Tonnetz* through the following *dual construction*: Map the faces of the *Tonnetz* in figure 4.16a or 4.16b to vertices. Now connect any two of these new vertices with an edge if their corresponding faces in the hexatonic *Tonnetz* share an edge. This construction results in one of the cubes discussed by Douthett and Steinbach; see figure 4.17. If the same construction is applied to the cube, the hexatonic *Tonnetz* is returned. This *dual* construction implies that the hexatonic *Tonnetz* and Douthett and Steinbach's cube are two ways of representing the same information.

Since the triad faces of the *Tonnetz* become vertices in the cube, the vertices in the cube represent triads. Similarly, the pc vertices in the *Tonnetz* become faces in the cube. Whence, the faces on the cube represent pcs. The pc name of each face is the pc in the intersection of the triads that define the face. For example, the name of the face defined by the triads C, c, A♭, and C aug is 0, which is the only pc common to all four triads (i.e., $\mathrm{C} \cap \mathrm{c} \cap \mathrm{A}^{\flat} \cap \mathrm{C\ aug} = \{0\}$). The names of the 6 faces of the cube are the members in the all-combinatorial

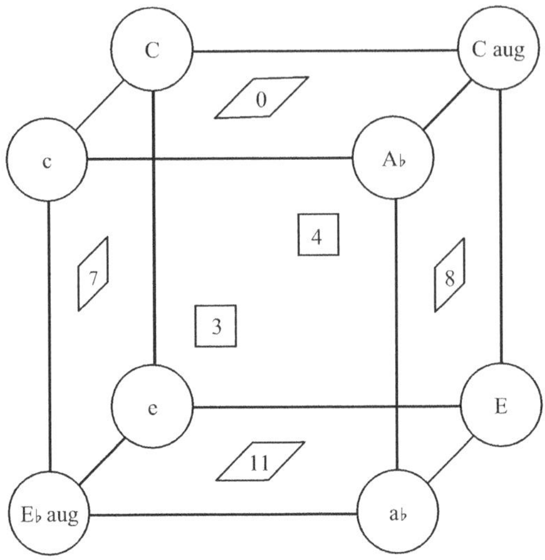

{0,3,4,7,8,11} Hexatonic Cube

Figure 4.17. Dual of the {0,3,4,7,8,11} Hexatonic *Tonnetz* (Hexatonic Cube).

hexachord $\{0, 3, 4, 7, 8, 11\}$. Moreover, the union of all the triads in the cube is also $\{0, 3, 4, 7, 8, 11\}$. Hence, we name this cube the $\{0, 3, 4, 7, 8, 11\}$ *hexatonic cube.* The bold edges of the cube in figure 4.17 trace the LP-cycle, and the union of this cube with the three cubes corresponding to the other hexachords in sc 6-20 yields Douthett and Steinbach's "Cube Dance," which is shown in figure 4.18. The cycles of bold edges in each of the cubes are the four LP-cycles; the dynamical data for these cycles are given in table 4.8.

The information in the bottom row of the table applies to all the LP-cycles. The triads in the cycles in Row 2 are given in the order in which they are generated by the dynamical representations in Row 5. For every 30° (clockwise) increase in the phase of the beacon ($\phi_1 = 21\frac{3}{7}°$, $51\frac{3}{7}°$, $81\frac{3}{7}°$, and $111\frac{3}{7}°$), a different LP-cycle is generated. If the phase of the beacon were to be increased by another 30° ($\phi_1 = 141\frac{3}{7}°$), the total increase from $\phi_1 = 21\frac{3}{7}°$ would be 120°. Without lamp labels, this configuration would be indistinguishable from the initial configuration ($\phi_1 = 21\frac{3}{7}°$). Whence, the LP-cycle in the second column of the first row would again be generated.

In his 1996 article, Cohn points out that one of these cycles appears in the first movement of Brahms's Concerto for Violin and Cello, Op. 102. In measures 270–78, the following sequence of triads appears:

A♭–a♭–E–e–C–c–A♭–a♭–E.

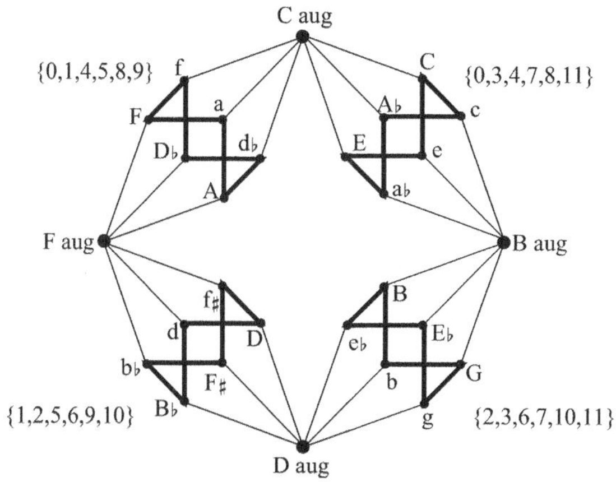

Cube Dance

Figure 4.18. The Union of the Four Hexatonic Cubes.

Table 4.8 Dynamical Information for the Hexatonic Systems (LP-Cycles)

Hexachord Parent	{0,3,4,7,8,11}	{0,1,4,5,8,9}	{1,2,5,6,9,10}	{0,3,4,7,8,11}
LP-Cycle	C–e–E–g♯–A♭–c–C	a–A–c♯–D♭–f–F–a	d–D–f♯–F♯–a♯–B♭–d	G–b–B–e♭–E♭–g–G
Second Mode Function	$m_2(t) = -0.5t$	$m_2(t) = -0.5t + 1.75$	$m_2(t) = -0.5t + 3.5$	$m_2(t) = -0.5t + 5.25$
Beacon Phase Angle	$\phi_2 = 21\,{}^3\!/_7\,°$	$\phi_2 = 51\,{}^3\!/_7\,°$	$\phi_2 = 81\,{}^3\!/_7\,°$	$\phi_2 = 111\,{}^3\!/_7\,°$
Dynamical Representation	$\left\langle J_{12,7,3}^{4t+5,-0.5t} \right\rangle_6$	$\left\langle J_{12,7,3}^{4t+5,-0.5t+1.75} \right\rangle_6$	$\left\langle J_{12,7,3}^{4t+5,-0.5t+3.5} \right\rangle_6$	$\left\langle J_{12,7,3}^{4t+5,-0.5t+5.25} \right\rangle_6$

$f_1 = {}^1\!/_{21}$ cps, $\phi_1 = 21\,{}^3\!/_7\,°$, $m_1(t) = 2t + 5$, and $f_2 = 0$ cps

If identifying the cycle in which the sequence appears is all that is needed, then $\left\langle J_{12,7,3}^{2t+5,-0.5t} \right\rangle_6$ is a satisfactory dynamical representation. However, because of their flexibility, dynamical representations can generate the sequence exactly as it appears here. First observe that the initial triad of the Brahms's sequence, A♭, appears at $t = 4$ in table 4.7. At $t = 4$, the *J*-representation of A♭ is $J_{12,7,3}^{13,-2}$ (see second row of table 4.7 or substitute 4 for *t* in $J_{12,7,3}^{2t+5,-0.5t}$). The superscripts of this representation now become the new phase numbers: $\hat{\phi}_1 = 13$ and $\hat{\phi}_2 = -2$, and the modified dynamical representation is $\left\langle J_{12,7,3}^{2t+13,-0.5t-2} \right\rangle_6$. The first nine triads generated by this representation (from $t = 0$ through $t = 8$) are

A♭–c–C–e–E–a♭–A♭—c–C.

The initial triad is correct, but the direction of the sequence is opposite to the sequence in the concerto. To reverse the direction of this sequence, one need only change the signs of the frequency numbers. This results in the dynamical representation $\left\langle J_{12,7,3}^{-2t+13,0.5t-2} \right\rangle_6$, which, for $t = 0$ through $t = 8$, corresponds to the exact order of the triads in the sequence in Brahms's concerto.

In addition to the LP-cycles, the other parsimonious cycles can also be represented dynamically. The dynamics of the octatonic sub-systems (PR-cycles) require that both the 7-hole filter and the beacon rotate. A 3-through-7-through-12 dynamical configuration for the $\{0, 1, 3, 4, 6, 7, 9, 10\}$ octatonic sub-system is shown in figure 4.19, and the pertinent dynamical information is given in table 4.9. As shown in table 4.9, the 7-hole filter rotates more slowly than the beacon ($f_1 = \frac{1}{56}$ cps and $f_2 = \frac{1}{24}$ cps), and the dynamical representation is $\left\langle J_{12,7,3}^{1.5t+2,0.5t-0.5} \right\rangle_8$. The stroboscopic diatonicism is given in row five of this table. At $t = 0$ the 7-hole filter is oriented in the E♭ major position. So, the E♭ triad is I of E♭ major; at $t = 1$ the 7-hole filter is oriented in the B♭ major position, and the triad c is ii of B♭ major; etc. Diatonic sets associated with adjacent triads in the PR-cycle are either $P_{1,0}$-related or $P_{2,0}$-related, and the dynamical representation that yields the cycle of diatonic sets in row 5 of table 4.9 is $\left\langle J_{12,7}^{1.5t+2} \right\rangle_8$. It is left to the reader to determine the continuous diatonicism of the cycle.

Recalling that there are three octatonic subsystems, there are two other PR-cycles that have dynamical representations in figure 4.6. The pertinent dynamical information for all three octatonic sub-systems is given in table 4.10. As with the LP-cycles, the frequencies for all three PR-cycles are the same, and it is the phases that differentiate one PR-cycle from another.

The final parsimonious cycle, the LR-cycle, contains all 24 harmonic triads. Table 4.11 gives the essential dynamical information for the LR-cycle shown in figure 4.20. In this cycle, the 7-hole filter rotates clockwise ($f_1 = \frac{1}{168}$ cps), and the beacon rotates counterclockwise ($f_2 = -\frac{1}{24}$ cps). The dynamical representation is $\left\langle J_{12,7,3}^{0.5t+5.5,-t-1} \right\rangle_{24}$, and the stroboscopic diatonicism is given in Row 6 of table 4.11. At $t = 0$ the 7-hole filter is oriented in the C-major position and the triad e is iii of C major. At $t = 1$ and $t = 2$ the 7-hole filter is oriented in the G-major position, and the triads G and b are I and iii of G major, respectively. At $t = 3$ and $t = 4$ the 7-hole filter is oriented in the D-major position, and the triads D and f♯ are I and iii of D major, respectively. The 7-hole filter continues to rotate through the cycle of fifths, generating the I and iii triads from each major scale. Adjacent triads from the same major scale in the LR-cycle are related by

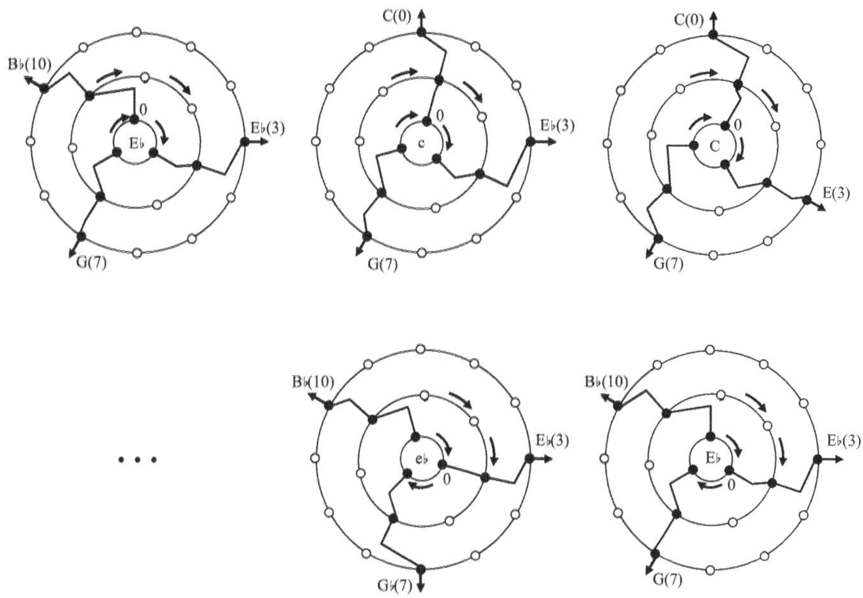

Figure 4.19. A 3-through-7-through-12 {0,1,3,4,6,7,10} Octatonic System (PR-Cycle).

Table 4.9 Stroboscopic Portrait of the $\{0, 1, 3, 4, 6, 7, 9, 10\}$ PR-Cycle $\langle J_{12,7,3}^{1.5t+2, 0.5t-0.5} \rangle_8$

t	0	1	2	3	4	5	6	7	8
$\langle J_{12,7,3}^{1.5t+2, 0.5t-0.5} \rangle_8$	$J_{12,7,3}^{2,-0.5}$	$J_{12,7,3}^{3.5,0}$	$J_{12,7,3}^{5,0.5}$	$J_{12,7,3}^{6.5,1}$	$J_{12,7,3}^{8,1.5}$	$J_{12,7,3}^{9.5,2}$	$J_{12,7,3}^{11,2.5}$	$J_{12,7,3}^{12.5,3}$	$J_{12,7,3}^{14,3.5}$
Triads	E♭	c	C	a	A	f♯	F♯	e♭	E♭
Hole 0 Angle On Circle 1	$8\frac{4}{7}°$	15°	$21\frac{3}{7}°$	$27\frac{6}{7}°$	$34\frac{2}{7}°$	$40\frac{5}{7}°$	$47\frac{1}{7}°$	$53\frac{4}{7}°$	60°
Lamp 0 Angle On Circle 2	15°	30°	45°	60°	75°	90°	105°	120°	135°
Triad Relation to Major Scale	I of E♭ Major	ii of B♭ Major	I of C Major	ii of G Major	I of A Major	ii of E Major	I of F♯ Major	ii of D♭ Major	I of E♭ Major

$f_1 = {}^{17}/_{840}$ cps $\left(6\frac{3}{7}°\text{ per sec.}\right)$ and $\phi_1 = 8\frac{4}{7}°$

$f_2 = {}^{1}/_{24}$ cps $(15°\text{ per sec.})$ and $\phi_2 = 0°$

the L transformation and, hence, are $P_{1,0}$-related. If the adjacent triads are from closely related scales, the triads are related by the R transformation, implying they are $P_{0,1}$-related.

Table 4.10 Dynamical Information for the Octatonic Systems (PR-Cycles)

Octachord Parent	{0,2,3,5,6,8,9,11}	{0,1,3,4,6,7,9,10}	{1,2,4,5,7,8,10,11}
PR-Cycle	A♭–f–F–d–D–b–B–g♯–A♭	E♭–c–C–a–A–f♯–F♯–e♭–E♭	B♭–g–G–e–E–c♯–D♭–b♭–B♭
First Mode Function	$m_1 = 1.5t + 1$	$m_1 = 1.5t + 2$	$m_1 = 1.5t + 3$
Second Mode Function	$m_2(t) = 0.5t + 1.5$	$m_2(t) = 0.5t - 0.5$	$m_2(t) = 0.5t - 2.5$
7-Hole Filter Phase Angle	$\phi_1 = 4\,^2\!/_7{}^\circ$	$\phi_1 = 8\,^4\!/_7{}^\circ$	$\phi_1 = 12\,^6\!/_7{}^\circ$
Beacon Phase Angle	$\phi_2 = 30^\circ$	$\phi_2 = 0^\circ$	$\phi_2 = -30^\circ$
Dynamical Representation	$\left\langle J_{12,7,3}^{1.5t+1,0.5t+1.5} \right\rangle_8$	$\left\langle J_{12,7,3}^{1.5t+2,0.5t-0.5} \right\rangle_8$	$\left\langle J_{12,7,3}^{1.5t+3,0.5t-2.5} \right\rangle_8$

$f_1 = {}^1\!/_{56}$ cps $\left(6\,^3\!/_7{}^\circ \text{ per sec.}\right)$ and $f_2 = {}^1\!/_{24}$ cps $(15^\circ$ per sec.)

Table 4.11 Stroboscopic Portrait of the LR-Cycle $\left\langle J_{12,7,3}^{0.5t+5,-t-1} \right\rangle_{24}$

t	0	1	2	3	4	. . .	22	23	24
$\left\langle J_{12,7,3}^{0.5t+5.5,-t-1} \right\rangle_{24}$	$J_{12,7,3}^{5.5,-1}$	$J_{12,7,3}^{6,-2}$	$J_{12,7,3}^{6.5,-3}$	$J_{12,7,3}^{7,-4}$	$J_{12,7,3}^{7.5,-5}$	. . .	$J_{12,7,3}^{16.5,-23}$	$J_{12,7,3}^{17,-24}$	$J_{12,7,3}^{17.5,-25}$
Triads	e	G	b	D	f♯	. . .	a	C	e
Hole 0 Angle On Circle 1	$6\,^3\!/_7{}^\circ$	$-8\,^4\!/_7{}^\circ$	$-23\,^4\!/_7{}^\circ$	$-38\,^4\!/_7{}^\circ$	$-53\,^4\!/_7{}^\circ$	. . .	$-323\,^4\!/_7{}^\circ$	$-338\,^4\!/_7{}^\circ$	$-353\,^4\!/_7{}^\circ$
Lamp 0 Angle On Circle 2	$23\,^4\!/_7{}^\circ$	$25\,^5\!/_7{}^\circ$	$27\,^6\!/_7{}^\circ$	30°	$32\,^1\!/_7{}^\circ$	. . .	$70\,^5\!/_7{}^\circ$	$72\,^6\!/_7{}^\circ$	75°
Triad Relation to Major Scale	iii of C Major	I of G Major	iii of G Major	I of D Major	iii of D Major	. . .	iii of F Major	I of C Major	iii of C Major

$f_1 = {}^1\!/_{168}$ cps $\left(2\,^1\!/_7{}^\circ \text{ per sec.}\right)$ and $\phi_1 = 23\,^4\!/_7{}^\circ$

$f_2 = {}^{-1}\!/_{24}$ cps $(-15^\circ$ per sec.) and $\phi_2 = 6\,^3\!/_7{}^\circ$

Cohn has examined a remarkable example of a sequence of triads from the LR-cycle in the second movement of Beethoven's Ninth Symphony, measures 143–76.[20] The sequence is

C–a–F–d–B♭–g–E♭–c–A♭–f–D♭–b♭–G♭–e♭–C♭–a♭—E–c♯–A.

This sequence is found in figure 4.7. As with the Brahms example, by choosing the appropriate control parameters, it is possible to generate this sequence beginning at $t = 0$. In table 4.11, the triad C appears at $t = 23$, and its J-representation is $J_{12,7,3}^{17,-24}$. So, for the sequence to begin on C, the phase numbers are $\hat{\phi}_1 = 17$ and $\hat{\phi}_2 = -24$. Note also that the Beethoven sequence proceeds in the opposite direction to that of the sequence in table 4.11. This means that the signs of the frequency numbers in the dynamical representation must be

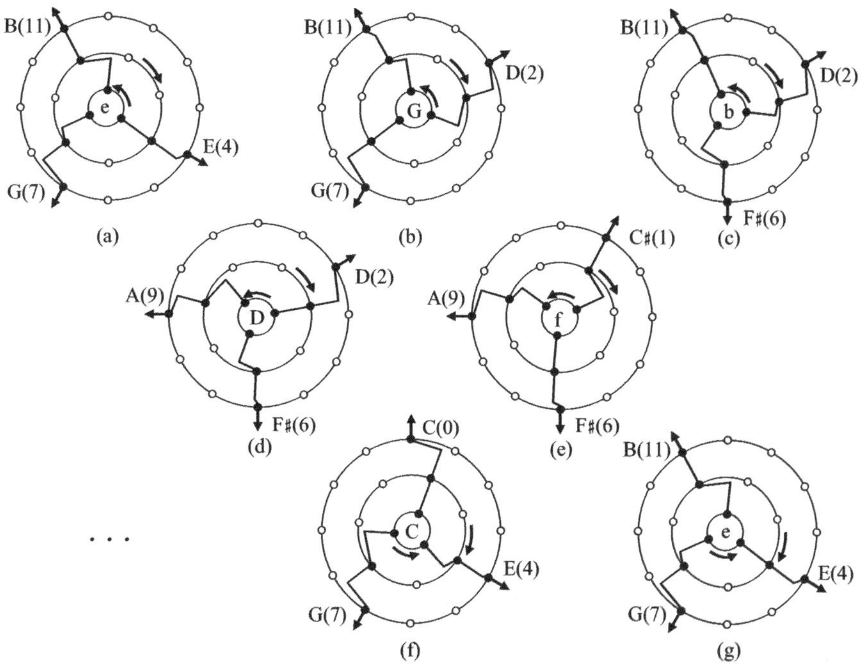

Figure 4.20. A 3-through-7-through-12 LR-Cycle.

reversed. Whence, for $0 \leq t \leq 18$, the dynamical representation that generates the sequence in this passage is $\left\langle J_{12,7,3}^{-0.5t+17,t-24}\right\rangle_{24}$.

4.6 Other Dynamical Cycles (What Next?)

We leave the reader with some observations and questions regarding cycles and sequences with other control parameters and dynamical configurations. For brevity, many of the details will be left to the reader to work out. Those who wish to experiment with dynamical configurations and control parameters might find the following website useful:

http://www.filteredpointsymmetry.com

This website allows the user to input configurations and control parameters and see the stroboscopic rotation of the circles.[21]

It is intriguing that, although the triadic cycles discussed above are very different in character and historical use, they can all be generated via the 3-through-7-through-12 dynamical configuration with simple changes in the control parameters. When viewed through this dynamical perspective, the fauxbourdon practice of the fifteenth century and the parsimonious chromatic approach of Brahms in his Double Concerto differ only by the control parameters of this 3-through-7-through-12 configuration. But what can be said of non-diatonic non-parsimonious sequences or other dynamical configurations?

In appendix 2-2 of her dissertation, So-Yung Ahn lists the triadic sequence

b♭–A–c♯–C–e–E♭–g–F♯–b♭–A–c♯–C–e–E♭–g–F♯–b♭–A,

which appears in Franz Liszt's *Grande Fantaisie Symphonique für Klavier und Orchester*, measures 185–99.[22] Clearly, this sequence of triads is not related to a diatonic triadic cycle. Moreover, since some of the adjacent triads are $P_{2,0}$-related, the sequence is not parsimonious either. However, it is still possible to find control parameters that model this sequence in the 3-through-7-through-12 dynamical configuration. By choosing $f_1 = -\frac{1}{56}$ cps ($-3\frac{3}{14}°$ per sec), $f_2 = -\frac{1}{24}$ cps ($-15°$ per sec), $\phi_1 = 47\frac{1}{7}°$, and $\phi_2 = 72\frac{6}{7}°$, the corresponding dynamical representation is $\left\langle J_{12,7,3}^{-1.5t+11,-0.5t+1.5} \right\rangle_8$, which generates the Liszt sequence for $0 \leq t \leq 17$. This is just one of many non-diatonic, non-parsimonious triadic cycles that can be generated by simple changes of the control parameters in the 3-through-7-through-12 dynamical configuration.

As noted at the end of section 4.3, some cycles of seventh chords embedded in a diatonic set have similar properties to the diatonic triadic cycles. This can be seen by fixing the 7-hole filter and rotating a 4-lamp beacon, which generates seventh chord cycles analogous to the three diatonic triadic cycles. In the C major scale, these cycles correspond to the dynamical representations $\left\langle J_{12,7,4}^{5,t} \right\rangle_7$, $\left\langle J_{12,7,4}^{5,2t} \right\rangle_7$, and $\left\langle J_{12,7,4}^{5,3t} \right\rangle_7$. In the cycle represented by $\left\langle J_{12,7,4}^{5,t} \right\rangle_7$, adjacent seventh chords are either $P_{1,0}$- or $P_{0,1}$-related; the pair of pcs not common to both seventh chords differ by 1 diatonic step. For $\left\langle J_{12,7,4}^{5,2t} \right\rangle_7$, adjacent seventh chords are $P_{1,1}$- or $P_{0,2}$-related and the pcs not common to both seventh chords can be paired so that the pcs differ by a diatonic step in each pair. For $\left\langle J_{12,7,4}^{5,3t} \right\rangle_7$, adjacent seventh chords are $P_{2,1}$-, $P_{1,2}$-, or $P_{0,3}$-related and the pcs not common to both can be paired so that the pcs in each pair differ by a diatonic step. Whence, if adjacent seventh chords are $P_{m,n}$-related, one can be obtained from the other by moving $m+n$ pcs by diatonic steps.

In their paper on parsimonious graphs, Douthett and Steinbach construct *starred parsimonious transformations* for seventh chords that mirror the parsimonious transformations for triads.[23] The transformation P_1^* exchanges half-diminished and minor seventh chords that have the same root, and the transformation L_1^* exchanges half-diminished and minor seventh chords that share an embedded minor triad (e.g., Eø7 and Gm7 share the triad g). This gives rise to $L_1^*P_1^*$-*cycles* for half-diminished and minor seventh chords analogous to the LP-cycles for triads; in both cases, adjacent chords are $P_{1,0}$-related. These cycles also have dynamical representations, but since seventh chords contain 4 pcs, the beacon must have 4 lamps. For example, the cycle

C ø7–Cm7–A ø7–Am7–F♯ø7–F♯m7–E♭ø7–E♭m7–C ø7

has a dynamical representation of $\left\langle J_{12,7,4}^{1.5t,-0.5t+3.5}\right\rangle_8$. Curiously, the dynamics of this cycle and the dynamics of the LP-cycles have one more thing in common; the 7-hole filter rotates while the beacon remains fixed (i.e., $f_2 = 0$ cps).

Douthett and Steinbach also introduce the transformations P_2^* and L_2^*. The transformation P_2^* exchanges dominant and minor seventh chords that have the same root, and L_2^* exchanges dominant and minor seventh chords that share a major triad (e.g., E7 and C♯m7 share E). This suggests $L_2^* P_2^*$*-cycles* of dominant and minor seventh chords. Adjacent seventh chords are also $P_{1,0}$-related and have dynamical representations. For example, the cycle

E♭m7–E♭7–Cm7–C7–Am7–A7–F♯m7–F♯7–E♭m7

has a dynamical representation of $\left\langle J_{12,7,4}^{1.5t,-0.5t+4.5}\right\rangle_8$. In this case too, the beacon remains fixed while circle 1 rotates. The control parameters of the $L_1^*P_1^*$- and $L_2^*P_2^*$-cycles are all identical except for the phases of their beacons. Note that, since the beacons of these cycles are fixed, it is possible to interpret the seventh chords in each cycle as a *perturbed fully diminished seventh chord*, similar to Zbikowski's observation regarding augmented triads and the LP-cycles. What *Tonnetz* might result from this observation and where this might lead is left to the reader.

Still another starred transformation introduced by Douthett and Steinbach is R^*, which, combined with the other starred transformations, yields other seventh chord cycles with dynamical representations.

In all the above, the cycles are filtered by a 7-hole filter. But the parsimonious triad cycles can also be generated when the 7-hole filter is replaced by an 8-hole filter. For example, an alternative dynamical representation for the $\{0, 3, 4, 7, 8, 11\}$ hexatonic cycle given in table 4.8 is $\left\langle J_{12,8,3}^{2t,-0.5t+1}\right\rangle_6$. Instead of picking harmonic triads out of diatonic sets (7 through 12), this dynamical system picks harmonic triads out of the all-combinatorial octachords from sc 8-28 (8 through 12). In addition to the 4-through-7-through-12 dynamical configuration, the $L_1^*P_1^*$- and $L_2^*P_2^*$-cycles can be generated with a 4-through-9-through-12 dynamical configuration. Dynamical representations for the $L_1^*P_1^*$- and $L_2^*P_2^*$-cycles above are $\left\langle J_{12,9,4}^{1.5t,-0.5t+4}\right\rangle_8$ and $\left\langle J_{12,9,4}^{1.5t,-0.5t+5}\right\rangle_8$, respectively. These dynamical systems "pluck" the seventh chords out of the all-combinatorial enneachord in sc 9-12 (9 through 12).

Still another application of dynamical systems relates to Cohn's *hyper-hexatonic systems*.[24] In his generalization of the Oettingen/Riemann *Tonnetz*, Cohn employs trichords whose interval vectors are $\langle x, x+1, x+2\rangle$ and $\langle x, x+2, x+1\rangle$, where x is any positive integer. It can be seen from the interval vectors that the cardinalities of the chromatic universes that support this generalized *Tonnetz* must be divisible by 3 (since $c = x+(x+1)+(x+2) = 3(x+1)$). When $x = 3$, the trichords reduce to harmonic triads and Cohn's generalized *Tonnetz* becomes the Oettingen/

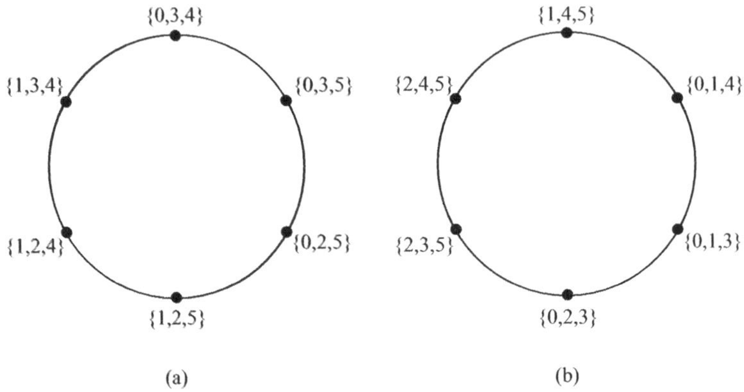

Figure 4.21. Hyper-Hexatonic Sub-Systems when $x = 1$ $(c = 6)$.

Riemann *Tonnetz*. This construction leads to *hyper-hexatonic sub-systems* analogous to the hexatonic sub-systems discussed in sections 4.2 and 4.4. These sub-systems consist of cycles of six trichords with the above interval vectors in which adjacent trichords are $P_{1,0}$-related. Figure 4.21 shows the two hyper-hexatonic sub-systems for $x = 1$ $(c = 6)$. A dynamical representation for the cycle in figure 4.21a is $\left\langle J_{6,4,3}^{t,-0,5t+3} \right\rangle_6$. For $x = 2$ $(c = 9)$, a dynamical representation for one of the hyper-hexatonic sub-systems is $\left\langle J_{9,6,4,3}^{0,t,-0,5t+3} \right\rangle_6$, and for $x = 3$ $(c = 12)$, a dynamical representation for the hexatonic sub-system in figure 4.5b is $\left\langle J_{12,9,6,4,3}^{0,0,t,-0,5t+3} \right\rangle_6$ (the other hyper-hexatonic sub-systems for these cardinalities can be generated by adjusting the phases). For any positive integer x, a dynamical representation for one of the hyper-hexatonic sub-systems is

$$\left\langle J_{3(x+1),3x,3(x-1),\cdots,6,4,3}^{\overbrace{0,0,0,\cdots,0}^{x-1 \text{ zeros}},t,-0,5t+3} \right\rangle_6 .$$

To conclude, we leave the reader with one final question. In all the above, the mode functions are linear functions. In the realm of real-valued functions, linear functions are relatively simple. But what cycles and sequences might result from non-linear mode functions (e.g., quadratic, exponential, or trigonometric functions), and are there musical questions for which dynamical systems with non-linear mode functions yield insight?

Notes

1. Those who wish to review the neo-Riemannian literature should refer to the *Journal of Music Theory* 42.2, which is a special issue on neo-Riemannian theory. In

addition to seminal papers on neo-Riemannian theory, this issue has a bibliography listing over 100 related publications.

2. Strogatz 1994, 2.

3. Douthett and Steinbach 1998, 243.

4. Agmon 1991 points out that any chord in the family of embedded triads and seventh chords can move to another chord in that family by step motion. Since the focus here is on triads, the interaction with seventh chords will be left out.

5. Cohn 1996, 9–11.

6. Klumpenhouwer 1994.

7. Cohn 1997.

8. In chapter 6, 137–60, of this volume, Hook relates figure 8 to his *signature group* by observing that if one of the lamps is designated as the "starting note of the scale," all 84 distinct modern modes (7 for each diatonic set) can be generated in one beacon revolution (1 for each 1/84th of a revolution).

9. Clough and Douthett 1991. Clough and Douthett's work on maximally even sets was, in part, inspired by, and is related to, two earlier publications by Clough and Myerson 1995 and 1996. Although the theory of maximally even sets was originally constructed to model scale structure, it has been shown by Douthett and Krantz 1996 and Krantz, Douthett, and Doty 1998 that this theory also relates to magnetic ordering in the two-state one-dimensional antiferrom agnetic Ising model. This interdisciplinary connection is also discussed by Krantz, Douthett, and Clough 2000. More recently and in a paper dedicated to the memory of John Clough, Douthett and Krantz 2007 present the scientific version of maximally even sets with applications in mathematics, physics, and music.

10. Lewin 1996.

11. Clampitt 1997.

12. For irrational frequencies, the generated sequence of sets is a *quasi-periodic sequence.* That is, the sequence is not periodic, but every finite contiguous subsequence of sets appears infinitely many times in the sequence. Carey and Clampitt 1996b have explored musical applications of quasi-periodic sequences, but it remains to be seen if dynamical systems with quasi-periodicity (irrational frequencies) have musical application.

13. The observation that small changes can have a dramatic effect has been expressed quite elegantly in a metaphor by the famous MIT meteorologist, Edward Lorenz. According to Hilborn 1994, Lorenz introduced his metaphor, known as the *butterfly effect,* at the December 1972 meeting of the American Association for the Advancement of Science in the title of his paper: "Predictability: Does the Flapping of a Butterfly's Wings in Brazil set off a Tornado in Texas."

14. Clough and Douthett 1991 and Clough, Douthett, Ramanathan, and Rowell 1993.

15. Clough, Cuciurean, and Douthett 1997.

16. Ibid., 89.

17. We will employ iterated maximally even sets to represent triads and seventh chords. However, any pcset can be expressed as an iterated maximally even set. For example, a *J*-representation for the tetrachord $\{0, 2, 3, 6\}$ is $J^{4,0,0,0}_{12,8,6,5,4}$; that is, $J^{4,0,0,0}_{12,8,6,5,4} = \{0, 2, 3, 6\}$. In general, a pcset does not have a unique *J*-representation. Even a *J*-representation that expresses a set in the fewest number of iterations is not necessarily unique; e.g., the *J*-representations $J^{4,0,0,0}_{12,8,6,5,4}$ and $J^{7,1,1,0}_{12,8,6,5,4}$ (4th-order maximally even sets) both express the set $\{0, 2, 3, 6\}$ in the fewest number of iterations.

18. Larry Zbikowski's suggestions came in the third of a series of seminars in music theory at the University of Chicago in 2003. It was at that seminar that I introduced the theory of dynamical voice leading.

19. Douthett and Steinbach 1998, 254.

20. Cohn 1991b, 1992, and 1997.

21. The graphics for the website were created by programmer, webmaster, and music theorist Richard Plotkin of the University of Chicago.

22. Ahn 2003.

23. Ibid., 1998.

24. Ibid., 1997.

Chapter Five

The "Over-Determined" Triad as a Source of Discord

Nascent Groups and the Emergent Chromatic Tonality in Nineteenth-Century German Harmonic Theory

Nora Engebretsen

Neo-Riemannian theory's demonstration of the susceptibility to group-theoretic interpretation of elements of Hugo Riemann's theories has sparked a reappraisal of nineteenth-century harmonic theory, focusing on its nascent group-theoretic content. This is evident in the resurgence of interest in Riemann's *Schritt/Wechsel* system (which can be interpreted as a formulation of a group isomorphic to the neo-Riemannian LPR group and familiar T_n/T_nI group), and in Richard Cohn's explorations of connections between neo-Riemannian theory and Carl Friedrich Weitzmann's 1853 monograph on the augmented triad.[1] In his essay "Introduction to Neo-Riemannian Theory: A Survey and a Historical Perspective," Cohn presents a broader view, suggesting that "the term 'neo-Riemannian' is most pertinently viewed as synecdochally appropriating the name of Riemann to represent a tradition of German harmonic theory which his writings culminated and perpetuated," and observing in particular that the neo-Riemannian approach "recuperates a number of concepts cultivated, often in isolation of each other, by individual nineteenth-century harmonic theorists."[2] Cohn identifies and locates historical precedents for six such concepts: triadic transformations, common-tone maximization, voice-leading parsimony, "mirror" or "dual" inversion, the *Tonnetz*, and enharmonic equivalence.

Neo-Riemannian theory has also fueled speculation about a group-theoretic orientation of theorists during the nineteenth century. This essay responds to Cohn's conjecture that the consonant triad might have contributed in the rise of a group-theoretic perspective.[3] Specifically, Cohn suggests that the consonant triad's "over-determined" nature—that is, its status as an optimal structure from the perspectives of both acoustic generability and voice-leading parsimony—may have led nineteenth-century theorists to inadvertently cultivate the group-theoretic perspective later made explicit by the neo-Riemannian movement.

In fact, two implicitly group-theoretic models of triadic relations, reflecting the over-determined triad's twofold nature, are suggested in works by Moritz Hauptmann, Arthur von Oettingen, and Riemann: one privileges maximal common-tone retention and incremental voice leading, and one privileges acoustically proximate (fifth- and third-based) root-interval relations. Though both approaches support a fully chromatic perspective (the former gives rise to the neo-Riemannian LPR group and the latter to the isomorphic *Schritt/Wechsel* group), nineteenth-century authors associated the common-tone approach with diatonic tonal space and the root-interval approach with chromatic tonal space, underscoring the over-determined triad as a source of conflict between diatonic and chromatic conceptions of tonal organization.

Though Hauptmann, Oettingen, and Riemann did not invoke the terminology of combinatorial group theory, their discussions of relationships among triads do appeal to a combinatorial perspective. All three privilege a few close relationships as "directly intelligible" generators, then account for the intelligibility of more distant relationships through the compounding or "composing" of these generators.[4] The combinatorial approach lends itself to the examination of group-theoretic content in the absence of an explicit group concept: the construction of a group from its generators corresponds to nineteenth-century theorists' tendency to build outward from local relations (leading, in several cases, to formulations of implicit groups)—rather than requiring recognition of the more global actions of these implicit groups on a particular set of triads. Moreover, the combinatorial approach resonates with aspects of nineteenth-century approaches to tonal coherence and structure, to be discussed below.

Rather than speaking of "generators" and their "composition," nineteenth-century theorists often characterized a relatively small number of chord-to-chord relationships as "directly intelligible," and then described all other relationships as "indirect," "mediated" or "elided" relationships, intelligible only when interpreted through reference to an implicit, underlying series of direct relationships. In choosing directly intelligible relationships, these authors appealed to Rameau's idea that the coherence of triadic successions stems from the nature of the consonant triad itself (though in nineteenth-century treatises these appeals drew upon both sides of the triad's "over-determined" nature). Their choices of generators were also shaped by their notions of tonality, though as the nineteenth century went on, the coherence of direct relationships increasingly came to be understood as being conceptually prior to and therefore independent of key. Each set of generators implied a specific hierarchy defined in terms of distance as a function of the number of combinatorial steps required to link triads. These hierarchies, which were often represented in spatial schemas, such as the *Tonnetz,* initially were easily reconciled with diatonically oriented models of tonal organization; but as chromatic relationships were incorporated into the generator sets, these hierarchies increasingly challenged key-based notions of distance and coherence, highlighting the structure supplied by the underlying group.

Hauptmann was by no means the first nineteenth-century theorist to emphasize common-tone retention and smooth voice leading in his treatment of harmonic succession. Indeed, as fundamental bass theory began to lose ground in Germany during the late eighteenth century, theorists began to turn to new models of harmonic succession based on melodic attraction and other voice-leading connections.[5] The emergence of this new perspective is widely associated with the work of Abbé Georg Joseph Vogler, whose treatment of harmonic succession influenced that of his student Gottfried Weber. A marked tendency toward the privileging of common-tone preservation and incremental voice leading, in particular, as determinants of harmonic relations is apparent in A. B. Marx's *Lehre von der musikalischen Komposition* (1837–41). What sets Hauptmann's views apart from those of earlier nineteenth-century authors is his view of common-tone retention as both a mechanism of succession and a measure of proximity, which brings to the fore the maximally parsimonious relationships modeled by the neo-Riemannian LPR operators.

Although Hauptmann's notion that common-tone connections determine triadic relations was not without precedent, his adaptation of this idea within a dialectical framework provided a catalyst for the developing group-theoretic perspective. Two aspects of Hauptmann's dialectical approach, in particular, contribute to the appearance of a group-theoretic perspective in *Die Natur der Harmonie und der Metrik* (1853): his notion that each member of a triad carries its own dialectical meaning, and the symmetry in his representation of tonal space.

Hauptmann begins his exposition on triadic succession by establishing the premise that "[t]he succession of two triads is again only intelligible in so far as both can be referred to a common element which changes meaning during the passage."[6] In a succession moving from a C-major triad to a G-major triad, for example, the common tone G passes from fifth meaning (associated with "contradiction" in Hauptmann's dialectical scheme) into root meaning (associated with the initial state of unity). The common-tone connection guarantees the succession's coherence, and the reinterpretation of that common tone's dialectical meaning makes the succession a dynamic event.

This notion of dialectical note quality also figures prominently in Hauptmann's discussion of the relationship between common-tone retention and harmonic proximity. So long as triads placed in succession share at least one common tone,

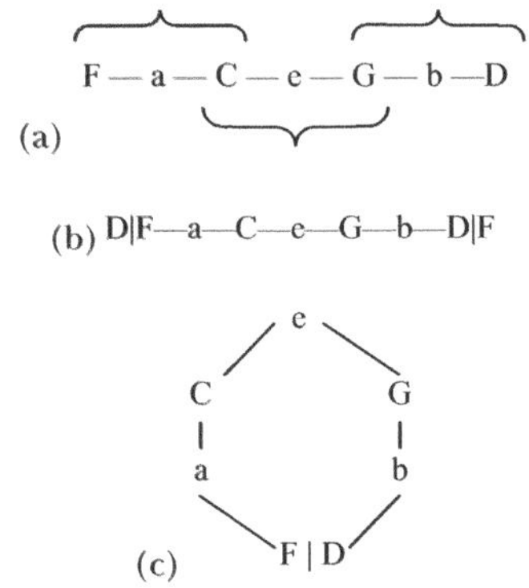

Figure 5.1 Hauptmann's representations of diatonic space.
5.1a The "triad of triads"
5.1b Extended version of the diatonic "triad of triads"
5.1c Diatonic space with its ends joined

Hauptmann considers their connection to be "self-evident"; however, he does assert that triads that share two notes are more closely related than those that share only one, at least within the strictly diatonic contexts he presents. Because a triad's third reconciles and therefore is invested with the dialectical qualities of that triad's root and fifth, a succession involving the reinterpretation of a third as a root or fifth (or vice versa) entails a closer relationship than one involving the reinterpretation of a root as its antithetical upper or lower fifth. Motion between the most closely related triads within a key, those sharing two common tones, such as between the C-major and A-minor or C-major and E-minor triads, involves no antithetical change of meaning, whereas motion between those sharing a single common tone does.

Hauptmann's representation of a major key as a "triad of triads"—with its tonic flanked by its upper and lower dominants, as shown in figure 5.1a, also illustrates the various common-tone relationships possible between the key's tonic triad and its other diatonic triads: starting from the middle of this diagram and moving outward, the tonic triad is shown to share two common tones each with the minor triads on the mediant and submediant, and one common tone each with the major triads on the dominant and subdominant (the case of the letter naming each pitch class reflects the origin of that pitch class as either a fifth [uppercase] or major third [lowercase]). As figure 5.1b illustrates, the tonic shares no common tones with either of the two remaining diatonic triads, the "diminished" triads on the leading tone and the supertonic.[7] This extended version of the key schema, which is presented in circular form in figure 5.1c, shows the common-tone relationship between any two triads within the key, and as such serves as a guide to harmonic proximity and succession.[8]

Having surveyed the common-tone relationships between a major tonic triad and the remaining diatonic triads in its key, Hauptmann turns to an examination of the voice-leading connections to be realized when these triads are placed in succession. Hauptmann notes that in a succession of two closely related triads (with two common tones), "only one voice will have to move melodically, while the other two remain, changing the harmonic meaning of their notes" (§88): in a progression from the root-position triad C–e–G to the triad a–C–e, he indicates that a–C–e should be placed in first inversion (C–e–G passing to C–e–a), and in a progression from the root-position triad C–e–G to the triad e–G–b, he indicates that e–G–b should be placed in second inversion (C–e–G passing to b–e–G) (§91). These examples, of course, model the parsimonious voice-leading relationships associated with the neo-Riemannian transformations R and L.[9]

Hauptmann displays the same concern for common-tone retention and incremental voice leading in his description of successions of triads linked by one common tone. He remarks that in these successions two parts move by step while the third is reinterpreted: C–e–G passes into C–F–a or into b–D–G. Interestingly, although Hauptmann recognizes these successions as intelligible in their own right, on the basis of the single common tone preserved, he later characterizes them as compound successions from which a third, mediating triad has been elided:

> The passage from C–e–G to F–a–C, which leads to the position C–F–a, is a compounded one, and consists of the progressions C–e–G. . .C–e–a. . .C–F–a. [. . .] Similarly with the succession from C–e–G to G–b–D which (in b–D–G) is compounded of the successions C–e–G. . .b–e–G. . .b–D–G.[10]

This view shows Hauptmann's concern for parsimonious voice-leading relationships.

Hauptmann also invokes an imagined mediator in his discussion of motion between triads not linked by a shared element. He holds that the intelligibility of a succession of this kind, in which all three parts move, emerges only through reference to a mediating triad with which both sounding triads share a common-tone connection. In particular, he specifies that the mediator should share two common tones (a "preponderance of community") with the first triad and one common tone with the second (§89). Hauptmann contends, for example, that the progression from the tonic triad C–e–G to the supertonic triad D|F–a is understood only through reference to the submediant a–C–e. In this context, however, Hauptmann does not conceive of the mediator as the middle element of a compound succession; rather, he substitutes the mediator for the initial triad in the succession. That is, instead of interpreting the succession from C–e–G to D|F–a as an elliptical version of the succession C–e–G. . .[C–e–a]. . .D|F–a, resulting in the voice-leading pattern C–e–G. . .D|F–a, Hauptmann claims that this succession from C–e–G to D|F–a must "be taken to be equivalent to the passage from a-C-e to D|F-a" and therefore substitutes the mediator, in root position, for the initial C-major triad and determines the voicing of the second sounding triad in relation to this substituted mediator, resulting in the voice-leading pattern C–e–G. . .a–D|F based on the succession C–e–G [=a–C–e]. . .a–D|F (§90). Although Hauptmann offers no explanation as to why the mediator should be understood as a substituted chord in this context but as an elided chord in others, his approach does provide for a common-tone connection between the disjunct triads without violating established voice-leading norms—whereas parallel fifths result if the C–e–G to D|F–a succession is accounted for as an ellipsis.[11]

The bulk of Hauptmann's treatment of triadic succession focuses on relationships within a strictly diatonic, major-mode context. To a degree, Hauptmann conceives of diatonic pitch-class space as a closed, cyclic space, as suggested above in figure 5.1c. Within this mod = 7 diatonic space, generic (024) triads are susceptible to a parsimonious cycle roughly akin to the <LR> cycle among consonant triads in chromatic space.[12] Although Hauptmann distinguishes between major, minor, and diminished triads in his examination of relationships among a major key's triads, he does treat the triads generically insofar as the rules of proximity and succession are concerned.[13] Even the diminished triads are treated as de facto consonant triads, governed by the same common-tone-based rules of succession as the major and minor triads, rather than by rules of dissonance resolution. The parsimonious cycle in which these "generic" triads participate underlies the representation of diatonic space given in figure 5.1c. Moving clockwise around the figure, starting from the tonic C–e–G, for example, successive motion between most closely related triads—those linked by two common tones—yields the cycle C–e–G . . . e–G–b . . . G–b–D . . . b–D|F . . . D|F–a . . . F–a–C . . . a–C–e . . . C–e–G, or following Hauptmann's rules of voice-leading, C–e–G . . . b–e–G . . . b–D–G . . . b–D|F . . . a–D|F . . . a–C–F . . . a–C–e . . . G–C–e.[14]

Hauptmann conceives of diatonic pitch-class space not only as a closed, cyclic space, but also as a segment of the infinite "chain of triads," which extends the "triad of triads" shown in figure 5.1a outward in both directions. Triadic successions that extend beyond the bounds of a given diatonic collection move outside the mod 7 system, and through the space of this justly intoned "chain of triads" system, which suggests the same conception of tonal space evident in the L/R alleys of the *Tonnetz*. Specifically, as Kevin Mooney has noted, "[t]he raising of the lowercase letters so that they form their own horizontal series (F ^{a} C ^{e} G ^{b} D) shows how small a step it is from Hauptmann's one-dimensional representation to the two-dimensional representation of the Table."[15] A single major key comprises only a small segment of an L/R alley, but the extended schema constitutes at least a partial map of chromatic, rather than diatonic, space.

Hauptmann's discussion of triadic succession focuses almost exclusively on diatonic successions, and even his forays outside of the established key tend to be diatonically predicated. Even in modulating passages, Hauptmann maintains "two chords in immediate succession shall belong to the same key" (§284). For example, if a C-major tonic triad is followed by a B-major triad, the succession would be rendered intelligible only by reinterpreting the first triad as the submediant of the closely related key of E minor, and then treating the succession as a diatonic succession in E minor (§282). Hauptmann, however, does eventually allow that, to effect a more direct modulation, "two chords belonging not to the same, but to very nearly related keys" might also be placed in succession, opening the door to the possibility of chromatic successions.[16]

Hauptmann examines a few instances of chromatic successions, which he treats as being roughly analogous to diatonic successions, at least insofar as common-tone connections are concerned.[17] He explains the close relationship between the triads C–e–G and C–e♭–G, for example, through reference to the triads' two shared common tones: in this case, the common tones preserve the interval of the fifth, which passes from positive to negative meaning, but which in both cases represents a contradiction between root and fifth, while the descent by semitone from e to e♭ is associated with a shift from positive to negative third meaning.[18] This, of course, is the voice-leading relationship associated with the neo-Riemannian transformation P.

As with diatonic triadic successions, Hauptmann seems to rely on common tones both as a gauge of harmonic proximity and as a source of coherence in his treatment of non-diatonic triadic successions. Though he never explicitly correlates maximal proximity with maximal common-tone retention, he does demonstrate that, with the exception of the parallel relation, most non-diatonic successions differ from their diatonic counterparts in at least one crucial respect: the chromatic successions that Hauptmann examines all require a more radical reinterpretation of the connecting common tones than was necessary in strictly diatonic contexts. The succession C–e–G . . . B–E–g♯, for example, not only requires a reassessment of the common tone's dialectical quality, as it passes from third meaning into root meaning, it also requires that e be reinterpreted as—in effect equated with—E, despite their acoustical difference of a comma under just intonation and their conceptual difference within Hauptmann's theory.[19]

The more pronounced change in meaning that a common tone undergoes in a chromatic succession such as this suggests a more distant relationship than that created by a common-tone link between diatonic or parallel triads, but Hauptmann is oddly noncommittal on this point.[20] In fact, Hauptmann makes no mention of the relative strength of relationships determined by common-tone connections in diatonic versus chromatic contexts. He never indicates, for example, whether the relationship between parallel triads is as close as that between diatonic triads similarly linked by two common tones, or even as close as that between any two diatonic triads. On the whole, his presentation suggests that while he conceives of diatonic and chromatic successions as being governed by the same principles, he does not conceive of them as being entirely equivalent. Specifically, while Hauptmann clearly views common tones as the agents

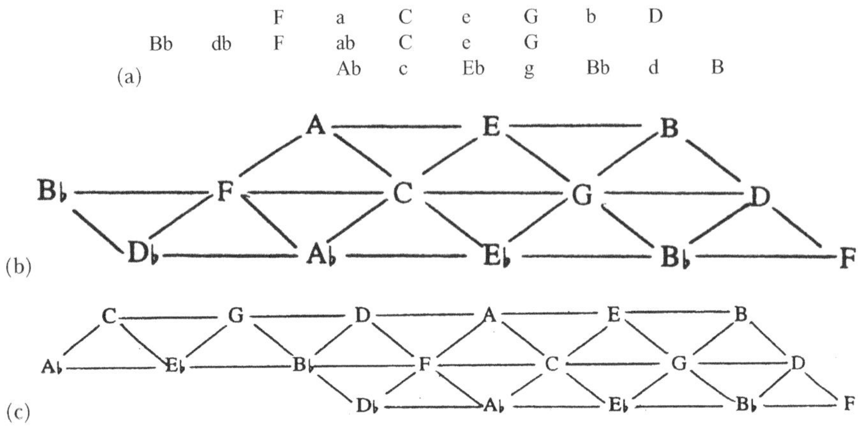

Figure 5.2 Hauptmann's implied model of chromatic space.
5.2a Common-tone connections between keys
5.2b Interpreted as a *Tonnetz* fragment
5.2c Extended *Tonnetz* fragment

of coherence in diatonic and chromatic successions alike, he seems to view common tones as an accurate measure of harmonic proximity only within a given diatonic or chromatic context—suggesting that, for Hauptmann, harmonic proximity is not just a function of voice-leading relationships, but involves key relationships as well. In other words, common-tone connections impart a sense of coherence independent of key, but harmonic proximity remains a function of both common-tone connection and key relations.

While the chain of triads serves admirably as a guide to relationships between successive fifth-related major keys and as a guide to common-tone connections among those keys' triads, it does not seem to accommodate relationships such as that between parallel major and minor keys, nor does it fully account for another interesting detail of Hauptmann's treatment of modulation: the shifts he makes between fifth-determined and third-determined pitch-classes. Although Hauptmann does not incorporate this additional information into a single, unified representation of tonal space, the descriptions he provides of these other relationships strongly suggest that he conceives of this space very much as the *Tonnetz* theorists later would.

The clearest encapsulation of Hauptmann's fuller view of chromatic space appears in the context of an explanation of a modulation from C major to E♭ major via the intermediary key of F minor (§275). As shown in figure 5.2a, Hauptmann illustrates the connections between these keys by arranging their "triad of triads" representations so as to align the pitch-classes that the keys hold in common. Hauptmann notes, for example, that the fifth C–G (later c–g) constitutes the "permanent, binding element" upon which "the rest of the intervals of the keys are formed."

Two details of this demonstration are particularly suggestive with regard to Hauptmann's conception of tonal space. First, the comparison of the three keys shows Hauptmann's notion of a double common-tone connection between parallel major and minor triads: parallel triads share the same fifth. Figure 5.2b reconfigures figure 5.2a to again show what a small step it is from Hauptmann's representation to that of the *Tonnetz*. Second, by emphasizing the role of the C–G/c–g fifth as the "permanent, binding element" connecting all three keys, Hauptmann draws a distinction between the key of E♭ major as reached through this modulation and the key of E♭ major one syntonic comma lower as reached by moving along the chain of triads toward the subdominant side of C major. The reinterpretation of the common tones, from the fifth-determined C–G to the third-determined c–g, creates a modulation in which "the new key springs right out of the middle of the first," as opposed to one that advances along a chain of triads, "leaving the seat of the first [key] and settling in the [essentially unrelated] region of the second."[21] Here too, it is but a small step from Hauptmann's description of these relationships to the *Tonnetz*. Figure 5.2c extends the fragment of the *Tonnetz* given in figure 5.2b to illustrate the difference between the two forms of E♭ major, but based on the parsimonious relationships that Hauptmann has defined, the fragment could in fact be extended infinitely—not only along its horizontal L/R alleys, but along its diagonal L/P and L/R alleys as well—to generate the full *Tonnetz*.

Hauptmann's discussions of chromatic successions and the conception of tonal space implied therein hint at a group presentation on the generators L, P, and R. As noted above, however, Hauptmann does not integrate the diatonically conceived L and R and the chromatically conceived P into a single system of relationships based on voice-leading parsimony, and his unflaggingly diatonic orientation precludes extensive exploration of the harmonic spaces represented by the P/L and P/R alleys of the *Tonnetz*. Moreover, Hauptmann does not explicitly define a role for the P relation as a mediator in chromatic successions featuring fewer than two common-tones in the way that he did for the L and R relations in diatonic contexts (the status of the P relation as a mediator is alluded to in Hauptmann's demonstration of a connection between C major and E♭ major on the basis of the preserved C–G/c–g fifth, but he does not pursue the matter). The appropriateness of asserting the P relation as an implicit generator within Hauptmann's system is uncertain. As with the shift from Hauptmann's representations of tonal space to the *Tonnetz*, the acceptance of the P relation as a generator alongside L and R is at once a small step (building on his correlation of harmonic proximity and coherence with maximal common-tone retention) and a giant leap (involving a shift toward a fully chromatic perspective). This shift in perspective and its group-theoretic implications are addressed more directly in Oettingen's work.

While Hauptmann's theory does not unequivocally imply a group presentation on P, L, and R, his treatment of P and other chromatic relations vis-à-vis the L and R relations and the series they generate both reinforces the primacy of the L/R series and anticipates the expanded collections of generators in both Oettingen's and Riemann's theories.[22] Specifically, in Hauptmann's discussion of relationships such as those between parallel major and minor triads or between chromatic mediants such as C major and E♭ major, we already encounter the notion that some relations are more directly intelligible than their relative positions within the L/R series suggests. This led Hauptmann to define a new category of relationships—among which P is the strongest—that move between distinct L/R series to establish a more direct link between the triads than that afforded by compounds of L and R alone. Oettingen and Riemann would likewise introduce direct relationships that allowed them to circumvent the notions of distance implicit in the L/R series. Unlike Hauptmann, however, Oettingen and Riemann would integrate these generative relationships into a coherent system including their own versions of R and L—reflecting the conceptions of tonal hierarchy in their theories.

The notion that chord-to-chord connections—whether established by common-tone retention, parsimonious voice leading, or some other means—impart a sense of tonal coherence independent of key grew increasingly prevalent during the latter half of the nineteenth century. Key-based notions of coherence were not wholly abandoned, but the view of coherence as a local phenomenon led to development of the nascent group-combinatorial approach evident in Hauptmann's treatment of triadic succession. Just as a group-theoretic approach based on parsimonious voice-leading connections seemed poised to emerge, however, German theorists rejected the common-tone model of succession in favor of one based primarily on consonant root relations, advancing a notion (based on Rameau's writings) that the coherence of triadic successions should derive from the acoustic coherence of the triad itself.

This shift from the common-tone model to the root-interval model was initiated somewhat inadvertently by Helmholtz and found confirmation in Oettingen's acoustically conceived approach to triadic relations. In most respects, Helmholtz's views on triadic succession, as presented in the final chapters of *Die Lehre von den Tonempfindungen* (1863), did not stray far from those of Hauptmann, but his justification of common-tone relations on acoustic grounds, coupled with his attempt to align the common-tone approach with Rameau's fundamental bass theory, paved the way for the emergence of the root-interval approach in Oettingen's work.

In his *Harmoniesystem in dualer Entwickelung* (1866), Oettingen questions Helmholtz's reliance on Hauptmann's common-tone measure of harmonic proximity, pointing to the conflict between the common-tone approach and Helmholtz's own views on pitch proximity—a conflict highlighted when Helmholtz turns to a fundamental-bass model of harmonic succession. Oettingen

$5^m \; 3^n.$

n:	—8	—7	—6	—5	—4	—3	—2	—1	0	1	2	3	4	5	6	7	8
m																	
2	*c*	*g*	*d*	*a*	*e*	*h*	*fis*	*cis*	*gis*	*dis*	*ais*	*eis*	*his*	*fisis*	*cisis*	*gisis*	*disis*
1	*as*	*es*	*b*	*f*	*c*	*g*	*d*	*a*	*e*	*h*	*fis*	*cis*	*gis*	*dis*	*ais*	*eis*	*his*
0	*fes*	*ces*	*ges*	*des*	*as*	*es*	*b*	*f*	*c*	*g*	*d*	*a*	*e*	*h*	*fis*	*cis*	*gis*
— 1	*deses*	*asas*	*eses*	*bb*	*fes*	*ces*	*ges*	*des*	*as*	*es*	*b*	*f*	*c*	*g*	*d*	*a*	*e*
— 2	*bbb*	*feses*	*ceses*	*geses*	*deses*	*asas*	*eses*	*bb*	*fes*	*ces*	*ges*	*des*	*as*	*es*	*b*	*f*	*c*

Figure 5.3 Oettingen's Table.

takes issue, for instance, with Helmholtz's assertion that triads whose roots stand in a diatonic third relation are more closely related than triads whose roots stand in a diatonic fifth relation, whereas the opposite would be true of third- and fifth-related pitches. He is even more dumbfounded by Hauptmann's claim that a fifth progression, such as that from a tonic to its dominant, should be understood as the compound of two diatonic third progressions from which the middle triad has been elided, and he indicts the common-tone approach for its failure to draw a distinction between diatonic and chromatic progressions, comparing the progression from C–E–G to C–F–A with that from C–E–G to C♯–E–G♯ and noting that both feature a single common-tone connection. While this is true, Hauptmann and the other common-tone theorists never abandoned all reference to key membership in their assessments of harmonic proximity. This final criticism is significant, however, insofar as it reflects Oettingen's awareness that the common-tone model presents a means of establishing a hierarchy of triadic relationships independent of tonic reference, and also insofar as it reflects Oettingen's discomfort with this notion. Ultimately, Oettingen rejects common tones as a guide to both triadic succession and proximity, and asserts that all matters of proximity and succession, among both pitches and chords, are governed by the same set of acoustically determined relationships.

Oettingen's views on triadic relations are deeply intertwined with the acoustically conceived model of tonal space he presents, which is reproduced here as figure 5.3. Oettingen's "Table"—which has come to be known as the *Tonnetz*—would be adopted in some form by a number of subsequent authors, most notably by Riemann, and would become a mainstay of German harmonic theory well into the twentieth century. Oettingen's Table features justly tuned pitches

Table 5.1 Oettingen's directly intelligible diatonic relations

name	from c+	analogous neo-Riemannian operation(s)
homonomic *Quintschritt*	c+–g+; c+–f+	D/D^{-1}
antinomic *Terzschritt*	c+–e°(a–)	R
antinomic *Leit-Schritt*	c+–b°(e–)	L

situated along intersecting horizontal and vertical axes: the horizontal axes are generated by acoustically pure (3:2) fifths and the vertical axes are generated by acoustically pure (5:4) major thirds. These fifths and thirds extend infinitely to capture the tuning differences that occur under just intonation between pitch classes associated with the same letter name.[23] Oettingen incorporates a system of *Horizontalstriche*, appropriated from Helmholtz, into his Table to indicate these slight tuning discrepancies.

In his treatment of relationships among consonant triads, Oettingen initially focuses on successions that remain within the bounds of a single key.[24] He indicates that all directly intelligible successions of consonant triads within a given key can be accounted for through reference to one of three relationships: the "homonomic" *Quintschritt*, the "antinomic" *Terzschritt*, and the antinomic *Leit-Schritt* (homonomic relationships preserve the chord quality [major or minor], while antinomic relationships reverse it).[25] The homonomic *Quintschritt* ("fifth step") connects the most closely related triads within a key—triads of the same quality whose roots are separated by a perfect fifth (here and throughout the remainder of this section, references to the "root" of a minor triad should be understood, in keeping with Oettingen's use, to indicate its "phonic" or dual root, the usual fifth of the triad). The antinomic *Terzschritt* ("third step") connects the next most closely related triads within a key—triads of opposing qualities whose roots are separated by a major third. Finally, the antinomic *Leit-Schritt* ("leading[-tone] step") connects the next most closely related triads within a key—triads of opposing qualities whose roots are separated by a minor second. Oettingen observes that the *Leit-Schritt* can be thought of as a *Quintschritt* plus a *Terzschritt*, though he is careful to note that he conceives of the *Leit-Schritt* as a single transformation.

Table 5.1 summarizes Oettingen's directly intelligible, diatonic relationships, illustrating each with respect to a C-major tonic triad (c+) and identifying the analogous neo-Riemannian operation. Oettingen's "phonic" labels are retained to make clear the root-interval relationship being invoked, but the more common "minor" label appears in parentheses immediately after the phonic label (for example, the antinomic *Terzschritt* connects a C-major triad (c+) with an A-minor triad (a–), which in Oettingen's terminology is an E-phonic triad (e°)).

Oettingen specifies only the size, not the direction, of the root interval underlying each of these relationships. Accordingly, the homonomic *Quintschritt* links the tonic c+ both to its dominant g+ (a perfect fifth upward) and to its subdominant f+ (a perfect fifth downward). Thus, the homonomic *Quintschritt* is equivalent in effect to either the neo-Riemannian operation D or the neo-Riemannian operation D^{-1}, depending upon the specific triads involved. Viewed as a transformation, the homonomic *Quintschritt* does not yield a one-to-one mapping; its inclusion within an implicit group of operations is possible only if we refine Oettingen's definition of the relationship to produce such a mapping.[26] Oettingen does not specify the direction of the root-intervals associated with the antinomic *Terzschritt* or *Leit-Schritt* either, but the constraint in effect at this point in his presentation—that these relationships obtain between consonant triads within the context of a diatonic collection—dictates that the *Terzschritt* is identical in its effect to the neo-Riemannian operation R and that the *Leit-Schritt* is identical in its effect to the neo-Riemannian operation L.

Having introduced his three basic transformations, Oettingen explores the ways in which they might be combined to navigate the Table (given above as figure 5.3).[27] Oettingen takes the c+ triad as his starting point, and observes that any other major triad with its root on the same horizontal series as c can be reached through repeated application of the homonomic *Quintschritt.* By using the antinomic *Terzschritt* and the antinomic *Leit-Schritt* in conjunction with the homonomic *Quintschritt,* he notes, all of the minor triads with phonic roots on the neighboring horizontal series to the north (the series with single upper *Striche*) can also be reached.

Here Oettingen uses a combinatorial approach, first taking the homonomic *Quintschritt* as a generator. In this context, he clearly conceives of the homonomic *Quintschritt* as a directed transformation, equivalent either to D or to D^{-1}, but not to both. The implicit infinite cyclic group, given by the presentation <D>, acts—as Oettingen notes—on the homonomic triads whose roots lie along a single fifth-series on the Table. The other two directly intelligible transformations, the antinomic *Terzschritt* (R) and *Leit-Schritt* (L), likewise generate an infinite group when taken in alternation, namely the same group that was produced by Hauptmann's theory.[28] This group acts on the major and minor triads whose roots lie along adjacent fifth series: the *Tonnetz* representation of this group is what Cohn terms an "<LR> chain" threading a pair of adjacent fifth axes.[29] That Oettingen does not recognize—or at least does not comment upon—the <LR> chain is certainly due to his preferring the *Quintschritt.* Although a *Terzschritt* and a *Leit-Schritt* yield the same result as a single *Quintschritt* when applied in succession (the direction of the *Terzschritt*-plus-*Leit-Schritt* transformation depends upon the order in which the transformations are applied), Oettingen clearly cannot characterize his closest transformational relationship as the compound of two more remote transformations—particularly not after chastising Hauptmann for asserting that the succession of fifth-related triads should be understood in this

Table 5.2 Oettingen's directly intelligible relations

name	from c+	analogous neo-Riemannian operation(s)
homonomic *Quintschritt*	c+–g+; c+–f+	D/D^{-1}
antinomic *Terzschritt*	c+–e°(a–)	R
antinomic *Leit-Schritt*	c+–b°(e–)	L
antinomic *Wechsel*	c+–c°(f–)	N

very way. Oettingen's privileging of three direct relationships involving motion along the Table's <LR> alleys suggests a group presentation on the three generators D, L, and R.[30]

While the three basic transformations can connect any two triads within a given horizontal alley on the Table, Oettingen notes that the remainder of the Table proves inaccessible. He therefore defines a fourth directly intelligible relationship, the antinomic *Wechsel* or *Octavschritt,* to introduce the possibility of motion between the horizontal <LR> alleys.[31] The antinomic *Wechsel* transforms a triad into its "reciprocal" (the triad of opposing quality with the same root), and as such is equivalent in effect to the neo-Riemannian operation N—which in turn derives from Weitzmann's "*nebenverwandt*" relation.[32] Insofar as it allows for motion between the Table's <LR> alleys, the antinomic *Wechsel* fulfills the same function as the neo-Riemannian operator P and Hauptmann's parallel relation. Table 5.2 summarizes Oettingen's collection of directly intelligible relations.

Oettingen remarks that by successively applying the *Wechsel* in alternation with one of the three other relationships used as generators—he suggests using the *Terzschritt,* which produces a chromatic third relation when used in conjunction with the *Wechsel*—even triads in remote fifth series can be reached. Using the four directly intelligible transformations in combination, it is possible to trace a path on the (infinitely extending) Table to establish a connection between any two triads. In terms of the familiar neo-Riemannian operations, Oettingen's system of chord connection suggests a group presentation on the generators D, N, L, and R.[33]

While Oettingen's views on the intelligibility of triadic relationships are shaped by the norms of diatonic tonality, the relationships he sees as directly intelligible—and ultimately treats as generative—offer a means of characterizing relationships between triads independent of reference to diatonic collections or tonal centers.[34] Oettingen also introduces three "shortcut" transformations: the antinomic *Quintschritt,* which reduces the two-step homonomic-*Quintschritt*-plus-antinomic-*Wechsel* compound to a one-step transformation (e.g., c+–g° replaces the succession c+–[g+]–g°, in which g+ is an implied mediator),[35] and the antinomic *kleine Ober-Terz Schritt* (g°–b♭+) and antinomic *kleine Unterterz Schritt* (f+–d°), which replace a homonomic *Quintschritt* plus an antinomic *Terzschritt* (in either order), the difference being the direction of the *Quintschritt*.[36]

Confronted with what he describes as "the chaos of possibilities" presented by the compounding of his direct relations, Oettingen also continues to rely on reference to key relations—"tone system" relations, in his terminology—as a source of order in his discussions of triadic successions.[37] Much like Hauptmann, Oettingen holds that the closest relationships between keys obtain when the tonic of one system is contained in another.[38] The C major system, for example, is most closely related to the F major, G major, A minor (E phonic), and E minor (B phonic) systems—precisely those with which its tonic stands in a *Quintschritt, Terzschritt,* or *Leit-Schritt* relation.[39] On the basis of the principle of mediated succession—the notion that certain successions of indirectly related triads can be understood through reference to some third, elided triad directly related to both—Oettingen also recognizes the coherence of motion between certain more remotely related tone systems, which he terms *Wechsel* tone-system relations.[40] Taking the c-tonic system as a starting point, the phonic triad c° (which is *Wechsel*-related to the tonic c+) can serve as a mediator between the c-tonic system and the a♭-tonic, d♭-tonic, g-phonic, and f-phonic systems. (From the mediating triad c°, a♭+ is reached by *Terzschritt,* d♭+ by *Leit-Schritt,* and g° and f° both by *Quintschritt.*) Similarly, by first moving to a direct relative of the c+ tonic—to f+ or g+ by *Quintschritt,* to e° by *Terzschritt,* or to b° by *Leit-Schritt*—and then applying the *Wechsel* transformation to each of those triads, the f-phonic, g-phonic, e-tonic, and b-tonic systems can be reached. In short, Oettingen conceives of tone-system relations in terms of transformational relations between the systems' tonic and/or phonic triads and deems coherent a range of key-system relationships on the basis of binary compositions of his basic transformations.[41]

Oettingen's theory of triadic relations points to the "over-determined" nature of the consonant triads. In his efforts to vanquish the common-tone approach to triadic succession and proximity and to present in its place an acoustically justified approach, Oettingen champions a set of transformations, based on consonant root relations within a diatonic context, which overlaps significantly with the set of relationships used in the common-tone model. In particular, Oettingen's view of the antinomic *Terzschritt* (R) and antinomic *Leit-Schritt* (L) as directly intelligible, generative relationships provides a point of contact between his system and Hauptmann's. The related connection between Oettingen's Table and Hauptmann's "chain of triads" schema further emphasizes the triads' acoustic and group-theoretic potential. Oettingen's views on triadic proximity do, of course, differ somewhat from those of Hauptmann. In classifying the *Quintschritt* (D/D^{-1}) as the closest possible tonal relationship, followed by the *Terzschritt* (R) and *Leit-Schritt* (L), Oettingen reverses Hauptmann's hierarchy, and Oettingen's definition of the *Quintschritt* as a discrete, generative transformation, rather than as a compound of the *Terzschritt* and *Leit-Schritt,* reflects the immediacy he attributes to the *Quintschritt* relations. Moreover, that Oettingen appends "short-cut" transformations to his system suggests that he considers some relationships—the parallel relation and certain chromatic-third relations—to be stronger than their derivations through composition of the generators would imply.

Between 1875 and 1880, Riemann published three treatises in which he put forward revised and extended versions of Oettingen's *Schritt/Wechsel* system of chord relations: *Die Hülfsmittel der Modulation* (1875), *Musikalische Syntaxis* (1877), and *Skizze einer neuen Methode der Harmonielehre* (1880). Although an implicit formulation of the finite *Schritt/Wechsel* group on the twenty-four consonant triads can already be detected in the earliest of these treatises, the relationships and the terminology used in the later *Skizze* would become the norm for Riemann's subsequent discussions of the *Schritt/Wechsel* system.[42] Accordingly, discussion of Riemann's *Schritt/Wechsel* system here will focus on the *Skizze.*

Riemann arrived at his implicit formulation of the *Schritt/Wechsel* group by exploring the combinatorial potential of a few directly intelligible relationships, much as Hauptmann and Oettingen had done. Like Oettingen, Riemann held relationships between triads to be a product of the acoustic relationships between their fundamentals, and he appropriated Oettingen's dually conceived terminology to describe these relationships. Whereas Oettingen classified relationships between consonant triads by indicating whether the chords were of the same or opposing qualities (yielding homonomic or antinomic relations) and by identifying the interval between the triads' roots, Riemann refined the system by treating the interval between the triads' roots as a directed interval, gauged from the direction of the initial triad's generation. Homonomic relationships are characterized as *Schritte* and antinomic relationships as *Wechsel,* and Riemann adds the modifiers "*schlicht*" (plain) and "*gegen-*" (contrary) to specify the direction of the root interval. "*Schlicht*" is assumed if the direction of a root motion is not specified. Oettingen's "homonomic *Quintschritt*" is thus replaced by two more specific relations, the "(*schlichter) Quintschritt*" and the "*Gegenquintschritt,*" his "antinomic *Terzschritt*" is replaced by the "*Terzwechsel,*" and so forth. The "antinomic *Wechsel*" is also renamed, becoming the "*Seitenwechsel.*"

By specifying the direction of the root-interval relationship, Riemann eliminated ambiguities of the sort encountered with Oettingen's *Quintschritte* and also eliminated the need to refer to a diatonic context in defining relationships between triads, as Oettingen did in defining his antinomic *Terzschritt* and antinomic *Leit-Schritt.* So, given a triad and a relationship defined under Riemann's system, there will be a unique triad that stands in the specified relationship to the original triad.

Like Oettingen, Riemann privileged triadic relationships involving no root motion or root motion through a perfect fifth or major third. Unlike Oettingen, however, Riemann did not limit himself to diatonic relationships when selecting the generators of the *Schritt/Wechsel* group. This change reflects a fundamental difference in the two theorists' conceptions of tonality: Riemann did not see tonality as linked to diatonic collections, but instead as a system of acoustically defined relationships around a given tonal center. While Oettingen's direct relationships proved to be predominantly diatonic, some of Riemann's direct relationships involved chords that lie beyond the bounds of a single diatonic collection yet have an acoustic connection based on the directly intelligible root intervals.

Riemann's differences with Oettingen in this respect are not immediately evident in the initial discussion of *Schritte* and *Wechsel* in the *Skizze.* Riemann's desire to present a readily accessible *Harmonielehre* led him to defer any substantive discussion of his own views on tonality until after an initial survey of root-interval relations within a diatonic context. Not surprisingly, Riemann's diatonically oriented system of *Schritte* and *Wechsel* bears a significant resemblance to Oettingen's original system, particularly so in its use of the diatonic *Terzwechsel* as a generative relationship.[43]

When Riemann does turn to his own views on tonality in the final chapter of the *Skizze,* he argues that diatonic and chromatic chords alike can relate directly to a tonic, so long as the primacy of the tonic—as the ultimate goal of cadential motion and as the chord in relation to which all other chords receive their individual effect and meaning—remains unshaken. Riemann offers a taxonomy of twenty-five "potentially intelligible" relationships. In this "*Systematik der Harmonieschritte,*" which is summarized in table 5.3, Riemann divides the twenty-five relations into seven categories on the basis of the root intervals involved. The categories are arranged by the relative immediacy of these root intervals. As the only progression involving a stationary root, the *Seitenwechsel* constitutes its own category, but each of the remaining six categories includes four relationships involving root motions of the same size—the *schlichter Schritt,* the *Gegenschritt,* the *schlichter Wechsel,* and the *Gegenwechsel.* Oddly, Riemann breaks this pattern without comment when he replaces the expected *Gegentritonuswechsel* with the *Doppelterzwechsel,* which does not share the root motion by tritone that links the other members of the category.[44]

Riemann's break with Oettingen is evident in his choice of generative relations. Table 5.3 includes a column showing the derivation of each of these relationships as a compound of the *Quintschritt* (Q) or *Gegenquintschritt* (Q^{-1}), *Terzschritt* (T) or *Gegenterzschritt* (T^{-1}), and/or *Seitenwechsel* (⊕) relationships.[45] Riemann's choice of the chromatic *Terzschritt,* rather than the diatonic *Terzwechsel,* as a generative relationship allows him to account for the chromatic third relations common to nineteenth-century practice.[46] The choice is also appealing from a systematizing standpoint, in that it allows Riemann to standardize his conception of all *Wechsel* relationships, other than the *Seitenwechsel* itself, as compounds of a *Schritt* followed by a *Seitenwechsel.* But the elevation of the chromatic *Terzschritt* also poses certain difficulties. Most notably, it leads Riemann to view the diatonic *Terzwechsel* as a compound relation (T⊕) in chromatic space. That is, the *Terzwechsel* (the neo-Riemannian R) is a derived relation. This derivation seems to contradict the perceptual immediacy—the direct intelligibility—of that progression.

Table 5.3 Riemann's *Systematik der Harmonieschritte* (from *Skizze*)

Categories of *Harmonieschritte*	Specific *Harmonieschritte*	from c$^+$	from $^\circ$e	Implied Derivation	Directly intelligible?
Seitenwechsel	*Seitenwechsel*	c$^+$–$^\circ$c	$^\circ$e–e$^+$	$\oplus$	•
Quintschritte	*schlichter Quintschritt*	c$^+$–g$^+$	$^\circ$e–$^\circ$a	Q	•
	Gegenquintschritt	c$^+$–f$^+$	$^\circ$e–$^\circ$h	Q^{-1}	•
	Quintwechsel	c$^+$–$^\circ$g	$^\circ$e–a$^+$	Q$\oplus$	•
	Gegenquintwechsel	c$^+$–$^\circ$f	$^\circ$e–h$^+$	Q$^{-1}\oplus$	
Terzschritte	*schlichter Terzschritt*	c$^+$–e$^+$	$^\circ$e–$^\circ$c	T	•
	Gegenterzschritt	c$^+$–as$^+$	$^\circ$e–$^\circ$gis	T^{-1}	•
	Terzwechsel	c$^+$–$^\circ$e	$^\circ$e–c$^+$	T$\oplus$	•
	Gegenterzwechsel	c$^+$–$^\circ$as	$^\circ$e–gis$^+$	T$^{-1}\oplus$	•
Kleinterzschritte	*schlichter Kleinterzschritt*	c$^+$–a$^+$	$^\circ$e–$^\circ$g	Q^{-1}T	•
	Gegen-Kleinterzschritt	c$^+$–es$^+$	$^\circ$e–$^\circ$cis	T^{-1}Q	•
	Kleinterzwechsel	c$^+$–$^\circ$a	$^\circ$e–g$^+$	Q^{-1}T$\oplus$	•
	Gegenkleinterzwechsel	c$^+$–$^\circ$es	$^\circ$e–cis$^+$	T^{-1}Q$\oplus$	•
Ganztonschritte	*schlichter Ganztonschritt*	c$^+$–d$^+$	$^\circ$e–$^\circ$d	Q^2	
	Gegenganztonschritt	c$^+$–b$^+$	$^\circ$e–$^\circ$fis	Q^{-2}	
	Ganztonwechsel	c$^+$–$^\circ$d	$^\circ$e–d$^+$	Q$^2\oplus$	
	Gegenganztonwechsel	c$^+$–$^\circ$b	$^\circ$e–fis$^+$	Q$^{-2}\oplus$	
Halbtonschritte	*schlichter Leittonschritt*	c$^+$–h$^+$	$^\circ$e–$^\circ$f	QT	•
	Gegenleittonschritt	c$^+$–des$^+$	$^\circ$e–$^\circ$dis	Q^{-1}T^{-1}	•
	Leittonwechsel	c$^+$–$^\circ$h	$^\circ$e–f$^+$	QT$\oplus$	•
	Gegenleittonwechsel	c$^+$–$^\circ$des	$^\circ$e–dis$^+$	Q^{-1}T$^{-1}\oplus$	
Tritonusschritte	*schlichter Tritonusschritt*	c$^+$–fis$^+$	$^\circ$e–$^\circ$b	Q^{2}T	
	Gegentritonusschritt	c$^+$–ges$^+$	$^\circ$e–$^\circ$ais	Q^{-2}T^{-1}	
	Tritonuswechsel	c$^+$–$^\circ$fis	$^\circ$e–b$^+$	Q^{2}T$\oplus$	
	Doppelterzwechsel	c$^+$–$^\circ$gis	$^\circ$e–as$^+$	T$^2\oplus$	•

This problem, of a conflict between derivation and intelligibility, was not unique to Riemann's theory. Oettingen had raised the issue in objecting to Hauptmann's common-tone approach, arguing against the notion that a progression between a tonic and either of its dominants must be mediated via diatonic third relations. Oettingen had encountered a similar problem with his own antinomic *Leit-Schritt* transformation, which he defined both as a fundamental, directly intelligible relationship and, contradictorily, as a compound *Quintschritt*-plus-*Terzschritt*, and which he eventually omitted from his final tally of direct relations. Riemann's reliance on the *Terzschritt* as a generator exacerbates this particular problem, as it leads him to represent the diatonic *Leittonwechsel* as an even more complex three-step compound (QT$\oplus$).

Riemann's reworking of Oettingen's *Schritt/Wechsel* system presents a more pronounced form of the problem, however, and one that seems to contradict the common perception of diatonic tonality. By privileging transposition by major third as a generative relation, and furthermore by expressing all inversions as compounds of transpositions and the *Seitenwechsel*, Riemann implies a tonal hierarchy in which diatonic relationships are represented not only as a subset of chromatic relationships, but in some cases even as subordinate to them.

The lack of consistent correlation between a relationship's intelligibility and the relative simplicity or complexity of its root-interval derivation is a recurring concern of Riemann's. The twenty-five relationships included in the *Systematik der Harmonieschritte* vary in strength and intelligibility. Riemann comments on the relative clarity of the relationships, and, in those cases in which the relationship is distant enough that it begins to call the supremacy of the tonic into question, he identifies appropriate mediators. He does not consider a succession from the tonic to the *Gegenquintwechsel* to be directly intelligible, for instance, and therefore requires that the progression c^{+}–^{o}f be followed by a mediating triad closely related to both c^{+} and ^{o}f—he suggests either ^{o}c or f^{+}.

Once his examination of all twenty-five relationships has been completed, Riemann compiles a roster of those relationships he considers to be directly intelligible, shown in the right-hand column of table 5.3.[47] Riemann does not indicate exactly why these sixteen *Schritte* and *Wechsel* are directly intelligible and the remaining nine are not, but the intelligibility of root-interval relationships involved seems to be the primary determinant: in addition to the *Seitenwechsel*, which involves no root motion, the direct relationships include three of the four relationships featuring root motion by perfect fifth, all four relationships featuring root motion by major third, all four relationships featuring root motion by minor third, and three of the four relationships featuring root motion by half step. Riemann consistently refers to the directly intelligible perfect fifth and major third throughout his discussion of root relations, but during his presentation of *Systematik der Harmonieschritte* he argues as well for the intelligibility of root motions by minor third and by half step.

The bases upon which Riemann argues for the intelligibility of these root relations suggest a point of contact with the common-tone model: though Riemann admits that neither the *Kleinterzschritt* (the difference between the *Quint* and *Terz*) nor the *Leittonschritt* (the sum of the *Quint* and *Terz*) is technically direct, he suggests that the former is nevertheless directly intelligible because both roots involved are representatives of the same *Klang*, and that the latter is nevertheless directly intelligible because of the strong melodic connection it affords. Yet the notion that the intelligibility of root-interval relationships serves as a guide to the intelligibility of triadic successions fails to explain why Riemann has excluded the *Gegenquintwechsel* and the *Gegenleittonwechsel* from his roster of directly intelligible relationships, and also fails to explain the inclusion on this roster of the *Doppelterzwechsel*, which involves root motion through an augmented fifth.[48]

Though Riemann never invokes the notion of a group, if an identity element is assumed and enharmonic equivalence is accepted, the twenty-five relationships registered in the *Skizze's Systematik der Harmonieschritte* constitute a group acting on the twenty-four consonant triads isomorphic to the neo-Riemannian PLR-group.[49] While Riemann continues to espouse a combinatorial approach in the *Skizze*, deriving relationships through composition of *Wechsel*, *Quint-* and *Terzschritte*, his increasingly inclusive views on the intelligibility of these various relationships are a subtle shift away from the stricter correlation of derivational simplicity and intelligibility seen in Hauptmann's and Oettingen's systems.

Riemann by no means abandons the notion that coherence is borne by the acoustically privileged fifth- and third-based generators, which is understandable, given his conception of tonality as the product of acoustically defined relationships. Rather, in some cases he relaxes the requirement that all relationships other than the generators must be understood through reference to mediators.[50]

While Riemann had explicitly rejected the diatonic bias of Oettingen's generators in developing his own version of the *Schritt/Wechsel* system, he openly embraced the diatonic underpinnings of Hauptmann's common-tone system in formulating his functional theory, which he introduced in the 1893 treatise *Vereinfachte Harmonielehre*. Through his reassertion of the primacy of the primary triads, however, and through his reassessments of the status of the secondary triads and the role of common-tone relations, Riemann would broach the major point of conflict between the common-tone and acoustic root-interval approaches, namely the discrepancy between their rankings of diatonic third and fifth relations. Moreover, the reintroduction of the diatonically oriented common-tone system forces Riemann to attempt a reconciliation of the common-tone and root-interval models. That Riemann appeals to the overdetermined *Tonnetz* as the vehicle of this reconciliation lends further support to Mooney's assertion that "the Table was the mise-en-scène for Riemann's struggle with his precursors."[51]

In formulating his functional scheme, Riemann returns to Hauptmann's conception of key as a "triad of triads" comprising the key's primary triads. Unlike Hauptmann, Riemann maintains that these three primary triads are the key's only "directly meaningful" triads. The remaining triads, the secondary triads, no longer mediate between the primary triads, as in Hauptmann's theory, but derive their functional meaning through their relationships to the primary triads. Among these relationships are, of course, the *Leittonwechsel* and *Parallel* relations, upon which the neo-Riemannian L and R operations are based, as well as the *Variante* relation, upon which the neo-Riemannian P operation is based. To avoid confusion, the *Parallel* and *Variante* relations will be referred to by their standard English names—that is, as the Relative and Parallel relations, respectively.

In Riemann's theory, as in Hauptmann's, both the *Leittonwechsel* and Relative relations are defined in terms of their parsimonious voice leading connection to the primary triads. The Relative triad of a primary triad shares the primary triad's root and third, but substitutes a sixth for the primary triad's fifth.[52]

Similarly, the *Leittonwechselklang* of a primary triad shares the primary triad's third and fifth, but replaces its root with its leading-tone (the note a half-step below the root of a major triad or a half-step above the "root" of a minor triad).[53] Again as in Hauptmann's presentation, Riemann does not grant the Parallel relation the same status as the other parsimonious relations: in his initial writings, the Parallel relation is not discussed alongside the Relative and *Leittonwechsel* relations and is not given its own symbol in the *Klangschlüssel* system that Riemann develops to identify triads' functional meanings.[54] In keeping with the diatonic orientation of his functional framework, Riemann categorizes the Parallel relation as the result of a chromatic alteration.

While Riemann's functional scheme follows Hauptmann's common-tone model in several respects, Hauptmann's focus on the changing dialectical meanings of the common tones is absent. As David Kopp has convincingly argued, Riemann's functional designations serve to fix chords' meanings within a key through reference to their associations with primary triads, whereas matters of chordal connection and harmonic coherence are the domain of the *Schritt/Wechsel* system, which continued to be used alongside the functional theory—primarily in the *Handbuch der Harmonielehre* series.[55] In effect, the functional system freed the implicit *Schritt/Wechsel* group from expressing meaning with respect to a given tonic and eased the way for it to act as a group of operations on a set of equally weighted triads.

Given this separation of duties, Riemann's adaptation of Hauptmann's diatonically oriented common-tone model does not compete directly with the chromatic, acoustically conceived *Schritt/Wechsel* group as an explanation of harmonic coherence, yet their coexistence invites a comparison. In many respects, the two systems align quite readily. Riemann's subjugation of the Relative and *Leittonwechsel* relations to the fifth-based system of primary triads—which reverses the hierarchy of Hauptmann's system—reflects the same understanding of tonal proximity that governs Oettingen's use of his *Dominantschritt, Terzschritt,* and *Leit-Schritt*: for Oettingen, moves horizontally across the Table are expressed in terms of the strongest relationship, the *Dominantschritt,* with a *Terzschritt* or *Leit-Schritt* as needed. If motion is restricted to a fragment of an L/R alley reflecting a given major key context, the relationships in Oettingen's system correspond exactly to those in Riemann's scheme of primary triads with their secondary Relative and *Leittonwechsel* associates.[56]

A comparison of the functional and *Schritt/Wechsel* systems also shows discrepancies, however. In particular, questions about the relative intelligibility and generative status of chromatic third relations suggest a need for reconciliation. Hints that Riemann was working toward such a reconciliation of the common-tone-based functional system and the acoustically conceived *Schritt/Wechsel* system appear in two of his later works, the 1914/15 essay "*Ideen*" and the 1917 edition of the *Handbuch der Harmonielehre.*

Riemann indicates in "*Ideen*" that the relationship between two triads is a function of the relationship between their roots, which can be calculated on the

Tonnetz and expressed in terms of the number and direction of *Quint* and/or *Terz* moves involved—with the proviso that the mind will always favor the shortest path. This corresponds to Riemann's reliance on the *Quintschritt* and *Terzschritt* as generators of the *Schritt/Wechsel* group. That the remaining *Schritt/Wechsel* generator—the *Wechsel*—is excluded from consideration in the *Quint/Terz* system is not particularly remarkable, as it involves no root motion.[57] Yet in this section of his essay—a section that addresses relationships not only between triads, but also between keys—Riemann clearly feels the need to distinguish between major and minor. Rather than invoking the *Wechsel* relationship, Riemann relies on the Relative and Parallel relations to account for change-of-mode relations, though he doesn't integrate them into the *Quint/Terz* mappings on the Table.

The combination of the *Quint* and *Terz* relations with the Relative and Parallel relations suggests that Riemann may have intended this new system as a rapprochement of sorts between the acoustic and common-tone models of tonal relations that had become aligned with the chromatic and the diatonic in his theories. Admittedly, however, Riemann's intent is somewhat unclear, largely because—although he purports to be discussing relations among both chords and keys—his discussion centers almost exclusively on key relations. Because key relationships are gauged in terms of the relationships between their tonic triads, it seems reasonable to assume that Riemann intended for his remarks to be understood to apply to both keys and triads. Yet, the focus on keys may well have dictated the use of the Relative and Parallel relations—in place of the *Wechsel* relation Riemann had previously used to account for change-of-mode relations between triads. Moreover, the derivation of the *Leittonwechsel* as a compound *Quint*-then-Relative relation suggests that Riemann's primary focus may not have been on voice-leading parsimony. On the other hand, it is possible that Riemann simply failed to privilege the parsimonious *Leittonwechsel* relation due to his focus on keys in this relatively brief passage.

An additional clue supporting the assertion that Riemann intended this system as a reconciliation of the two approaches appears in the edition of the *Handbuch* that appeared following the publication of "*Ideen.*" Kopp has observed that in the sixth edition of the *Handbuch*, Riemann for the first time incorporates strong mediant relations into his functional theory by identifying the *Terzklänge* (the triads lying a *Terzschritt* to either side of the tonic) in their own symbols, rather than in relation to the primary triads—in effect identifying the *Terzklänge* as a distinct functional category.[58] This elevation of the *Terzklänge* eases the diatonic strictures Riemann had originally placed on his functional theory and brings the functional side of his theory more into accord with the *Schritt/Wechsel* system.

Kopp suggests that the inclusion of the *Terz* relation in "*Ideen*" paved the way for the elevation of the *Terzklänge* within Riemann's functional system, in that Riemann's treatment of the *Terz* relation in "*Ideen*" constitutes a formalization of the independent mediant relation, and in that "[i]ts expression as mathematically-determined moves on a harmonic grid, rather than a set of diatonic archetypes, neutralizes Riemann's reservations about the mediants' chromatic content."[59] As shown above, the independence of these chromatic mediant relations had been formalized in Riemann's works long before "*Ideen*"—as early as *Hülfsmittel*, in which a version of the Table also appeared—but the reassertion of the *Quint/Terz* representation, in conjunction with Riemann's newly articulated ideas about the tonal imagination and perception, may well have prompted him to address the acknowledged immediacy of these chromatic third relations within his functional theory as well. That the Table should assume a central role in Riemann's attempts to harmonize, if not synthesize, the two competing approaches is not surprising.

The issue of tonal coherence motivated the development of both the neo-Riemannian approach and the nascent group-theoretic approaches of the nineteenth century. Faced with an ever-growing repertoire of music in which chromatic relationships resisted reconciliation to a diatonic basis, Riemann, like the neo-Riemannians, locates the coherence of these relationships in the combinatorial logic of the group: Riemann turns to the logic inherent in his implicit, acoustically conceived *Schritt/Wechsel* group, while neo-Riemannian theory turns most often to the logic inherent in its explicit, parsimonious PLR group.

Riemann was by no means the first nineteenth-century theorist to advocate a combinatorial and nascently group-theoretic approach to triadic relations—in this respect, neo-Riemannian theory invokes his name synecdochally—nor were his predecessors unanimously concerned with tonal coherence in chromatic contexts. The nascent group-theoretic approach developed not as a response to the exigencies of an increasingly chromatic practice, but rather, seemingly fortuitously, out of elements which, as Cohn has asserted, "had been long present, if not fully articulated or mobilized."[60] As the emergence of the group concept was deeply intertwined with a shift away from a diatonic, scale-based conception of tonality, toward a chromatic conception, the over-determined triad and *Tonnetz* are implicated in this regard as well.

That the eventual formulation of a complete group on the twenty-four consonant triads involved *Schritt* and *Wechsel* relations, rather than the parsimonious PLR relations, is to an extent unremarkable: the consonant triad's over-determined nature clearly supports two distinct (but isomorphic) group-theoretic systems—one linked to the triad's acoustic generability and the other linked to its voice-leading potential—both of which are modeled by the *Tonnetz*. This outcome is, however, of interest from the perspective of the historical development of tonality. Though both the common-tone and root-interval approaches are capable of supporting a fully chromatic orientation, the historical alignment of the common-tone model with diatonic space and the *Schritt/Wechsel* model with chromatic space manifests a conflict between the diatonic and chromatic that comes to the fore in Riemann's works. That Oettingen, in initiating the shift between the approaches, took issue not with Hauptmann's diatonic conception of tonality but with his hierarchy of diatonic relations lends further credence to Cohn's conjecture that the seeds of tonality's destruction were sown from within.[61] In fact, an incipient group-theoretic orientation entered nineteenth-century German harmonic theory not only under the guise of the acoustic relations in general (as Cohn speculates) but also under the guise of tonality in particular. In this respect—in his dissolution of a diatonically predicated conception of tonal relations—as much as in his actual formulation of the *Schritt/Wechsel* group, Riemann's theories represent the culmination of a nascent group-theoretic tradition.

Notes

1. The group-theoretic potential of Riemann's *Schritt/Wechsel* system was first identified in Klumpenhouwer 1994, though for reasons to be discussed below Klumpenhouwer suggests that he has had to add transformations to Riemann's system to complete the group. Kopp 1995 provides a more accurate overview of Riemann's *Schritt/Wechsel* system, but does not comment on its implicit group structure. Both Mooney 1996 and Gollin 2000 credit Riemann with outlining the complete *Schritt/Wechsel* group and examine its structure. On the structure of the group, see also Gollin 1998 and Clough 1998—the latter explores the isomorphism between the *Schritt/Wechsel* group and T_n/T_nI group. Cohn 1998a cites the *Schritt/Wechsel* group in his survey of neo-Riemannian theory's historical precedents. Cohn's discussions of Weitzmann's work appear in Cohn 1998c and 2000.

2. Cohn 1998a, 175 and 169.

3. Cohn 1997, 59.

4. Combinatorial group theory offers a convenient way of describing a group via what is known as a group presentation. A group presentation—which is generally given in the form $G=\langle f_1, \ldots, f_n; r_1, \ldots, r_q \rangle$—describes a group in terms of a system of "generators" (given above as "$f_1, \ldots, f_n$") and a set of "defining relators" on those generators ("$r_1, \ldots, r_q$"). The system of generators is a collection of group elements from which all other elements can be derived as compounds. Relators place constraints on the ways in which the generators compose by showing how certain combinations of the generators align with the group's identity element. A set of defining relators on a given set of generators is a set of relators from which all other relators on those generators can be derived.

5. Shirlaw [1917] 1970, 325–26, notes a similar abandonment of the fundamental bass approach in France during this period, which he attributes in part to fundamental bass theory's inability to account for novel harmonic progressions appearing in practice and also in part to an increasing emphasis on practical rather than theoretical approaches to harmony. See, for example, Jérôme-Joseph de Momigny's *Cours complet d'harmonie et de composition* (1803–5).

6. Hauptmann [1853] 1991, §87. All translations of passages from Hauptmann's treatise are by Heathcote.

7. Hauptmann's characterization of the supertonic triad as "diminished" is linked to his espousal of just intonation. Hauptmann conceives of the supertonic triad in C major not as d–F–a—that is, not as a minor triad—but as D|F–a, a triad with a flatted ("diminished") fifth spanning a pure major third and a Pythagorean third (32/27, a syntonic comma smaller than a pure minor third). (The "d" would be the next element to the left of the schema in figure 5.1a, but "D"—the fifth above "G"—preempts "d" in the context of C major, leading to the juxtaposition of two fifth-series elements

["D|F"] and the "diminished" fifth in the triad D|F–a.) In the context of his dialectical scheme, Hauptmann counts both diminished triads as dissonant, in the sense that these triads do not constitute unities, but instead embody an initial contradiction that is never reconciled (§46). Hauptmann makes no mention of the diminished triad's dialectical character in his discussion of relationships among the diatonic triads; as will be discussed below, he treats the diminished triad as if it were a consonant triad.

8. Kopp 1995, 89 and McCune 1986, 15 make the same observation about the schema's role as a guide to harmonic proximity and succession. Although Hauptmann does not actually include this circular representation of the key in his treatise, he explicitly states that the joining of the limits of the key—the "passing into self" of the key—should be conceived in this manner: "We can picture the idea of something passing into self by thinking of a finite straight line bent into a circle with its beginning and end united: *finite* as *infinite*, or *infinite* as *finite*" (§45). The circular representation of the key does appear in Heathcote's appendix to his translation of Hauptmann's treatise, Hauptmann [1853] 1991, 351.

9. The mapping of these successions onto figure 5.1c would seem to suggest that the triads are in each case related by inversion about a shared major or minor third, and would therefore seem to suggest an additional link between Hauptmann's approach and the R and L transformations as contextual inversion operators. Hauptmann's characterization of major and minor triads as being generated through opposing means—whereas the root of a major triad *has* both a perfect fifth and a major third above it, the fifth of a minor triad *is* both the perfect fifth above the triad's root and the major third above the triad's third—seems to support the idea that the triads in these successions are related by inversion; however, his descriptions of the various possible relationships between triads emphasize the proximate voice-leading of common-tone connections rather than the transformational voice-leading associated with inversion.

10. §116. Cohn 1997, 64, cites this passage as one example of a historical treatment of elision suggestive of the notion of compound operators. Hauptmann's claim that motion from a tonic to either of the two remaining primary triads—the upper or lower dominants—should be understood to be mediated via a secondary triad will be rejected by Oettingen, who finds this to be untenable given the acoustic relationships between the triads' roots.

11. Though the D-a fifth of the D|F-a triad is not a perfect fifth in Hauptmann's just system (hence Hauptmann's classification of the supertonic triad as "diminished"), normative voice-leading practices would favor the less parsimonious C–e–G. . .a–D|F voicing (with two voices moving by step and one skipping by third) over the smoother C–e–G. . .D|F–a voicing (with all voices moving by step) for this reason. Kopp 1995, 91, also comments on Hauptmann's curious explanation of disjunct successions, remarking in particular on Hauptmann's apparent disregard for dialectical note qualities in this context.

12. The capacity of the generic triad to enter into "maximally smooth" or parsimonious cycles has been remarked upon in the neo-Riemannian literature by Cohn 1996, 32 and 1997, 58 and Lewin 1996, 182. Agmon 1991 does not specifically refer to this sort of cycle, but he does note the generic triad's potential with respect to what he terms "efficient" voice leading, and he also notes that generic triads can be ordered hierarchically on the basis of the number of common tones they share with a given triad.

In terms of group presentations, the cyclic group introduced by the parsimonious cycle might be given by the presentation $\langle T_1; (T_1)^7 \rangle$, where T_1 is the operation mapping a generic (024) triad to the next member of the parsimonious cycle, moving in clockwise motion. In its effect, this T operation is reminiscent of Lewin's 1982 SHIFT operation, but differs in that T acts on generic triads within the given mod-7 cyclic space, whereas SHIFT acts on what Lewin terms "Riemann systems" in a justly tuned environment.

13. Hauptmann does not fully embrace this generic, mod-7 perspective in his discussions of triadic successions within minor key systems. Hauptmann clearly conceives of minor diatonic space as a cyclic space, insofar as he refers to a joining of a minor key's limits in his derivations of its diminished triads. (Hauptmann's minor-key systems feature a minor tonic triad flanked by its minor subdominant and major dominant: F a♭ C e♭ G b D. The diminished triads that result from a joining of the system's ends are D|F–a♭ and b–D|F.) Specific—rather than generic—interval sizes figure in his discussion of triadic succession, however, as he addresses the voice-leading problem posed by the melodic augmented second found in harmonic minor. Specifically, Hauptmann observes that in the C-minor succession from C–F–a♭ to b–D|F, the latter chord should appear as D–F–b to emphasize the common-tone connection between the triads, but he notes that this introduces a melodic discontinuity in the form of the augmented second a♭–b. A move from C–F–a♭ to b–D|F would avoid this melodic error, but would also abandon the common-tone connection—the very source of the succession's coherence (§106).

14. Hauptmann presents one example of this parsimonious cycle, in reverse order (i.e., moving counterclockwise around figure 5.1c), during a discussion of the details of voice-leading connections in triadic successions: "in the continued series each subordinate triad too may appear in primary form, as, e.g., in the series C–e–G . . . C–e–a . . . C–F–a . . . D|F–a . . . D|F–b . . . D–G–b . . . e–G–b . . . e–G–C . . . e–a–C . . . F–a–C, and so on" (§105). Here Hauptmann's concern is with the way in which voice leading determines the position of the triads in the succession, and this version of the cycle will circle figure 5.1c three times before closing back at C–e–G.

15. Mooney 1996, 48.

16. §285. Hauptmann does not specify exactly how "very nearly related" these two keys must be. He describes two kinds of relationships between keys: those in which keys are linked by shared triads (specifically those in which the tonic triad of at least one of the keys appears in the other key), and those in which the keys' tonics are linked by at least one common tone. Yet, Hauptmann provides an example of a chromatic succession in which triads he describes as belonging to C major and D minor are juxtaposed, and these two keys do not seem to be related in either respect (since the triad built on D, in C major, is regarded as diminished). In fact, in §258, Hauptmann makes a point of reminding the reader that because of the structure of the diminished triad built on the second degree of a major scale, that triad appears only in one particular key system.

17. The most explicit discussion of direct successions between non-diatonic triads occurs in §§285–86. It is tempting to consider much of the discussion of key relations and modulation in §§271–88 applicable as well, as Hauptmann generally characterizes the relationship between two keys in terms of the common-tone connections between their tonic triads (though he describes key relations more precisely in terms of their shared primary triads) and also as he notes somewhat vaguely that extended chord progressions leading from one tonic to the next "might certainly be very much contracted without loss of clearness" (§274; see also Kopp 1995, 95).

18. §248, §253. The major triad presents a "positive" unity through its active relationships of "having a third and fifth above its root," while the minor triad represents a "negative" unity through its fifth's passive status as "being" a fifth and third (see note 9).

19. The conceptual difference between these two pitch classes can be illustrated by extending the schema of pure major thirds within pure perfect fifths that Hauptmann uses to represent the major key:

F a C e G b D f♯ A c♯ E.

While "e" belongs to C major, "E" appears as part of the dominant of D major. Hauptmann discusses this notion of changing note quality in §§271–72 and 274–75.

20. Hauptmann's reticence might stem in part from the fact that similar adjustments appear, by necessity, in almost all modulatory passages, even in those in which each pair of successive triads can be accounted for as a diatonic succession. In §275, for example, Hauptmann presents the following succession with the accompanying Roman numeral analysis:

* *

C-e-G ... B♭–C–e–G ... a♭–C-F ... A♭–D♭–f... A♭–B♭–D♭–f... g–B♭–D♭–E♭... A♭–c–E♭
C: I
f: V — V_7 — I

A♭: VI— IV — II_7 — V_7 — I

Hauptmann's interest in this passage lies primarily in the connection it establishes between the C in the initial triad and the c in the final triad, but in the midst of the succession a♭–C–F (the tonic of F minor) has been implicitly reinterpreted as A♭–c–f (the submediant of A♭ major). Once this switch from F minor to A♭ major is understood to be in effect, however, the succession from A♭–c–f to A♭–D♭–f is strictly diatonic. This aspect of Hauptmann's system holds interesting implications in terms of his conception of tonal space, as will be considered below.

21. Hauptmann [1853] 1991, §271, 266. Like the earlier common-tone theorist Marx, Hauptmann relies on pitch-class intersection as a test for relatedness among keys, but Hauptmann focuses on common-tone connections between the keys' tonic triads rather than on common-tone connections between complete diatonic collections. Whereas Marx conceives of relationships between triads in terms of the relationships between the keys they represent, Hauptmann conceives of relationships between keys in terms of the relationships between their tonic triads.

22. While Hauptmann's theory does not fully support a group presentation on P, L, and R, it does imply a finitely generated infinite group on L and R, given by the presentation $\langle L, R; L^2, R^2 \rangle$. This group is considered to be "free" on the generators L and R in the sense that there does not exist a non-trivial relator governing their interaction with one another. On the primacy of the L/R series, see Cohn 1997, 36–37.

23. Although the lowercase note names that appear in the Table indicate a specific octave within Oettingen's notational scheme, the Table should be understood to model relationships among pitch classes, rather than among pitches. Oettingen 1866, 15, emphasizes that this is the case by demonstrating various octave-specific realizations of relationships shown in the Table.

24. Oettingen 1866, 137ff.

25. Like Hauptmann, Oettingen considers the supertonic triad to be a dissonant chord.

26. As will be discussed below, Riemann introduces a refinement to Oettingen's root-interval system that has precisely this effect.

27. Oettingen 1866, 141–42. Gollin 2000, 162–67, argues that Oettingen conceives of the Table as a static table, akin to the periodic table of elements, demonstrating the derivation of pitch classes, rather than as a "map" of a "traversable space." The latter understanding of Oettingen's Table is suggested by Mooney 1996, 72 and echoed in Cohn 1997 and 1998a. While Oettingen's references to the Table prior to this point in his treatise do suggest a static conception, the passage at hand indicates that he also conceives of the Table as a navigable map of triadic space.

28. That is, the group given by the presentation <L, R; L^2, R^2>.

29. Cohn 1997, 29.

30. <D, L, R; L^2, R^2, DL=LD, DR=RD>

31. Riemann would later standardize the use of the term "*Schritt*" ("step") for transposition operators and the term "*Wechsel*" ("exchange") for inversion operators. Oettingen does not draw as clear a distinction, however: while he always refers to relationships based on transposition as "*Schritte*," he refers to those based on inversion as both "*Schritte*" and "*Wechsel*."

32. Oettingen 1866, 145, observes that this relationship is a familiar one, encountered in dominant-to-tonic progressions in minor and also in tonic-to-subdominant progressions in Hauptmann's "major minor" key.

33. <D, N, L, R; N^2, L^2, R^2, DN=ND, DL=LD, DR=RD>

34. Oettingen stresses acoustically determined root relations as the principle guiding his choice of transformations, but diatonic reference plays a large role as well. Based solely on root relations, the progressions c+–e°, c+–e+, c+–a♭° and c+–a♭+ appear to be equivalent relationships, but Oettingen singles out the first relationship as the fundamental transformation based on root motion by a third. Riemann will embrace a somewhat wider range of relationships between triads on the basis of the root-intervals involved, yet he does not rely solely on root intervals as a guide to triadic relations, either.

35. Oettingen 1866, 149–52. The characterization of this transformation as a "shortcut" transformation comes from Mooney 1996, 81. Mooney observes that "Shortcut transformations are Oettingen's response to commonplace harmonic relations that happen to be remote in his theory."

36. The examples given here are Oettingen's. The *klein Oberterz Schritt* differs from the *klein Oberterz Wechsel* (as the transformation would more accurately be called) given in the summary of the shortcut transformations in Mooney 1996, 82.

37. Oettingen 1866, 156. Mooney 1996, 75–76 and 79, remarks upon this shortcoming of Oettingen's system, noting that it would soon be addressed by the functional component of Riemann's theory.

38. Oettingen 1866, 125ff.

39. At this point in the treatise, Oettingen has not yet defined these transformational relationships. Even once he has introduced the transformations, however, he does not explicitly note the connection, although it comes into play during his discussion of mediated relationships between key systems, to be discussed below.

40. Oettingen 1866, 145.

41. This conception of tone-system relations is not made clear in his discussions of tone-system relations elsewhere in the treatise, including in the more detailed examination in his chapter 4. A good introduction to the approach taken is provided in Mooney 1996, 82–94.

42. The group-theoretic potential of Riemann's discussion of the *Schritt/Wechsel* system in *Die Hülfsmittel der Modulation* was first noted in Gollin 2000, 217–24. In subsequent editions of the *Skizze* (which, beginning with the second edition, was retitled *Handbuch der Harmonielehre*), Riemann slightly revised his presentation to include additional relationships that are redundant under equal temperament.

43. One particularly telling difference between Riemann's and Oettingen's discussions involves Riemann's apparent elevation of the *Seitenwechsel* relation—a later addition to Oettingen's diatonically conceived system—to a status roughly equivalent to that of the *Quintschritte*.

44. Riemann 1880, 80–81. Riemann's inclusion of the *Gegentritonusschritt* is curious in that it introduces a redundancy if his system is viewed as a group of transformations on the equal-tempered triads: under equal temperament, the *Tritonusschritt* is its own inverse. Yet Riemann clearly conceives of the *Gegentritonusschritt* as the inverse of the *Tritonusschritt,* as is evident from his discussion of their use: Riemann stresses that the two *Tritonusschritte* move in the opposite directions and therefore play different harmonic roles, noting in particular that the *Gegentritonusschritt* serves well in modulations to the tonic's "contrary side" (*Gegen-Seite*), whereas the *Tritonusschritt* leads in the opposite direction and as such is particularly effective in modulations back from the contrary side. Riemann's omission of the expected *Gegentritonuswechsel*—which he indicates is simply not found in practice—is not problematic in this regard, as the *Tritonuswechsel* serves as its own inverse. Riemann would, however, include the *Gegentritonuswechsel* in subsequent presentations of the *Systematik der Harmonieschritte* (see Mooney 1996, 246).

45. Riemann does not provide symbolic representations of the relationships' derivations in the *Skizze,* but they are readily inferred from his presentation. Gollin 2000, 233, provides a similar tally. Riemann does present a similar symbolic schema in the earlier *Hülfsmittel.*

46. Klumpenhouwer's 1994 presentation of the *Schritt/Wechsel* group in fact involves the extension of Riemann's diatonic *Schritt/Wechsel* system, but curiously privileges the *Terzschritt* instead of the *Terzwechsel.* Klumpenhouwer's departure from Riemann's exposition seems to stem from his desire to treat all *Schritte* as a class before turning to the *Wechsel,* which in turn leads him to conceive of the *Wechsel* as composite *Schritt*-then-*Seitenwechsel* relations. As such, Klumpenhouwer's conception of the group proves to be much more in keeping with Riemann's second, "chromatic" presentation of the *Schritt/Wechsel* system than with the initial diatonic presentation.

47. Even among "direct" relations, not all successions are equally admissible in all contexts. Riemann 1880, 84, indicates that successions such as those through a *Gegenterzwechsel, Gegenkleinterzwechsel,* or *Doppelterzwechsel* are better used in instrumental, rather than vocal music. Kopp 1995, 126, notes that Riemann immediately backs away from his claims of widespread intelligibility: during his discussion of key relations (which are described in terms of root relations between tonics), he reverts to the more limited list of relationships shown in table 5.3.

48. Kopp 1995, 120, suggests that Riemann relies on common tones as a guide to the intelligibility of successions. For example, the *Gegenquintwechsel* is only member of *Quintschritte* category to lack a common-tone connection, and it is the only

member of the *Quintschritte* category not identified as directly intelligible. Yet the *Kleinterzwechsel* and *Gegenkleinterzwechsel* both lack common-tone connections but are directly intelligible. Likewise, common-tone connections alone do not explain why the two *Leittonschritte* are direct, but the *Gegenleittonwechsel* is not. The presence of a common-tone connection does seem a reasonable explanation for Riemann's inclusion of the *Doppelterzwechsel* among the direct relationships, though.

49. The specific presentation is as $\langle \oplus, Q, T; \oplus^2, (Q\oplus)^2, (T\oplus)^2, Q^4 = T \rangle$.

50. Riemann's strategy is not without precedent: Oettingen characterizes the *Leit-Schritt* as a composite relationship, but immediately argues that it is directly intelligible and treats it as a generator.

51. Mooney 1996, 146.

52. Riemann 1893, 71. Here the usual fifth of a minor triad should be understood as its "root."

53. Riemann 1893, 76 and 80. In this context, Riemann initially introduces the *Leittonwechsel* as the result of motion between a major tonic and the relative of its dominant or between a major key's subdominant and the relative of the tonic. He initially admits *Leittonwechselklänge* defined only with respect to functions for which the leading tone involved will be diatonic. That is, in major key contexts, he recognizes *Leittonwechsel* relations with respect to the tonic and subdominant, but not with respect to the dominant, as the note a half-step below the dominant's root would be outside the diatonic collection.

54. Riemann 1893, 44, does include a passing reference to the tonic's Parallel in discussing the relationship between a major tonic and its minor subdominant (the *Quintklang* of the tonic Parallel). The relationship is not defined explicitly in terms of common-tone connections, however, and no indication is given that this is a standard relationship linking primary and secondary triads. As Mooney 1996, 234–35, has remarked, Riemann would not introduce a *Klangschlüssel* symbol unique to the Parallel relation until 1918, in the sixth edition of the *Handbuch der Harmonielehre.*

55. Kopp 1995, 181–84. Mooney 1996, 229–36, however, argues in favor of a transformational reading of Riemann's functional system.

56. Connections between Riemann's system of functions and the *Tonnetz* have been explored in greater detail in Imig 1970, Mooney 1996, and Smith 2001.

57. As noted in the previous section, Riemann's discussions of harmonic proximity reflect some ambivalence as to the extent to which the *Wechsel* affects perceptions of tonal distance. Certainly, it does not seem to figure into Riemann's calculations of distance in terms of what have been characterized here as number of steps in the generators $\oplus$, Q, and T, yet Riemann does acknowledge a difference in the intelligibility of the *Gegenquintschritt* versus the *Gegenquintwechsel.*

58. Kopp 1995, 185.

59. Kopp 1995, 191. Although Kopp explores Riemann's *Schritt/Wechsel* system as given in *Skizze,* he does not comment on the *Terzschritt's* privileged status as a generator, which is admittedly much less pronounced in presentations after *Hülfsmittel.*

60. Cohn 1997, 59.

61. Cohn 1997, 59.

Chapter Six

Signature Transformations

Julian Hook

Among the many distinctive features of the Schubert *Valse sentimentale* in A major reproduced in figure 6.1, the various adventures of the four-eighth-note pickup figure (first appearing as C♯–E–D–C♯ in measure 2) merit special attention, as does the sudden appearance of the key of C♯ major in the second strain.[1] We consider the latter first. This chromatic mediant C♯ major could be explained in traditional terms as a product of mixture, a sort of parallel substitution for the diatonic mediant, C♯ minor. This explanation could be modeled using neo-Riemannian transformations as shown in the simple network of figure 6.2, in which the Riemannian *Leittonwechsel L* describes the relationship between A major and C♯ minor, and *P* is the parallel relation between C♯ minor and C♯ major.

It might be objected that figure 6.2 is only marginally relevant to Schubert's waltz, inasmuch as C♯ minor never actually appears in the piece: there is neither a C♯ minor triad nor a tonicization of the key of C♯ minor. There is, however, at least a suggestion of C♯ minor in the form taken by the four-note melodic fragment at the point where the mixture first appears, in measure 18: C♯–E–D♯–C♯. Two measures later, the figure takes its C♯ major form, C♯–E♯–D♯–C♯. The latter, it should be noted, is not simply measure 2 transposed from A major to C♯ major; rather, both forms start on the same note, C-sharp, which is scale degree $\hat{3}$ in A major but $\hat{1}$ in C♯ major.

All three of these forms of the four-note motive are shown in figure 6.3. The melodic transformations shown here precisely reflect the harmonic structure outlined in figure 6.2, but in a way that is more apparent in the waltz. The variants of the melodic figure are represented here by changes of key signature rather than the accidentals that Schubert used. When shifts of diatonic collection are implied, appropriate key signatures are a convenient means of representation for analytical purposes, and will be used thus throughout this paper, sometimes without regard to a composer's original notation.

The transformations relating the melodic forms in figure 6.3 are *signature transformations.* The first signature transformation, s_1, adds one sharp to the

Figure 6.1. Schubert, *Valse sentimentale,* D. 779 (Op. 50), No. 13.

key signature, and s_3 adds three more sharps, all while leaving the written notes unchanged.

The three melodic fragments in figure 6.3 exhibit three different patterns of exact interval sizes (m3–M2–m2, m3–m2–M2, and M3–M2–M2). In the language of diatonic set theory, they represent three different chromatic *species* of one diatonic *genus.* In fact, because the genus in this case consists of only three different notes, three is the maximum number of species that can be formed from it, in accordance with the principle "cardinality equals variety."[2]

$$A \xrightarrow{L} c\sharp \xrightarrow{P} C\sharp$$

Figure 6.2. Appearance of C-sharp major in Schubert's waltz: neo-Riemannian analysis.

Figure 6.3. Signature transformations among melodic fragments in Schubert's waltz.

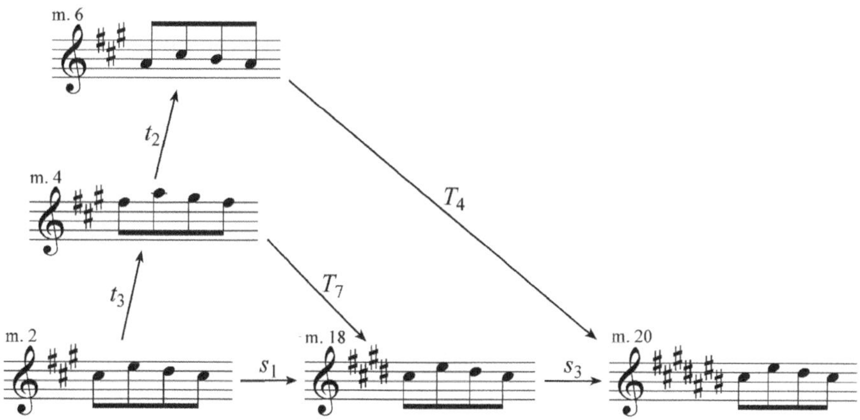

Figure 6.4. Expanded network of signature transformations and transposition operators in Schubert's waltz.

Although the fragments at measures 18 and 20 do not inhabit the A major diatonic collection, both of their *species* have been realized previously in an A major context. In fact, the three species are represented by the first three occurrences of the four-note motive in the piece. The expanded network of figure 6.4 incorporates these additional variants. In addition to the signature transformations s_1 and s_3, two kinds of transformations appear in figure 6.4. The three A major fragments (mm. 2, 4, 6) are related by the *diatonic transposition operators* t_3 and t_2 (note the lower-case *t*'s); that is, every note of the figure in measure 2 is raised three A major scale-steps, yielding the figure at measure 4, then another two scale-steps, yielding (under the assumption of octave equivalence) the figure at measure 6. Diatonic transposition operates in a mod-7 universe; that is, t_7 is the identity.

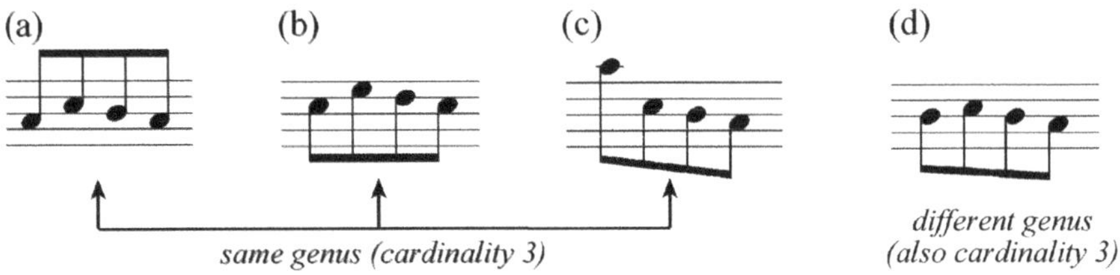

Figure 6.5. Four floating diatonic forms.

Upper-case T_4 and T_7, meanwhile, are the familiar chromatic mod-12 transposition operators, but with a slight twist: they are applied here to diatonic objects, and each T_n is understood to transpose the underlying diatonic scale (and therefore the key signature) along with the notes. The fragment in measure 6, for example, is in A major and starts on $\hat{1}$; in measure 20, transposed by T_4, it becomes a figure in C♯ major, also starting on $\hat{1}$.

Diatonic transpositions t_n preserve diatonic context, but not necessarily species: A major remains A major under any t_n. Chromatic transpositions T_n do not preserve diatonic context, but do preserve species: a major third remains a major third under any T_n. Signature transformations s_n, in general, preserve neither species nor diatonic context, but do preserve letter names: an E remains an E under s_n, though it may acquire a sharp.[3]

As this analysis is beginning to reveal, Schubert's varied presentations of this seemingly inconsequential four-note figure participate in a rich network of foreshadowings and reminiscences while mirroring significant aspects of the harmonic structure of the waltz. These points would be reinforced by expanding the analysis to encompass the remaining variants of the motive, such as the tonicization of D major at measure 30 and the modified-contour figures of measures 10, 12, and 14. Rather than pursuing this thread, we shall turn our attention to the development of a theoretical framework for the study of signature transformations.

A hurdle to be overcome in the formalization of signature transformations is a precise definition of the objects on which these transformations act. Each of the melodic fragments in figure 6.4 is not simply a pitch-class set or string, for several reasons. First, it must be diatonic, and the diatonic collection to which it belongs must be specified (as indicated by the attached key signature). For the chromatic transpositions T_n to form a closed mod-12 system both octave equivalence and enharmonic equivalence must be assumed; the latter, as we shall discover, adds interest to the behavior of signature transformations.

We start by defining a *floating diatonic form* as a fragment of diatonic music written on a staff with no clef, key signature, or accidentals.[4] Four examples of float-

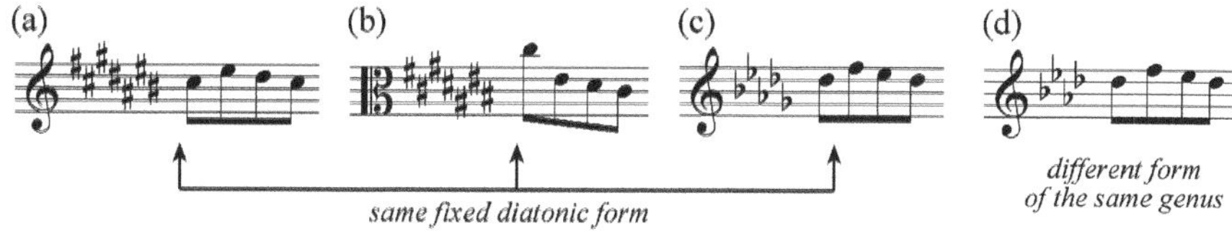

Figure 6.6. Four realizations of floating diatonic forms.

ing diatonic forms are shown in figure 6.5. Confronted with one of these forms, we don't know what the notes are; we imagine that any theoretically possible clef could be assigned to the staff, so the first note could in principle be any of the seven letter names (we may imagine the notes fixed on the staff, with a C or G clef allowed to "float" up or down to any position, or alternatively we may imagine a clef fixed on the staff, with the notes allowed to "float" up or down as a group).

Some properties of a floating diatonic form can be known despite the indeterminacy in the definition. We know, for instance, that forms (a) and (b) of figure 6.5 both follow the pattern "up a third, down a step, down a step"; both, that is, belong to the same *genus* (we do not know the species, since the seconds and thirds may be either major or minor). Form (c) belongs to this genus also: because of octave equivalence, "down a sixth" is the same as "up a third." But (d), which begins "up a step," belongs to a different genus. We may define the *cardinality* of a floating diatonic form as the number of different notes in it, assuming octave equivalence; all the forms in figure 6.5 are of cardinality 3. Two forms in the same genus always have the same cardinality, so we may speak of the "cardinality of a genus" without ambiguity.

A *realization* of a floating diatonic form assigns a clef and standard key signature to it. Figure 6.6 shows realizations of four floating diatonic forms in the same genus. We consider the first three of these to be *equivalent*, in the sense that the following two conditions are satisfied: (1) their pitch-class content is identical (in the usual mod-12 universe), and (2) their key signatures are equivalent according to the usual rules of enharmonic equivalence (i.e., they determine the same diatonic collection). By this definition, realization (d) in figure 6.6 is not equivalent to the others, because the second condition is not satisfied (the key signature does not match). The *notes* of (d) are precisely the same as those of (c), as the flat missing from its key signature does not affect any of the notes of the form; nevertheless, we consider the realizations nonequivalent in the sense that they inhabit different diatonic worlds.

Equivalence classes of realizations of floating diatonic forms will be called *fixed diatonic forms*, or simply *forms*. Examples (a), (b), and (c) in figure 6.6, therefore, represent one fixed diatonic form, while (d) is a different form entirely (though it belongs to the same genus as the others).

Figure 6.7. The action of transposition operators and signature transformations.

(a) Chromatic transposition operators (T_n)
(b) Diatonic transposition operators (t_n)
(c) Sharpwise signature transformations (s_n)
(d) Flatwise signature transformations (f_n)

Fixed diatonic forms are the objects on which signature transformations act. Figure 6.7 illustrates the action of the various transposition operators and signature transformations on forms. The *chromatic transposition operator* T_n transposes both the key signature and the notes of a form up n semitones. The *diatonic transposition operator* t_n transposes the notes of a form up n steps within its diatonic scale, leaving the key signature unchanged. The *sharpwise signature transformation* s_n shifts the key signature of a form n positions in the sharpwise direction on the circle of fifths (adding sharps or removing flats as appropriate), leaving the written notes unchanged. Similarly, the *flatwise signature transformation* f_n shifts the key signature n positions in the flatwise direction (adding flats or removing sharps), again leaving the written notes unchanged.

Theorem. The transformation operators and signature transformations satisfy the following properties:

(a) $T_n = (T_1)^n$; $t_n = (t_1)^n$; $s_n = (s_1)^n$; $f_n = (f_1)^n$.
(b) $T_{12} = T_0$ = the identity transformation.
(c) $t_7 = t_0$ = the identity transformation.
(d) $s_7 = T_1$; $f_7 = T_{-1} = T_{11}$.
(e) $s_{12} = t_1$; $f_{12} = t_{-1} = t_6$.
(f) $s_{84} = s_0 = f_{84} = f_0$ = the identity transformation.

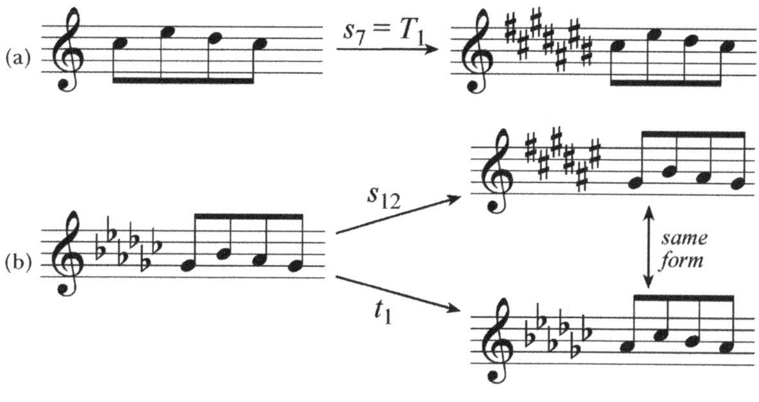

Figure 6.8. Transposition operators as signature transformations.
(a) $s_7 = T_1$
(b) $s_{12} = t_1$

(g) All transformations T_n, t_n, s_n, and f_n commute, and all belong to a cyclic group of order 84, the *signature group*, which is generated by s_1 (or equivalently by $f_1 = (s_1)^{-1}$).

(h) Every signature transformation s_n may be represented as the product of some t_i and T_j. Specifically, $s_n = t_i T_j$, where $i = 3n \pmod 7$ and $j = 7n \pmod{12}$.

(i) Every diatonic genus (with trivial exceptions) has exactly 84 forms, each of which may be obtained from any other by applying some signature transformation s_n. The signature group acts in simply transitive fashion on these 84 forms.

Part (a) of this theorem says simply that applying T_n is equivalent to applying T_1 n times (and likewise for t_n, s_n, and f_n). Parts (b) and (c) describe the mod-12 and mod-7 behavior of T_n and t_n respectively, observed previously in conjunction with figure 6.4.

The first nontrivial part of the theorem is (d). Adding seven sharps to a key signature effectively transposes the implied diatonic collection up a semitone, and any notes along with it; that is, s_7 has the same effect as T_1 (see figure 6.8(a)). Similarly f_7, which adds seven flats, is equivalent to $T_{-1} = T_{11}$.

Part (e) is the heart of the matter. The equivalence of the signature transformation s_{12} and the diatonic transposition t_1 amounts to an explicit mathematical statement of the principle of enharmonic equivalence. This is most easily visualized in the application of these transformations to a form with a six-flat key signature, as in figure 6.8(b). A shift of twelve positions in the sharpwise direction converts this six-flat signature to six sharps. But this six-sharp realization is enharmonically equivalent to a six-*flat* realization whose written notes are one scale-step higher than the notes of the original form—a transformation that could have been accomplished via t_1. The two realizations on the right side of figure 6.8(b) are equivalent, and are therefore considered the same fixed diatonic form; as operations on fixed diatonic forms, s_{12} and t_1 are identical. Figures 6.9–6.11 will provide additional illustrations of statement (e), offering skeptical readers further opportunities to persuade themselves of its validity. The flatwise counterpart of the statement $s_{12} = t_1$ is the statement $f_{12} = t_{-1} = t_6$.

The remaining parts of the theorem are direct consequences of those just discussed. Because seven applications of s_1 give T_1, and twelve applications of T_1 give the identity, it follows that 84 (= 7 × 12) applications of s_1 give the identity, as stated in part (f). Of course, no one is about to add 84 sharps to a key signature; the point is that if we add sharps one at a time and adjust for enharmonic equivalence whenever it becomes convenient to do so, after 84 iterations the original form will reappear.

We have shown that every transposition operator, chromatic or diatonic, can be expressed as some s_n, and that s_{84} is the identity. In the language of group theory, these transformations form a cyclic group of order 84 generated by s_1, as stated in (g). Equivalently, the same group is generated by f_1, the inverse of s_1. Cyclic groups are always commutative, so it follows that a combination of transpositions and signature transformations can be written in any order with the same result; for example, $T_5t_3 = t_3T_5$, $t_2s_9 = s_9t_2$, and so forth.[5]

These properties yield new insights about the transformation network of figure 6.4. It may be observed in that network that application of t_3 followed by T_7 gives the same result as a single application of s_1. We can now see mathematically why this is so:

$$t_3T_7 = (t_1)^3(T_1)^7 = (s_{12})^3(s_7)^7 = s_{36}s_{49} = s_{85} = s_{84}s_1 = s_1.$$

(In this and all subsequent formulas, transformations are combined in left-to-right order.) This equation demonstrates a transformational identity: t_3 followed by T_7 is the same as s_1, not merely on these particular forms from the Schubert waltz, but always, on any fixed diatonic forms. A similar calculation applied to the larger triangle in figure 6.4 gives the identity $t_5T_4 = s_4$.[6]

Parts (d) and (e) of the theorem imply that every t_i or T_j may be written as some s_n. Part (h) is a sort of converse; namely, every s_n can be written not necessarily as some t_i or T_j, but as a product of some t_i and T_j. The identity $t_3T_7 = s_1$, established above, is an example of this property, and in fact the general formula given in (h) for s_n follows directly from this formula for s_1, simply by applying each transformation n times.[7]

The 84 forms of one genus, described in part (i), are the result of all possible combinations of the seven possible clefs (up to octave equivalence) and

Figure 6.9. The 84 forms of one diatonic genus, and the T_n, t_n, and s_n relations among them.

the twelve possible key signatures (up to enharmonic equivalence).[8] Figure 6.9 shows the 84 forms of the genus from the Schubert waltz, arranged in twelve rows and seven columns (there appear to be thirteen rows, but only because the diagram includes two enharmonically equivalent realizations of the six-sharp/six-flat forms, which are considered the same). Each row consists of seven forms with the same key signature, arranged in diatonic circle-of-fifths order (t_4). In any one column, the forms are chromatic transpositions of each other, also in circle-of-fifths order (T_7). The signature transformation s_1, in this representation, is a diagonal move one space to the southwest. The chain of five s_1 arrows shown here could be continued: when it runs off the left side of the figure, it reappears at the right side, and when it runs off the bottom, it reappears at the top. In other words, the figure is topologically a torus, like the Riemannian *Tonnetz* and many other graphical representations of musical spaces. Readers may find it instructive to trace a chain of s_1 arrows through seven iterations (verifying that $s_7 = T_1$), twelve iterations ($s_{12} = t_1$), or even 84 iterations (a complete cycle back to the original form). Any two of the 84 forms in figure 6.9 are linked by a unique s_n; this is what is meant by the *simply transitive* group action asserted in the theorem's last sentence.[9]

The various oval shapes drawn in figure 6.9 illustrate several types of equivalence relations. Recall that the entire figure represents one *genus*. The three *species* of this genus, in the usual diatonic-set-theory sense, are visible as tall dashed rectangles, the full twelve rows high; the first is three columns wide, the others two. As mentioned previously, the number of species is always equal to the cardinality of the genus. The forms in one species are always adjacent along the circle of fifths, so any species in a diagram like figure 6.9 will always be a grouping of adjacent columns.[10]

The forms in any one column are related by chromatic transpositions, so we call them *T_n-equivalent*, and we call the columns *T_n-classes*. The forms in each row are related by diatonic transpositions, so we call them *t_n-equivalent*, and we call the rows *t_n-classes*. The rows may also be called *signature classes*, since each row consists of all the forms with the same key signature. There are always twelve signature classes, and always (with trivial exceptions) seven T_n-classes.

The intersection of a signature class and a species is called a *signature subspecies*. A few of these are shown as light dotted ovals in the bottom two rows of figure 6.9. In this case there are 36 signature subspecies altogether; in general this number is $12c$, where c is the cardinality of the genus. Two forms identical in pitch-class content (but possibly differing in key signature) are called *pitch-class equivalent* and are said to belong to the same *pitch-class subspecies*; we encountered an example in parts (c) and (d) of figure 6.6. Several pitch-class subspecies appear as light dashed outlines, diagonally oriented, across the middle of figure 6.9; there are again 36 (again, generally $12c$) of these. The species of a fixed diatonic form consists of all forms in its pitch-class subspecies and all other forms which are T_n-equivalent to them.

Occasionally when a signature transformation is applied to a form, the resulting form is pitch-class equivalent to the original; that is, the key signature is changed but the notes are not. (Refer again to figure 6.6: application of s_1 to form (c) would yield (d), a different form with the same notes. The two leftmost s_1 arrows in figure 6.9 are also examples.) In such cases we say that the signature transformation *acts inertly* on the form to which it is applied. The idea of pitch-class equivalent forms—one set of notes that functions in two different diatonic worlds—is a familiar one: this is the characteristic of a pivot chord in a modulation. Such a modulation can be modeled by a signature transformation acting inertly on the pivot chord, adjusting the diatonic context of the chord without changing the chord itself.

The genus of diatonic scales offers an instructive illustration. The 84 forms of a diatonic scale correspond to the seven modes on the twelve possible tonic notes. Figure 6.10 illustrates that starting with C major, repeated application of the signature transformation s_1 yields C Lydian, C♯ Locrian, C♯ Phrygian, C♯ Aeolian, C♯ Dorian, C♯ Mixolydian, and after seven iterations C♯ major (T_1 of C major). Respelling in flats for convenience, we continue through D♭ Lydian, D Locrian, D Phrygian, D Aeolian, and then, after the twelfth iteration, D Dorian (t_1 of C major). Every seventh scale is major; every twelfth is a white-key scale. The major scales and the white-key scales coincide at the 84th iteration, when the entire cycle has been circumnavigated and C major reappears.

This process is illustrated musically by a passage from *The Yellow Pages* by Michael Torke, partially reproduced in figure 6.11.[11] Starting at measure 9, a two-measure segment is played thirteen times, with one additional sharp in the music at each successive iteration. The effect is that the music heard initially in G major (step 0) is in A-flat major by measure 23 (step 7) and A Dorian (the original diatonic scale but a step higher) by measure 33 (step 12). At this point Torke breaks the cycle with a move of $t_{-1} = t_6 = f_{12}$, undoing in a single blow the previous twelve s_1 transformations and thereby returning to the original pitch level for a final hearing, a restatement of step 0. (There are other processes at work in the upper parts of this passage: the flute and clarinet play more notes with each iteration, while the violin plays fewer. The arrival at step 7—A-flat major, T_1 of the original G major—is marked by complete silence in the violin, which then switches to sustained pitches. The cello and piano parts, however, apply the signature transformations strictly.)

Recent work by Jack Douthett, presented elsewhere in this volume, describes the same process another way. In figures 4.8(a) and 4.8(b), Douthett models diatonic scales by a pair of concentric circles; seven lamps in the inner circle (the "beacon") shine beams of light outward, where they navigate their way through twelve holes in the outer circle (the "filter"). A small rotation of the beacon brings about a shift of diatonic scale—from D-flat major to A-flat major, in Douthett's description. If, however, we always take beam number 1 as the starting note of the scale, then the "A-flat major" of figure 4.8(b) is really

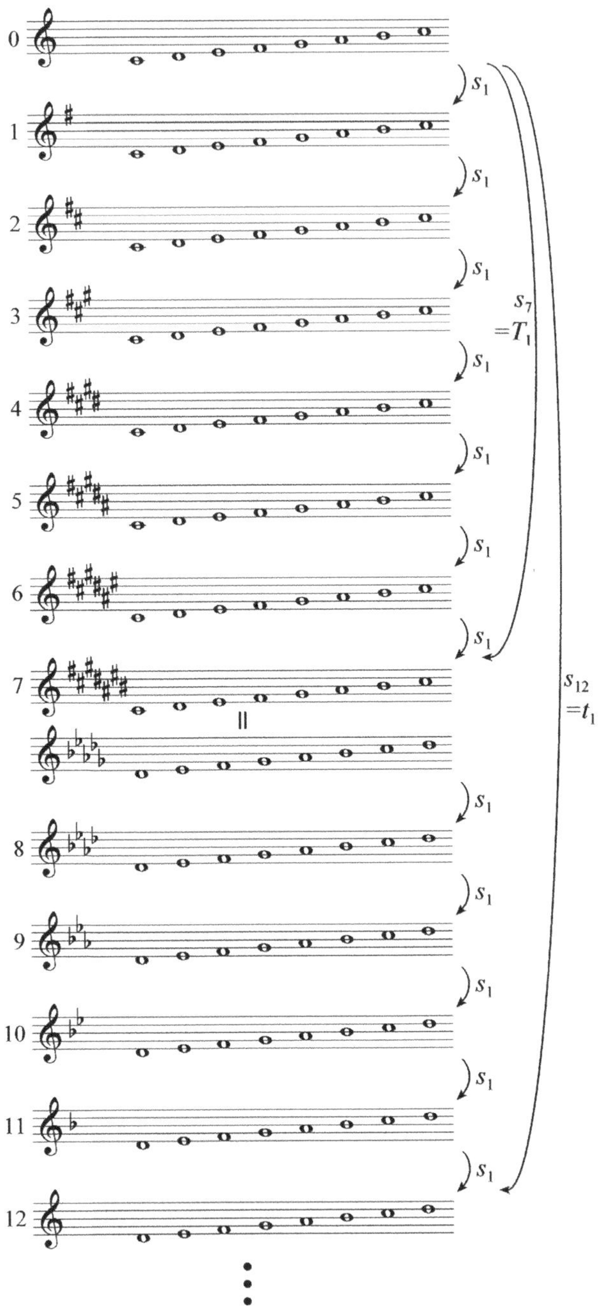

Figure 6.10. The 84 forms of a diatonic scale.

Figure 6.11. Cycle of signature transformations in Michael Torke, *The Yellow Pages*.

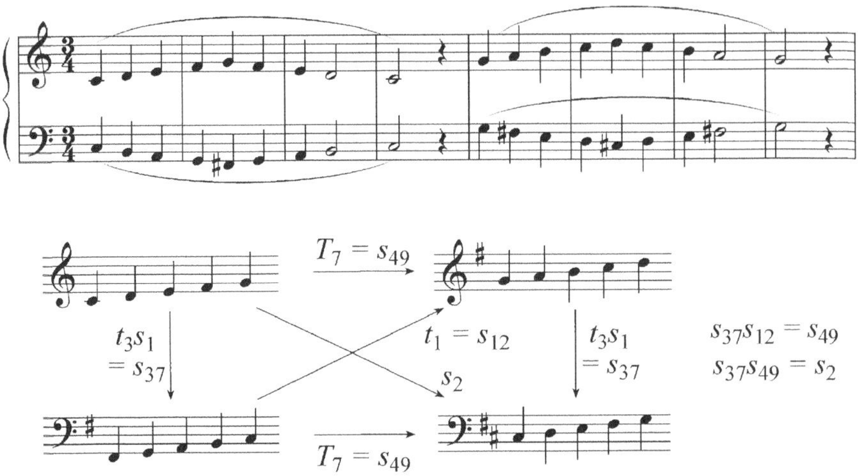

Figure 6.12. Signature transformations in Bartók, "Contrary Motion" (*Mikrokosmos,* vol. 1, no. 17).

D-flat Lydian, and the shift is just what we saw previously in figure 6.10 as the s_1 move from step 7 to step 8. The structure of the signature group as described above is precisely reflected in the angle of rotation calculated by Douthett: $4^2/_7$ degrees, exactly 1/84 of a complete circle.[12]

Douthett shows that the structures generated by his beacon-and-filter systems correspond closely to maximally even sets.[13] It therefore seems likely that signature transformations generalize in interesting ways to other cardinalities of scales, as the property of maximal evenness does. An investigation of this possibility would be complicated, however, by the lack of a standardized system of notation for accidentals and key signatures in scales of arbitrary cardinalities. Rather than exploring these ideas here, we shall conclude with several additional analytical examples, intended to convey something of the flavor and variety of the musical settings in which signature transformations may be profitably employed.

Signature transformations may shed light on any music that makes use of shifts from one diatonic collection to another. The short excerpt in figure 6.12, from one of the simple pieces in Volume 1 of Bartók's *Mikrokosmos,* illustrates a passage in which five-note segments of three different diatonic collections appear in close proximity. The transformations identified as T_7, t_1, and s_2 are straightforward. The two vertical arrows describe identical transformations, each of which can be thought of as a combination of a three-steps-up diatonic transposition and the addition of one sharp, in other words t_3s_1. By the theory developed above, this is equivalent to the signature transformation $(s_{12})^3s_1 = s_{37}$ (the same transformation

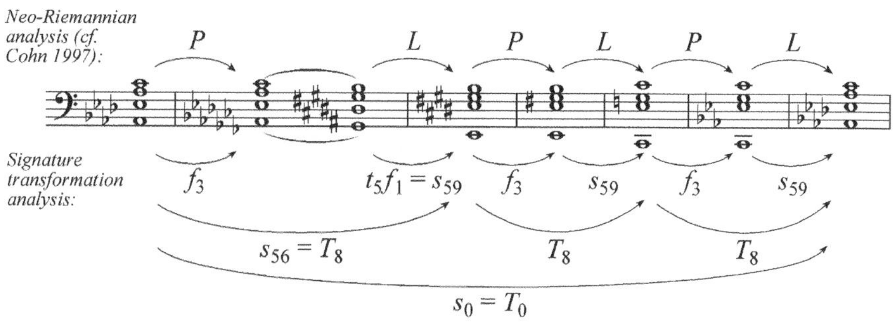

Figure 6.13. Two analyses of a hexatonic cycle (Brahms, Double Concerto, Op. 102, mvt. I, mm. 270–76).

could be conceived as a chromatic T_5 followed by the signature transformation s_2; because $T_5 = s_{35}$, this again works out to s_{37}). Displayed to the right of the graph are two transformational identities reflected in the structure of the network, analogous to those discussed earlier in conjunction with figure 6.4.

Many of the chromatic chord progressions that are staples of the neo-Riemannian literature are amenable to analysis by signature transformations, as figure 6.13 illustrates in the case of a hexatonic *PL*-cycle. The signature transformation analysis requires assigning a complete diatonic context to each triad, which the neo-Riemannian analysis does not—an additional layer of interpretation that may be taken as either an advantage or a drawback.

Our final series of examples is based on the excerpt in figure 6.14, an extended passage from the middle of the first movement of Debussy's Violin Sonata. This excerpt serves as the development section of this rather free sonata form. For analytical purposes it may be divided into five segments:

Segment 1 (mm. 84–97): piano arpeggiations in E, violin melody
Segment 2 (mm. 98–105): arpeggiations of planing perfect-fifth dyads
Segment 3 (mm. 106–19): piano arpeggiations in C, violin melody
Segment 4 (mm. 120–27): transposed restatement of segment 2
Segment 5 (mm. 128–49): quintuplets in piano, static melody in violin, transition to recapitulation at measure 150

Segments 1 and 3 are similar, and both are purely diatonic, but while the former is in E major, the latter is in C Lydian. The violin melody is not "transposed"; in fact, apart from the sharps, it is written with exactly the same notes both times. Here is a situation tailor-made for a signature transformation, modeled in figure 6.15 as f_3 because of the disappearance of three sharps.

The piano part in the same measures, however, is not transformed by f_3 at all; it travels from E major to C Lydian by an entirely different

Figure 6.14. Debussy, Sonata for Violin and Piano, mvt. I, mm. 84–155.

Figure 6.14 (*continued*).

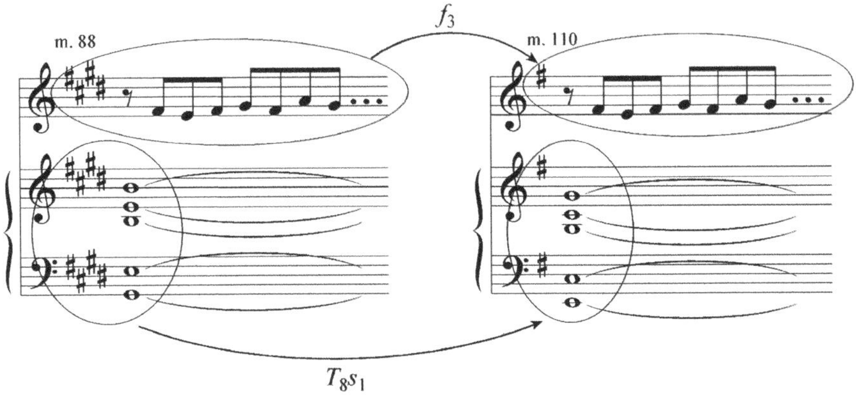

Figure 6.15. Signature transformations in Debussy, mm. 88–97 and 110–19.

transformational route. The arpeggiations, in fact, are simply transposed from E to C. We could call this T_8 but for the fact that T_8 applied to E-major music would yield C-major music, whereas we know from the violin's activities that the correct diatonic collection in segment 3 is C Lydian. (The piano does nothing to contradict the violin's C Lydian, since it plays no F of either kind; asserting that the piano plays in C major while the violin plays in C Lydian would be ascribing to Debussy's music a bimodality that is surely not appropriate here.) The transformation in the piano in figure 6.15 is therefore shown as T_8s_1: T_8 to transpose the notes, and s_1 to adjust the key signature. Because there are no F's in the piano, this s_1 is an example of a signature transformation acting inertly, as previously defined. It does not actually change any notes in the arpeggiations; it merely changes our understanding of the diatonic context in which those arpeggiations reside.[14]

In the first part of segment 5, measures 128–39, the violin hovers around G-flat while the piano plays quintuplets. The mode is E-flat Dorian (notice the C-naturals in the piano in measures 137 and 139). The echo at measure 140 introduces a different modal context: the bass E-flat drops to D, the violin's G-flat is respelled as F-sharp, and the mode is D Lydian. This too is a simple signature transformation, namely f_4, made more apparent in figure 6.16 by respelling E-flat Dorian as D-sharp Dorian. In principle this transformation applies to the entire violin-and-piano texture as a unit, in contrast to the f_3 discussed above; it is worth noting, however, that while f_4 changes all the notes in the piano part, it does not affect either of the two notes played by the violin.

The violin enters in all four parts of figures 6.15 and 6.16 (mm. 88, 110, 133, and 140) on the same F-sharp/G-flat, but this note is heard in four different

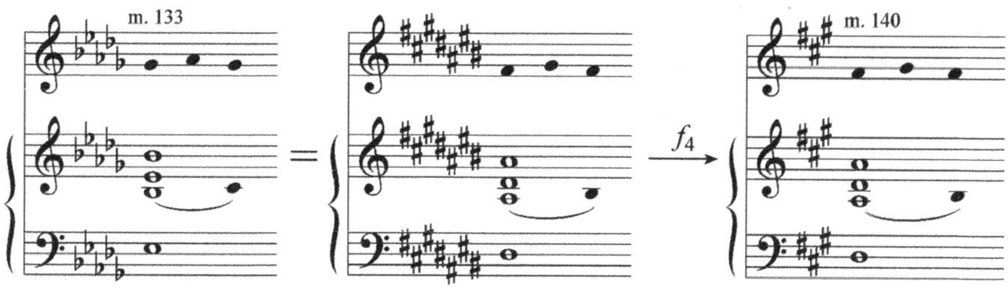

Figure 6.16. The signature transformation f_4 in Debussy, mm. 133–41.

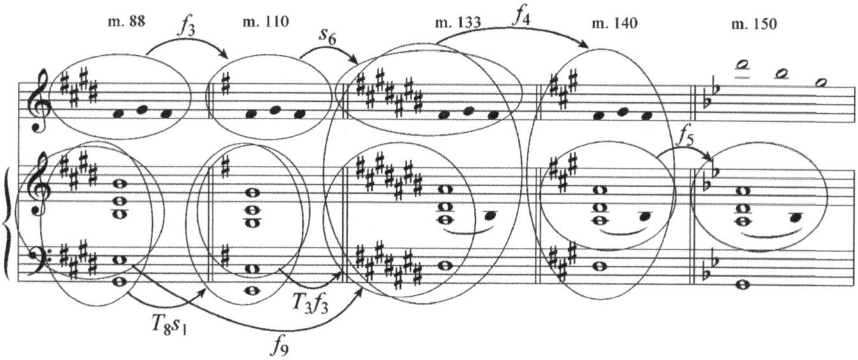

Figure 6.17. "Background" signature transformations in Debussy, mm. 88–150.

modal contexts. Figure 6.17 incorporates this observation into a graph of the "background" transformational structure of the entire excerpt. The reader is encouraged to play all the notes of figure 6.17 at the keyboard, listening to the transformations and their coloristic effects. The first two "measures" in this figure are a shorthand form of figure 6.15, while the next two repeat figure 6.16. Transformations from the second measure (C Lydian) to the third (D-sharp Dorian) are shown also. Here again a simple signature transformation (this time s_6) is operative in the violin, while the piano combines a chromatic transposition (T_3) with an inert signature transformation (f_3) in order to match the mode implied by the violin.

The note-grouping ovals in figure 6.17 vary in size in accordance with small shifts in the extent of the diatonic forms. Thus in measures 88 and 110 the piano arpeggiates first-inversion triads, but in measure 133 the bass drops out, leaving open fifths. At the same time, however, B-sharp (C-natural in the score) appears in the piano as an upper neighbor to A-sharp, and this note participates significantly in the transformational activity thereafter: changing first to B-natural in the global f_4 that leads into measure 140, then to B-flat in a remarkable f_5 into measure 150. The recapitulation is signaled first by the G in the bass, then by the descending G minor triad melody (actually in the piano, notwithstanding its placement for convenience in the violin part of figure 6.17). Meanwhile the quintuplet arpeggiations continue, migrating to the violin simultaneously with the f_5 transformation from the three-sharp signature of D Lydian to the two flats of G minor. The A and D of these arpeggiations are the only two notes of the D Lydian scale that are *not* affected by f_5. As B gives way to B-flat, however, an ingenious role reversal takes place: in the G minor context of measure 150, B-flat is no longer an upper neighbor to A but instead the stable note to which the A resolves.

The transformations in figure 6.17 bridge the intervening spans of segments 2 and 4. These two segments are unlike the others in that neither one is restricted to the notes of a single diatonic collection; segment 2, for instance, contains both B-naturals and B-flats. Nevertheless, at any given moment, a listener readily assimilates the music into an implied diatonic context. Figure 6.18 models a listener's perception of these changing diatonic contexts in quasi-phenomenological fashion, situating each dyad within the diatonic scale in which, I believe, it is most likely to be heard, and modeling the relationships between successive dyads with signature transformations. The top half of the figure shows segment 2, while the bottom half shows segment 4.

At measure 98, the ten-measure violin melody in E major has just ended, so we naturally interpret the E–B dyad in a four-sharp context. The D–A dyad in the very next measure, however, contradicts that context, so we fit it into the nearest available diatonic context, given by a three-sharp signature. Retrospectively, we realize that the E–B dyad would have fit the three-sharp signature as well, so it can be reinterpreted that way, as a sort of pivot chord. In figure 6.18 an inert f_1 relates the two interpretations of E–B; then t_6 relates measure 98 to measure 99.

The following F–C dyad requires a more drastic adjustment, leapfrogging two sharps to the all-white-note signature. Retrospectively, however, the entire segment (so far) fits this new key signature, so it can be accommodated via inert f_3's applied to the first two dyads. This signature then holds sway for three more measures. The fifth and sixth dyads repeat the first two, but are now unlikely to be heard in reference to the original four-sharp signature because of what has intervened. The B-flat in the seventh measure forces a final adjustment, to a one-flat signature. The previous dyad can again be heard as a pivot, but for the first time we reach a point where the entire segment cannot be accommodated by one signature.

Segment 2 ends at measure 105 with the C–G dyad that becomes the backdrop for segment 3. Because segment 3 is in C Lydian as discussed above, an inert s_2

Figure 6.18. Signature transformations acting on perfect-fifth dyads in Debussy, mm. 98–128.

is required to get there. The relationship between E major at measure 98 and C Lydian at measure 106 can be conceived in two ways. It was described earlier, in conjunction with figure 6.15, as $T_8 s_1$, the combination of a chromatic transposition with an inert signature adjustment. On the other hand, from figure 6.18 we can calculate that the composite of all the diatonic transposition operators is $t_6 t_2 t_1 t_5 t_6 t_5 t_1 = t_5$ (add the subscripts mod 7), and the net signature change is f_3 (from four sharps to one), so the overall transformation should be $t_5 f_3$. The algebra of signature transformations developed above confirms that $T_8 s_1$ and $t_5 f_3$ are indeed the same; both work out to s_{57}.

Segment 4 is a chromatic transposition of segment 2, as indicated by the two largest bubbles in figure 6.18 and the T_8 arrow linking them, but the two passages are in different relationships with the surrounding measures. The music preceding the C–G dyad of measure 120 is not in C major but in C Lydian, so the first signature adjustment in segment 4 is f_2 rather than f_1. This f_2 undoes the s_2 with which segment 2 ended, leading back to a one-flat signature. The first two dyads of segment 4 are simply the retrograde of the last two dyads of segment 2 (in fact, the first *four* dyads of segment 4 retrograde the last four dyads of segment 2, all transformed to a four-flat signature).

The five-flat signature in which segment 4 concludes fits the E-flat Dorian (D-sharp Dorian) music that follows in segment 5. The diatonic transposition t_4 into measure 128, however, is something new. Diatonic t_3 and t_4 are fifth relations, the most fundamental root progressions in tonal music. A quick survey of figure 6.18 reveals that Debussy has avoided these throughout the excerpt, using t_1, t_2, t_5, and t_6 in abundance but no t_3's at all and only this single t_4, a "plagal cadence" setting the stage for segment 5.

Figure 6.18 begins and ends with E–B dyads, the first in a four-sharp signature, the last in a five-flat signature. The net transformational trajectory from segment 1 to the beginning of segment 5 is therefore f_9; this appears as the large arrow looping around the left side of figure 6.18, and also at the lower left of figure 6.17.

Key signatures, diatonic scales, and enharmonic equivalence may seem simple, and many musicians, students, and even music theorists would probably be surprised that there might still be things to be learned about them. But the profound consequences of David Lewin's ruminations about getting from one musical entity to another, or of John Clough's speculations about the logic behind the arrangement of the black and white keys of the piano, offer ample evidence that sometimes the simple questions can be the deepest.

Notes

1. This waltz has been widely analyzed. Schachter 1999, 70–76, explores such features as the two-part counterpoint in the right hand, the persistent hemiola, and

several subtle aspects of meter on a larger scale. Lerdahl and Jackendoff 1983, 250–53, cite the piece as an unusually complex example of the interaction of meter and grouping structures, asserting a conflict between a regular four-bar hypermeter and the several ten-bar phrase groupings. Neumeyer (forthcoming) devotes an entire chapter to a series of alternative readings of this waltz.

2. See Clough and Myerson 1985.

3. The letter-name-preserving property of signature transformations is valid only as long as no enharmonic spelling adjustment is made, as will be discussed below.

4. Clough 1979–80 represented several diatonic fragments this way. The term "fragment of diatonic music" is deliberately vague; it could be a chord (set), a melody (line), or a combination of these—essentially anything that can be written on a staff as specified.

5. For background in group theory the reader may consult any abstract algebra text, such as Dummit and Foote 2003.

6. In the terminology introduced in Hook 2007, these identities imply that the network of figure 6.4 is *path-consistent.*

7. Mathematically, statement (h) is tantamount to the observation that every number mod 84 can be written in the form $7x + 12y$; the integer x is uniquely determined mod 12, and y is uniquely determined mod 7. Readers who have studied number theory will recognize here a special case of the "Chinese remainder theorem"; see Stillwell 2003, 171–75.

8. The "trivial exceptions" mentioned in part (i) of the theorem are cases in which the notes of a form make no distinctions among the seven diatonic pitch classes: a seven-note cluster chord, for example, or even more trivially, the "empty form" with no notes at all. Such a genus has only 12 forms, distinguished by key signature, rather than 84.

9. See chapter 7 of Lewin 1987 for a discussion of simply transitive group actions.

10. This adjacency follows from the proof of "cardinality equals variety" in Clough and Myerson 1985; see particularly the illustration of species in their figure 2.

11. Figure 6.11 is condensed from the original score for flute, clarinet, violin, cello, and piano. I thank John Roeder and Ian Quinn for calling this example to my attention.

12. Cyclic groups of order 84 have appeared in the music theory literature, at least implicitly, on several previous occasions. Brinkman 1986 and Agmon 1989 both represent pitch structures by ordered pairs consisting of an integer mod 12 and an integer mod 7. These ordered pairs do form a cyclic group of order 84, isomorphic to the signature group. Neither author studies the group or the transformations as such: Brinkman's objective is a representation of musical data in a form useful for computer processing, while Agmon is concerned with formulating a general definition of "diatonic system," encompassing the familiar 7-in-12 case and other structures. In a recent conference paper, Agmon 2003 revisits his ordered-pair representation but without the assumption of octave equivalence, so that both the mod-12 and mod-7 aspects of the structure are effectively abandoned.

13. See Clough and Douthett 1991.

14. The sonata excerpt moves in and out of several different modes, and the present analysis is intended to show that signature transformations are an appropriate tool for studying Debussy's changes of modal coloration. It should perhaps be

clarified, however, that the transformations do not recognize tonal centers, and therefore do not distinguish among the various modes with the same key signature. In principle, the right side of figure 6.15 could be C Lydian, G major, E minor, or any of four other tonic-mode combinations; the choice has no bearing on the signature transformations appearing there. As analysts, of course, we are free to assert a specific mode whenever we find compelling evidence for a tonal center.

Chapter Seven

Some Pedagogical Implications of Diatonic and Neo-Riemannian Theory

Timothy A. Johnson

Introduction

John Clough's contributions have tremendous potential for the teaching of music theory at all levels. He used a rigorous style more inviting to professional music theorists than to students, but his work can be relevant and interesting even to beginners. Beginning students need some help, however—not because the work of Clough and his collaborators lacks clarity, but because the students lack the necessary background.

This chapter introduces several of Clough's ideas to beginning students in music theory. Some of my prior work suggests methods and materials for introducing certain aspects of diatonic theory to beginning music students.[1] My textbook and accompanying *Instructor Resources* focus primarily on three essential features of the diatonic collection—cardinality equals variety (CV), structure implies multiplicity (SM), and maximal evenness (ME)—identified in two articles, one by Clough and Gerald Myerson and another by Clough and Jack Douthett.[2] Building on this prior work, here I suggest pedagogical approaches for concepts developed in two more recent essays by John Clough, as discussed below.

Clough, Nora Engebretsen, and Jonathan Kochavi examine earlier scholars' work in diatonic theory to reveal important interrelationships.[3] Apparently disparate ways of identifying and understanding scales—and more generally sets (or collections, the term I prefer)—turn out to be related if one looks at interval cycles and other associated constructs—including CV, SM, and ME.

Clough's chapter in this book, "Flip-Flop Circles and their Groups," also uses interval cycles, and ultimately the diatonic collection, to explore neo-Riemannian theory. Although Clough's two essays seem to deal with quite different topics, they are related in both their cyclic approaches and their diatonic goals. My aim here is to shape Clough's musical and mathematical approaches into a form suitable for beginning students, without trivializing these approaches. Although I am assuming the role of an interpreter here, most students will need another interpreter, their teacher, to guide them through the activities I suggest and to discuss the various outcomes. Although the ultimate test of this chapter's ideas is whether they succeed with students, teachers are my immediate audience, especially teachers of introductory music theory or music fundamentals.

Features and Interval Cycles

The principal aim of Clough, Engebretsen, and Kochavi is to develop a taxonomy of scales (specifically) or pitch-class sets (generally)—or, more familiarly, collections—based on certain features they possess as related to interval cycles. Some of these features are found in the work of individual scholars, some are common to different scholars, and others are presented for the first time in the authors' article. Each of the features may be explored advantageously by introductory students.[4] By means of the following suggested methodology, students will be able to develop mathematical skills—including working through a simple, but elegant and far-reaching, algebraic proof—in a familiar (musical) environment, thus providing them with cross-disciplinary experience. They will discover certain features for themselves, rather than just receiving information; they will learn many important attributes of the diatonic collection, the primary musical context in which they are likely to work; and they will explore some intriguing extensions of these diatonic notions to microtonal and other non-12-based musical systems, stretching their imaginations and suggesting areas for future investigation. Although all of the material to be discussed here is latent in the scholarly essays, this chapter provides a pedagogically oriented focus. I start with a discussion of some terms from Clough, Engebretsen, and Kochavi 1999.

Generated

"Generated" (G) is part of the definition of several other terms and in fact is an implied aspect of each of the other eight features identified in the article. Elsewhere I have presented pedagogically oriented explanations of certain generated collections; however, this prior work does not provide a student-initiated approach to the concept.[5] A companion *Instructor Resources* presents a few exercises involving a generator with diatonic and other maximally even collections, but because of the primacy of the generator, and especially its relation to the other features, a more investigative pedagogical approach is warranted here.[6]

A generator is simply an interval cycle. A generated collection is produced by employing a single specific interval, in succession. Generated collections may be identified succinctly using the notation developed by Marc Wooldridge in which the number of notes in the universe (c), the number of notes in the collection (d), and the generator (g) are given as an ordered triple: G(c, d, g).[7] For example, the familiar diatonic collection (d = 7) generated by the circle of perfect fourths (g = 5) in the usual 12-note chromatic universe (c = 12) is represented as G(12, 7, 5). A productive activity for beginning students who are learning intervals for the first time in a fundamentals course, or reviewing intervals in basic music theory, would be to build various generated collections. In addition to practicing working with intervals, students may be led to explore universes of other sizes, including the diatonic collection (where c = 7) or microtonal collections (where c > 12). Students can label their results using Wooldridge's notation. If desired, and especially with microtonal universes, a mathematical approach to the application of the generator may be introduced: "G(c, d, g) = {0, g, 2g, . . . , (d – 1)g}, where products are reduced mod c."[8] The formula is relatively simple, presents musical intervallic work in a mathematical context, and can be used by students even at the introductory level.

Well-Formed

The "well-formed" (WF) feature is also introduced in my earlier work for selected collections, though not through pedagogical activities.[9] Defined by Norman Carey and David Clampitt, a WF collection's generator consists of consistent specific and generic intervals.[10] Creating WF collections is considerably more difficult for beginning students than creating generated collections. However, each of the generated collections from the previously described exercises may be examined for WF. By plotting the collections using dots on a circle, both the generic intervals (the distance in terms of the number of dots) and specific intervals (the distance in terms of the number of lines) of the generator may be observed easily, as illustrated in figure 7.1. The circles show the gradual generation (starting from F) of the diatonic collection by a g of 7. Inspection of the final circle reveals that each specific interval of 7 half steps corresponds to a generic distance of 3 dots. Circles displaying generated collections with universes of any size may be constructed in a similar manner by students, and then may be examined for the WF feature, or lack thereof.

Features Involving the Spectrum

Three features defined by Clough, Engebretsen, and Kochavi might best be introduced to beginning students simultaneously. "Myhill's property," "distributionally

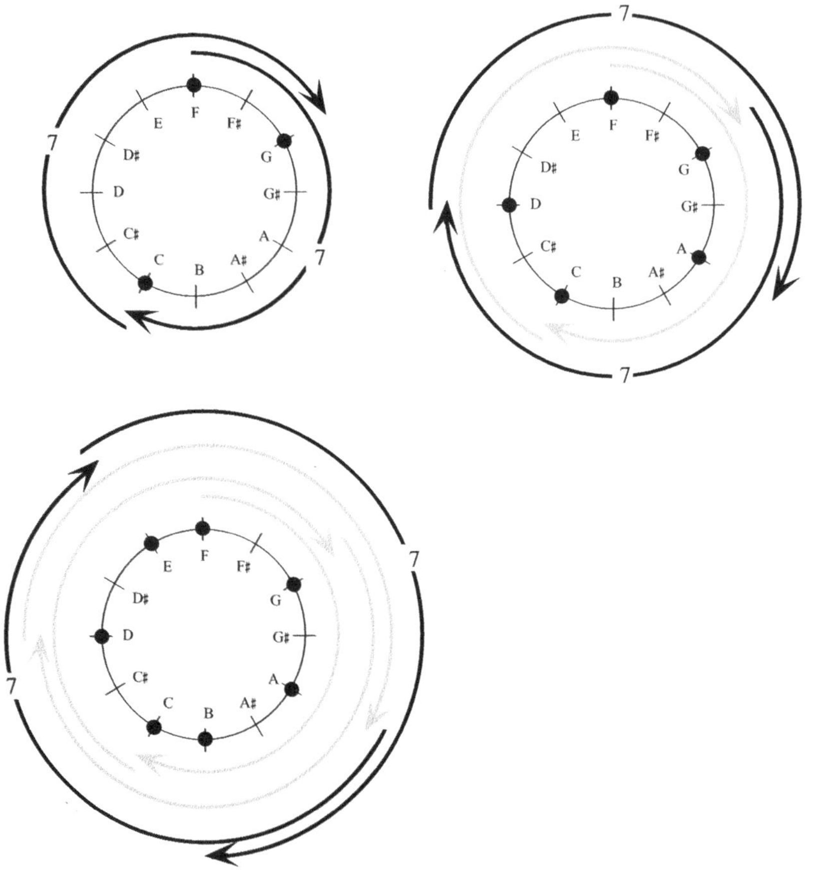

Figure 7.1. Demonstration of a well-formed collection. Adapted from Timothy A. Johnson, *Foundations of Diatonic Theory: A Mathematically Based Approach to Music Fundamentals.* ©2003 Key College Publishing, 82. All rights reserved.

even," and "maximally even" (MP, DE, and ME respectively) are defined in reference to the spectrum of a generic interval. The spectrum of a generic interval is the set of specific intervals that corresponds to it. Using the familiar diatonic collection in a 12-note universe as an example, the possible number of half steps that correspond to each diatonic interval is the spectrum for that interval. Perhaps the easiest way for introductory-level students to grasp the concept of the spectrum is to complete a table for a given collection, as shown in figure 7.2. Any collection of d notes may be plotted on a circle of c lines, and the intervals may be observed and

D harmonic minor scale

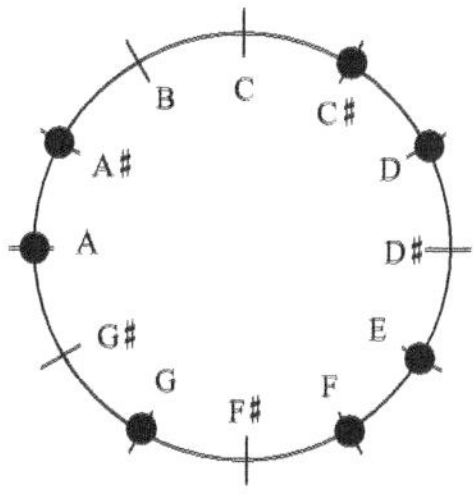

Clockwise distance between dots	
d distance	c distance
1	1, 2, 3
2	3, 4
3	4, 5, 6
4	6, 7, 8
5	8, 9
6	9, 10, 11

Figure 7.2. Calculating spectra (c distances) for generic intervals (d distances) in the D harmonic minor scale. Adapted from Timothy A. Johnson, *Instructor Resources: Foundations of Diatonic Theory*. ©2003 Key College Publishing, 39.

noted in the table as d distances and c distances. For example, the D harmonic minor scale, shown in figure 7.2, presents a variety of c distances corresponding to each d distance, or specific intervals corresponding to each generic interval. The c distances in the table are the spectrum for each generic interval. Once students are comfortable with this counting process, they will find the definitions of the three spectrum-oriented features easy to handle.

For the MP feature—first developed by Clough and Myerson (1985) but named after a colleague in mathematics, John Myhill, who sparked their investigation of this idea—the spectrum for each generic interval has exactly two members.[11] A less restricted feature, DE—developed initially by Clough, Engebretsen, and Kochavi—stipulates that the spectrum may have either one or two members. Finally, tightening the restriction placed on DE, the ME feature allows the spectrum for each generic interval to have either one or two members, but any two members of a spectrum must be consecutive numbers. Because some spectra of the D harmonic minor scale, displayed in figure 7.2, have three members, this scale possesses none of these features. The diatonic collection possesses all three features. A simple but fruitful problem for beginning students to explore would be to examine familiar collections for these features. Because the process of determining if these features apply to a collection is essentially the same for each feature, students can explore their chosen collections easily. Further, incorporating more creativity into the task, students may be asked to invent their own collections, which possess one or more of these features—in universes of any size, not just the familiar 12-note chromatic universe. Students may discover ME for themselves by means of a graduated series of exercises found in my earlier work, which includes some further discussion of MP geared toward beginning students.[12]

Deep

The definition of the "deep" (DP) feature relies upon the interval class vector, a staple of pitch-class set theory, but not usually included as part of the curriculum at the introductory level. Nevertheless, beginning students find the interval class vector easy to calculate, especially when collections are mapped onto circles, and they easily can apply the definition of DP: every interval class appears a unique number of times. The diatonic collection <254361> and the Guidonian hexachord <143250> are perhaps the most obvious examples of DP, but here again exploration of collections in universes that do not have 12 notes can spark the curiosity of students at any level. One example—provided by Clough, Engebretsen, and Kochavi—is a 5-note collection in an 11-note universe (with "half steps" just a bit larger than the familiar chromatic), generated by the specific interval 4, or G(11, 5, 4), with an interval class vector of <20341>.[13]

Diatonic

Although the term "diatonic" is often taken for granted as referring to transpositions of the pattern of whole and half steps that correspond to the white keys of a piano, and indeed the term has been used loosely in that sense in this chapter, the diatonic (DT) *feature* has a more general definition that includes the usual diatonic collection, as well as similarly shaped collections in universes that contain other than 12 notes. Eytan Agmon was the first to define diatonic in this more general sense, and later Clough and Douthett independently defined an essentially identical feature, which they termed "hyperdiatonic."[14] This chapter—following Clough, Engebretsen, and Kochavi—uses Agmon's term and Clough and Douthett's definition.

DT collections are always ME. In addition, DT collections only occur in universes sized in multiples of 4. Finally, with these two restrictions in place, DT collections always have one note more than half of the number of notes in the universe. In algebraic terms, DT is an "ME-set with $c = 2(d - 1)$ and $c \equiv 0$, mod 4."[15] The challenge for beginning students is to form diatonic collections in universes that do not contain 12 notes. In this way they can review the ME feature as they employ it to create DT collections in universes of varying sizes, they can observe commonalities between these DT collections, and they can practice using a simple algebraic formula (which is employed later in this chapter in a proof).

All of the features studied thus far may be reviewed along with DT, because any DT collection possesses each of the features described above, some of which imply others.[16] For example, because all DT collections are ME by definition, they are also DE. Clough, Engebretsen, and Kochavi prove each of these implicative relationships, but an interesting challenge for beginning students might be to try to find a DT collection that does not possess all of the other features (though none exist).[17] Or, they may be asked to demonstrate that all of the other features apply to any DT collection.

Balzano

A collection with the Balzano (BZ) feature—named after its originator, Gerald Balzano—has a generator (g) that is equivalent to the number of notes in the collection (d).[18] The generator is also equivalent to the sum of the bisectors of the collection, or the two intervals that divide the number of notes in the collection (d) approximately in half. And the universe (c) is equivalent to the product of the bisectors. Finally, the smaller of the bisectors must be greater than or equal to 3, and all Balzano collections are also ME. In algebraic terms, with the bisectors labeled as k and $k + 1$, BZ is an "ME-set with $c = k(k + 1)$ and $d = g = 2k + 1$, $k \geq 3$."[19]

This adaptation of the term "bisector," a term originally introduced by Jay Rahn, is not used in the definition of BZ either by Balzano or in the article under discussion; however, a similar use of the term is made in an article by Clough.[20] Clough, Engebretsen, and Kochavi identify what I am calling the bisector simply as "an interval k and the next larger interval $k + 1$, whose sum is equal to the size of the generator."[21] Because the sizes of the generator and the collection are equivalent under BZ, their interval (k) and the next larger interval $(k + 1)$ are effectively the bisectors of a BZ collection. The bisector is explored in more detail through pedagogical exercises designed for beginning students in my earlier work.[22]

As with DT, because of the straightforward algebraic formula used to define the BZ feature, students can easily create BZ collections simply by plugging in various values for k into the formula, beginning with 3, which produces the usual diatonic collection in a 12-note universe, and continuing through microtonal universes. Once they have worked with the formulas associated with both of these concepts (BZ and DT), students should be ready to work through an elegant proof (given below) that shows that there is only one BZ collection that is also DT, which is the familiar diatonic collection in a 12-note universe. Although many students, not just music students, freeze up when asked to deal with proofs, working through a simple but beautiful proof like this one can help them increase their mathematical skills and gain confidence in their abilities.

The proof begins with a review of definitions. For DT collections $c = 2(d - 1)$ and $c \equiv 0$, mod 4. Unfortunately, the Clough, Engebretsen, and Kochavi article contains a typographical error in its proof by including $d = g$ under the definition of DT in the proof but not in the original definition.[23] This equivalence would be true if BZ implied DT, but it does not, as demonstrated by the authors' example of generating the usual diatonic collection in a 12-note universe by means of the circle of perfect fourths, or G(12, 7, 5), where $d \neq g$. Therefore d does not necessarily equal g as implied by the equivalence of d and g in the formula.

For BZ collections, $c = k(k + 1)$ and $d = g = 2k + 1$, where $k \geq 3$. This time the article harmlessly alters the definition, using $k > 2$ rather than $k \geq 3$, presumably for mathematical correctness; however, for introductory-level students, maintaining the original definition is more "pc"—in this case, "pedagogically correct."

As students begin to work through the proof, the commonalities between the two definitions should be noted. DT solves for c in terms of the variable d, whereas BZ solves for both c and d in terms of the variable k. Therefore, students may assert that the formulas that are indicated as equivalent to c for each feature must be equivalent to each other (the proof in the article uses d as the common variable by rearranging the DT formula to solve for d, but the proof offered here relates more directly to the original definitions and thus is easier to grasp for students). Since $c = 2(d - 1)$ for the DT feature and $c = k(k + 1)$ for the BZ feature, then $2(d - 1) = k\ (k + 1)$ by substitution for c. Similarly, by substituting the other portion of the definition of BZ into this formula, where $d = 2k + 1$, to replace the variable d, students may assert that $2((2k + 1) - 1) = k\ (k + 1)$. Then, simplifying each portion of the formula, $2(2k + 0) = k^2 + k$, or $4k = k^2 + k$. Dividing each side of the formula by k yields $4 = k + 1$, and subtracting 1 from each side reveals that $3 = k$. Thus the only collection that is both DT and BZ has a value of $k = 3$, and using this integer with the BZ formula, as done by the students themselves in the exercises above, produces *only* the usual diatonic collection in a 12-note universe. Finally, because all of the other features are implied by the DT feature, as asserted earlier in this chapter, and as proved by Clough, Engebretsen, and Kochavi, the only collection that possesses all eight of these features is the familiar 7-note diatonic collection in a universe of 12 notes.[24]

The article includes extensive discussion of other aspects of the features, their implicative relationships, and the collections that have them. However, these issues might best be left for more advanced study. The methods outlined above provide an introduction to the features and to the creation of collections that possess these features, led by student inquiry, rather than delivered in a lecture format, and suitable for study at the introductory level. The remainder of this chapter will introduce ways to achieve the same goals with a different approach to the theoretical application of interval cycles in John Clough's chapter in this book, "Flip-Flop Circles and their Groups."

Flip-Flop Circles

Whereas interval cycles generate the pitch classes of feature sets, Clough employs interval cycles with triads to form "Flip-Flop Circles," or cycles of alternating major and minor triads connected by motion through pairs of interval cycles.

More generally, and in the spirit of the dualistic approach of Riemannian theory on which these ideas are based, Clough examines musical objects that appear in binary states (not just major and minor triads but any musical objects that can appear in two states, including pairs of individual pitch classes) built by pairs of interval cycles. In this way Clough's work in neo-Riemannian theory intersects with his customary diatonically oriented goals through his application of non-diatonic groups and procedures to diatonic situations. For students of music theory at the introductory level, both the non-diatonic and diatonic contexts may be explored fruitfully. Although diatonicism constitutes the principal material studied by beginning students, the simple major and minor triads involved in non-diatonic groups are a standard part of the music fundamentals curriculum, and using interval cycles lets students explore interesting aspects of these familiar musical objects through a different approach than is usually taken at this level.

Clough's Flip-Flop Circle is a generalization of the group of major and minor triads typified by the hexatonic system, as presented by Richard Cohn.[25] A hexatonic system is a series of alternating major and minor triads (for example, C, c, A♭, a♭, E, e) that is formed by a series of alternating transformations—here, parallel (P) followed by leading-tone exchange (L)—and whose pitch classes correspond to the hexatonic collection (in this example, the pitch classes C, E♭, E, G, A♭, and B). Other similar systems may be formed following related procedures, using different combinations of transformations and corresponding to different collections, as described by Cohn and other scholars, but Clough generalizes both the systems formed and the transformations that form them. Each of the transformations combines transposition and inversion, which may be construed productively in neo-Riemannian, dualistic terms.

Wechsel and *Schritte*

Although much of the neo-Riemannian literature has been dominated by the primary standard transformations, P, L, and R (relative), these transformations are just three members of a group of 24 that can transform any major or minor triad into any of the 12 major and minor triads, as described by Henry Klumpenhouwer.[26] In Riemannian theory a *Wechsel,* or contextual inversion, transforms a triad into the opposite mode based on some member of the triad or interval (or combination of intervals) in the triad. A *Schritt,* or contextual transposition, transposes a triad based on some member of the triad or interval (or combination of intervals) in the triad (keeping the mode the same). As shown in figure 7.3, *Wechsel* and *Schritte* act on major and minor triads in opposite ways, "flipping" minor triads up or transposing them down, and "flopping" major triads down or transposing them up. The combination of a *Wechsel* and a *Schritt* simplifies the transformation between triads of different modes significantly, as

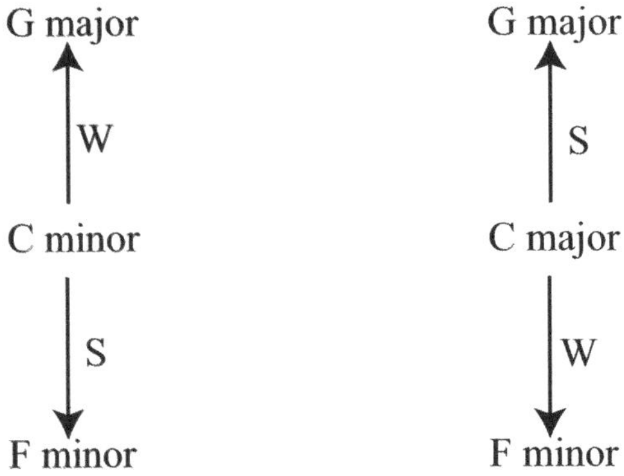

Figure 7.3. Beginning with a minor triad (for example, the C minor triad in the middle of the first diagram), a *Wechsel* flips it up to form a major triad, and a *Schritt* transposes it down to form a minor triad. Beginning with a major triad (for example, the C major triad in the middle of the second diagram), a *Wechsel* flops it down to form a minor triad, and a *Schritt* transposes it up to form a major triad.

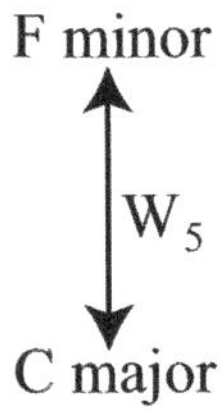

Figure 7.4. A combination *Wechsel* (W_5) transposes a major triad (the bottom triad in the example) up by 5 semitones and contextually inverts it into a minor triad. The same combination *Wechsel* (W_5) transposes a minor triad (the top triad in the example) down by 5 semitones and contextually inverts it into a major triad.

Clough pointed out, and thus provides an easy entry point into the concept for beginning students.

A simple way to conceive of the *Schritt/Wechsel* group is to define a *Schritt* (S) as a transposition by a given number of semitones up for major triads or down for minor triads, shown as S_n, and to define a *Wechsel* as a combination of transposition with contextual inversion, shown as W_n. The easiest *Wechsel* for beginning students to handle, and the one selected by Clough for this purpose, is the one that transforms a major triad into its parallel minor triad and vice versa (also known as P), a contextual inversion around the interval of a fifth of the chord. This *Wechsel* in combination with all of the *Schritte* can produce all of the major-minor triad relationships, though, as Clough showed, the combination is easier to navigate if the *Wechsel* follows the *Schritt.* Thus, the W_n transformation transposes first, then inverts, rather than the other way around, as in the traditional T_nI or I_n operation. Beginning students can learn to apply these contextual transformations quickly; the only problem will be to remember to transpose major triads up and minor triads down. The payoff of this approach based on harmonic dualism, however, is that the same transposition level applies to a relationship between two chords, regardless of which appears first, as shown in figure 7.4.

Creating *Wechsel* Patterns

Once students are familiar with the W_n transformation, they can create their own flip-flop circles simply by alternating between a pair of *Wechsel* (called a "pattern") to create a circle of alternating major and minor triads. Clough's chapter contains a number of examples that may be used for reference, but perhaps the best pedagogical procedure would be to let students discover flip-flop circles for themselves. Clough's question ("what is the space of all such circles?") is a good place to begin, though insisting that students write out every pattern is probably too much. As students navigate through flip-flop circles using the W_n transformation, they will gain valuable experience with intervals and triads, basic components of the typical fundamentals curriculum.

Students' work with flip-flop circles can be related to their earlier work with interval cycles, as outlined previously in this chapter, by asking them to observe the transpositional relationships between chords of the same quality in the circles they create. For this exercise, either traditional transposition (T_n) or contextual transposition (S_n) may be used effectively. In either case, what students will discover is that each chord quality in the circle is transposed by a consistent interval, creating two interlocking cycles of major and minor triads.

Diatonic Extensions

Another way to relate students' earlier work with interval cycles to flip-flop circles and the W_n transformation may be found in the last part of Clough's chapter, where he extends these concepts to a diatonic context. Rather than restricting himself to major and minor triads, Clough generalizes the idea to any two contrasting musical structures, for example the $^5_3$ and $^6_3$ inversions of diatonic triads, or two different rhythmic values of pitch classes. Both of these extensions may be grasped fairly easily by introductory students, again following a self-discovery approach.

First, the basic idea of W_n as a transformation that changes a musical entity into a contrasting music entity must be established. Because of this generalization, moving from contextual inversion (or W) to undefined flips or flops, a new transformation (J_n) replaces it, as described in detail in the latter part of Clough's chapter, though his theoretical explanation is not pedagogically necessary. Like W_n, J_n transposes a musical entity by n half steps and flips (or flops) it to another musical entity that is in some way different. Unlike W_n, however, for J_n it does not matter if the transposition occurs before or after the flip/flop, again as explained in detail by Clough. From the perspective of beginning students, a brief explanation is that the direction and order of contextual transposition and inversion for W_n was defined by the quality of the triad, whereas for J_n the quality is immaterial.

Second, a method of diatonic transposition, or transposition by scale steps rather than half steps, must be established. J_n is a transformation that applies to diatonic contexts, rather than chromatic ones. Students may need a few moments of adjustment in order to translate the concept of the interval of a second to n = 1 and the interval of a third to n = 2, but due to their previous work with d distances, there likely will be little concern.

With their new vocabulary, students may be asked to create patterns that form diatonic flip-flop circles using musical entities in any two states. Beginning with a diatonic sequence, as in Clough's chapter, may be appropriate or even necessary for some students, depending on their level of creativity and willingness to explore this new approach. They might also be challenged to find examples of flip-flop circles in the music that they are studying on their instrument—including method and etude books, where flip-flop circles are rather common, and their own musical literature, where sequences often abound.

Finally, the idea of interval cycles can be applied to these new flip-flop circles created or discovered by the students. Each flip-flop circle displays an interval cycle between every other musical entity. These interval cycles can be labeled using the standard transposition operation (T_n), and students may observe the limited variety of these cycles, mod 7, despite the comparatively large variety of pairs of the J_n transformation. In so doing students may discover what Clough called "one of the more underappreciated facts of music theory," that there are only three diatonic interval cycles and that each of them includes all seven notes of the diatonic collection.

Conclusion

This chapter suggests ways for beginning students to become acquainted with concepts of advanced music theory, while mastering the fundamentals. Clough's ideas provide myriad opportunities for creativity at a level of music theory that far too often is concerned only with memorizing information. Through these methods, students not only will have a chance to master fundamentals, but they will do so in a musically stimulating manner, preparing them for more advanced work and giving them a new appreciation of intervals, triads, and the diatonic collection.

Notes

1. Johnson 2003a and 2003b.
2. Clough and Myerson 1985, and Clough and Douthett 1991.
3. Clough, Engebretsen, and Kochavi 1999. The authors draw primarily on the following work: Agmon 1989, 1991, and 1996; Balzano 1980; Carey and Clampitt 1989; Clough and Douthett 1991; Clough and Myerson 1985 and 1986; and Gamer 1967.
4. Clough, Engebretsen, and Kochavi 1999, 74.
5. Johnson 2003a, 83–84, 94–97, 103–8.
6. Johnson 2003b, 44, 47–48.
7. Wooldridge 1992.
8. Clough, Engebretsen, and Kochavi 1999, 77.
9. Johnson 2003a, 83–84, 94–95, 103–8.
10. Carey and Clampitt 1989.
11. Clough and Myerson 1985.
12. Johnson 2003a, 105–8, presents MP. Johnson 2003a, 40–43, and Johnson 2003b, 25, present DP.
13. Clough, Engebretsen, and Kochavi 1999, 76.
14. Agmon 1989; Clough and Douthett 1991.
15. Clough, Engebretsen, and Kochavi 1999, 76.
16. Clough, Engebretsen, and Kochavi 1999, 77–78.
17. Clough, Engebretsen, and Kochavi 1999, 78–84.
18. Balzano 1980.
19. Clough, Engebretsen, and Kochavi 1999, 76.
20. Rahn 1977; Clough 1994, 235.
21. Clough, Engebretsen, and Kochavi 1999, 77.
22. Johnson 2003a, 97–105.
23. Clough, Engebretsen, and Kochavi 1999, 86, 76.
24. Clough, Engebretsen, and Kochavi 1999, 86.
25. Cohn 1996.
26. Klumpenhouwer 1994.

Chapter Eight

A Parsimony Metric for Diatonic Sequences

Jonathan Kochavi

Sequences use familiar harmonies but combine them with a grammar quite separate from functional harmony. Let us assume that we are working in diatonic pitch-class space, and suppose each sequential unit is to be made of up exactly two chords. Building on preliminary work from 1996, John Clough explored the application and adaptation of neo-Riemannian theory to such a context.[1] Using the nomenclature of Hook's uniform triadic transformations, Clough proposed that the binary opposition of major and minor triads that underlies so much neo-Riemannian theory could be extended to chord types found in such sequences.[2] Instead of dichotomizing harmonic triads based on transposition class, chord classes in sequences can be based on chordal inversion, outer voice intervals, or other configural distinctions. Using this new conception of dualism, Clough went on to define transformations analogous to the neo-Riemannian *Schritte* and *Wechsel* to describe progressions within the diatonic space traversed by particular harmonic sequences.

The present study, building upon earlier work,[3] investigates the sequence types explored by Clough from the perspective of parsimony. What types of diatonic sequences are possible under the constraint of parsimony? Is there some way to classify sequences based on their level of voice-leading parsimony? To answer these questions, of course, there must be a clear idea of what parsimony means in this context. To this end, we will identify two basic types of parsimony in such a sequence: *chord-to-chord parsimony*, as measured by the voice-leading connections between any two adjacent chords in the sequence, and *unit-to-unit parsimony*, determined by the relationship between corresponding chords in adjacent sequential units.[4] Limiting the scope of our investigation to sequences made up entirely of triadic harmonies, we will explore the meaning and implication of both types of parsimony. In order to undertake this exploration in such a restricted context, the sense of the term "parsimony" must first be made precise in the general setting.[5]

Part I. Toward a Definition of Parsimony

The general definition of parsimonious voice leading has been developed through a series of recent writings by Richard Cohn.[6] In his 1997 article, Cohn relates the notion of parsimony to the P, L, and R operators, which are generalized to apply to all trichords by isolating their double common-tone properties: "a trichord is parsimonious if its voice-leading is non-zero but minimal under L, P, and R."[7] For our purposes, however, we will work with a definition of parsimony that is independent of any particular operator or set-class cardinality. To that end, we can turn to Cohn's earlier work, where the term is defined in terms of *maximal common-tone retention* (excluding repetition) and *incremental voice leading*, whereby "the single moving voice travels, between adjacent harmonies, by the smallest possible interval."[8] While an explicit comparative measure is not given, inherent in this definition is a preference for maximizing the number of common tones over minimizing the size of the interval by which the non-retained voice(s) move. Presumably, a transition from a C-major triad to an A-minor triad exhibits a greater degree of parsimony in its voice leading than does that from C major to E major (figure 8.1.1), despite a total displacement of two semitones in each. Figure 8.1.2 presents a less clear-cut situation. The first transition, C major to D♭ major, is simply a T_1 transposition and retains no common tones while encompassing a total displacement of three semitones. On the other hand, the transition from C major to G minor does retain a common tone but requires voice-leading intervals totaling four semitones in the other two parts. Further confounding this situation, the relative displacement of these two progressions reverses if considered in diatonic space: the first remains as three steps, while the second reduces to two.

Figure 8.1.1. C major to A minor/E major.

Figure 8.1.2. C major to D♭ major/G minor.

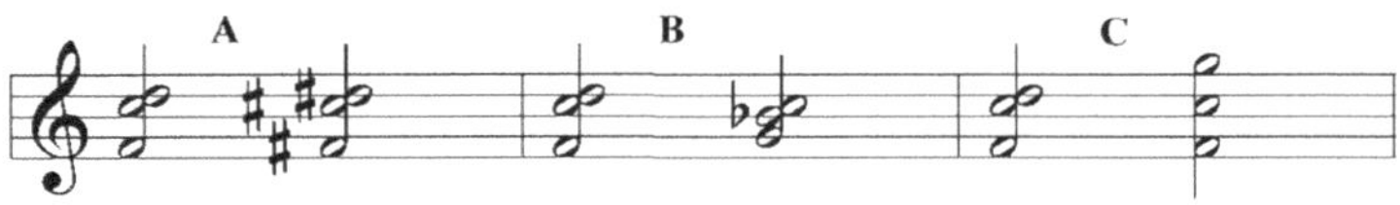

Figure 8.1.3. Three progressions from {0, 2, 5}.

In order to properly sort through the concept of "relative parsimony," we must recall that the applicability of parsimony is not limited to the harmonic triads. Figure 8.1.3 shows three progressions from the pc set {0, 2, 5}, each preserving set-class identity. To compare each progression with respect to parsimony, we need to propose a working definition of *total voice-leading displacement.* In figures 8.1.1 and 8.1.2, the measurement of this displacement was unambiguous, but in figure 8.1.3 a precise definition is required. We start by defining the concept of distance in octave-equivalent space. Note that modular distance as defined below is identical to our notion of interval class, although abstracted from the usual pitch-class (or beat class) context.

Definition 8.1.1. $\boldsymbol{x\ \widetilde{\mathbf{mod}}\ n}$ represents the specific integer between 0 and $n - 1$ that is equivalent to x mod n.

Definition 8.1.2. Suppose x and y are integer classes mod n. The **modular distance between x and y** is given by $\min\{(x-y)\ \widetilde{\bmod}\ n,\ (y-x)\ \widetilde{\bmod}\ n\}$. We represent this distance with the expression $|y - x|$. Furthermore, the **magnitude of an integer class x** is the modular distance between x and 0, represented with the expression $|x|$.

Definition 8.1.3. Suppose X and Y are pc sets of equal cardinality, n, with $X = \{x_1, x_2, \ldots, x_n\}$ and $Y = \{y_1, y_2, \ldots, y_n\}$ where $x_1 = y_1$, $x_2 = y_2, \ldots, x_k = y_k$, and $x_i \neq y_j$ for any $i,j > k$. Then the **total voice-leading displacement** (or, **VL-shift**) between X and Y is given by

$$\min\left\{\sum_{r=k+1}^{n}\left|y_r - x_{\sigma(r)}\right|\right\},$$

where σ is a permutation of $\{k+1, k+2, \ldots, n\}$.

Definition 8.1.3 essentially measures the sum of all the voice-leading intervals between two chords assuming that the voices are paired off in the most efficient way, i.e., the pairing is chosen to minimize the displacement. The only constraint on the choice of pairing is that if the two chords share a pc, then the voices that contain this pc must be paired.[9] This constraint makes good musical sense and is found in other writings on parsimony. In their Relation Definition, Douthett and Steinbach identify a range of possible parsimonious relations based on individual voices remaining fixed, moving by half step, and moving by whole step.[10] However, their goal is to explore webs of harmonic relations which result from applying different measures of parsimony, each of which is dichotomous—that is, each identifies a progression as being parsimonious or non-parsimonious—as opposed to generating relative measures of parsimony.

Applying Definition 8.1.3 to our {0, 2, 5} examples, we find that figure 8.1.3A shifts three semitones and retains no common tones, figure 8.1.3B shifts six semitones and retains one common tone, and figure 8.1.3C shifts five semitones and retains two common tones. It is clear from these examples that the maximization of the number of common tones does not necessarily lead to a minimization of the VL-shift. Therefore, any parsimony metric placed on voice-leading space must incorporate and weigh both components.

Considering the level of generality of the current study, it would be foolish to insist on a single ordering on VL space to provide a measure of relative parsimony in all possible contexts. Instead, two partial orderings are proposed here, one which favors common-tone retention, and one which favors minimal VL-shifts.

Definition 8.1.4a. (Favoring common tones). For all integers i, suppose X_i and Y_i are pc sets of equal cardinality sharing c_i common tones and exhibiting a VL-shift of v_i. Then the progression from X_j to Y_j is said to have a **greater degree of CT-parsimony** than X_k to Y_k iff:

(1) $c_j > c_k$, or
(2) $c_j = c_k$ and $v_j < v_i$.

If $c_j = c_k$ and $v_j = v_k$, then the two progressions are said to have **equal degrees of parsimony**.

Definition 8.1.4b. (Favoring minimal VL-shifts). For all integers i, suppose X_i and Y_i are pc sets of equal cardinality sharing c_i common tones and exhibiting a VL-shift of v_i. Then the progression from X_j to Y_j is said to have a **greater degree of VL-parsimony** than X_k to Y_k iff:

(1) $v_j < v_i$, or
(2) $v_j = v_i$, and $c_j > c_k$.

If $c_j = c_k$ and $v_j = v_k$, then the two progressions are said to have **equal degrees of parsimony**.

Definition 8.1.4c. If the progression from X_j to Y_j has a greater degree of CT- and VL-parsimony than X_k to Y_k then X_j to Y_j is **more parsimonious** than X_k to Y_k.

Turning back to our earlier examples, the first progression in figure 8.1.1 is more parsimonious than the second, whereas in figure 8.1.2, the first progression is more VL-parsimonious and the second is more CT-parsimonious. Ordering the figure 8.1.3 progressions by CT-parsimony from most to least yields CBA, but the VL-parsimony ordering is ACB. As the positions of progressions C and B are unchanged relative to one another in these two orderings, we can tell that C is more parsimonious that B.

The following theorem, preceded by an elementary lemma, shows us that the notion of parsimony is significantly simplified for triads and seventh chords in a diatonic setting.

Lemma 8.1.5. Suppose X is a pc set in c-space. If X is transformed by two inversely related transposition operators and each resultant set is paired with X to form a progression, the two progressions yield the same number of common tones and the same VL-shift.

Proof. Consider the progression from X to Y with t common tones and a VL-shift of v semitones. Note that these values are left unchanged after the application of a transposition operator T_n: the progression from $T_n(X)$ to $T_n(Y)$ still has t common tones and VL-shift v. Therefore, the number of common tones and the VL-shift of the progression from $T_{-n}(X)$ to X are the same as the corresponding values for the progression from $X = T_n(T_{-n}(X))$ to $T_n(X)$.

Theorem 8.1.6. In diatonic space, relative measures of CT-parsimony and VL-parsimony are equivalent on progressions involving elements of the diatonic set class $\{0, 2, 4\}_7$ (the generic triads) and on progressions involving elements of $\{0, 2, 4, 6\}_7$ (the generic seventh chords).

Proof. For all i, suppose X_i and Y_i are pc sets of equal cardinality sharing c_i common tones and exhibiting a VL-shift of v_i. For each of the two cases, we must show that if $c_1 > c_2$ then $v_1 \leq v_2$, (i.e., greater CT-parsimony ⇒ greater VL-parsimony), and if $v_1 < v_2$, then $c_1 \geq c_2$ (i.e., greater VL-parsimony ⇒ greater CT-parsimony).

(a) Let X_1 and Y_1, and X_2 and Y_2 be two pairs of diatonic pc sets that are members of the set class $\{0, 2, 4\}_7$. Since the members of the set class are inversionally symmetric, all of these pc sets are transpositionally related. Furthermore, because the transposition of a progression does not change its VL-shift or number of common tones, we can assume without loss of generality that $X_1 = X_2$ by transposing both X_2 and Y_2 at the appropriate level, if necessary. Find k_1 and k_2 such that $T_{k_1}(X_1) = Y_1$ and $T_{k_2}(X_2) = Y_2$. By Lemma 8.1.5, it suffices to show the desired relationships hold for $0 \leq k_1, k_2 \leq 3$. (Any k_i between 4 and 6 yields parsimony values identical to that of its inverse between 1 and 3). It is easy to determine the possibilities for k_i and the resultant values for v_i given the value of c_i. In fact, this computation reveals that each c_i value allows exactly one value for v_i:

(i) If $c_i = 3$, then $k_i = 0$ and therefore $v_i = 0$.
(ii) If $c_i = 2$, then $k_i = 2$ and therefore $v_i = 1$.
(iii) If $c_i = 1$, then $k_i = 3$ and therefore $v_i = 2$.
(iv) If $c_i = 0$, then $k_i = 1$ and therefore $v_i = 3$.

As this summary exhausts all possible values for k_i, the converse of each statement holds as well:

(i') If $v_i = 0$, then $k_i = 0$ and therefore $c_i = 3$.
(ii') If $v_i = 1$, then $k_i = 2$ and therefore $c_i = 2$.
(iii') If $v_i = 2$, then $k_i = 3$ and therefore $c_i = 1$.
(iv') If $v_i = 3$, then $k_i = 1$ and therefore $c_i = 0$.

The first set of statements shows that if $c_1 > c_2$, then $v_1 < v_2$, and the second set shows that if $v_1 < v_2$, then $c_1 > c_2$.

(b) Now let X_1 and Y_1, and X_2 and Y_2 be two pairs of diatonic pc sets that are members of the set class $\{0, 2, 4, 6\}_7$. Again, the members of the set class are inversionally symmetric, so all of these pc sets are transpositionally related, and again, we can assume without loss of generality that $X_1 = X_2$. Find k_1 and k_2 such that $T_{k_1}(X_1) = Y_1$ and $T_{k_2}(X_2) = Y_2$. As above, it suffices to show the desired relationships hold for $0 \leq k_1, k_2 \leq 3$. Computations analogous to the generic triad case yield:[11]

(i) If $c_i = 4$, then $k_i = 0$ and therefore $v_i = 0$.
(ii) If $c_i = 3$, then $k_i = 2$ and therefore $v_i = 1$.
(iii) If $c_i = 2$, then $k_i = 3$ and therefore $v_i = 2$.
(iv) If $c_i = 1$, then $k_i = 1$ and therefore $v_i = 3$.
(v) $c_i \neq 0$ (since the set cardinality exceeds ½ the space cardinality).

And similarly:

(i') If $v_i = 0$, then $k_i = 0$ and therefore $c_i = 4$.
(ii') If $v_i = 1$, then $k_i = 2$ and therefore $c_i = 3$.
(iii') If $v_i = 2$, then $k_i = 3$ and therefore $c_i = 2$.
(iv') If $v_i = 3$, then $k_i = 1$ and therefore $c_i = 1$.

Therefore, in this case as well, the first set of statements shows that if $c_1 > c_2$, then $v_1 < v_2$, and the second set shows that if $v_1 < v_2$, then $c_1 > c_2$.

Theorem 8.1.6 provides us with a definition of parsimony for the familiar triads and seventh chords. Therefore, we can say without ambiguity that the

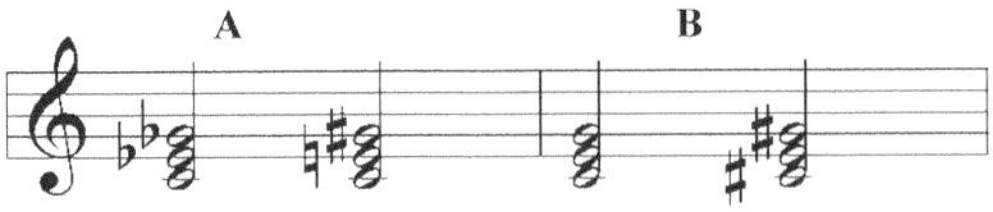

Figure 8.1.4. Equally parsimonious?

progression from a triad to its relative (an R-progression) is more parsimonious that the progression from a triad X to $T_1(X)$ in diatonic space. However, the theorem stipulates that these objects be viewed in diatonic space, and by doing so, tells only half the story. While there are contexts in which viewing R-progressions and L-progressions (each having a VL-shift of one in diatonic space) as equally parsimonious could be musically justified, and even in viewing P-progressions (VL-shift of zero) as more parsimonious than L-progressions, no one hears the progression from a diminished to an augmented triad given in figure 8.1.4A as equally parsimonious to that given in figure 8.1.4B.[12] This counterintuitive aspect of parsimony in the diatonic setting arises from the fact that multiple 12-space set classes are encompassed by a single set class in diatonic space.[13] Of course, the other problematic feature of measuring parsimony in diatonic space is the inability to register chromatic shifts that alter a chord's quality. The progressions in figure 8.1.4 are indistinguishable not only from each other, but also from a C major to C-major triad "progression" in diatonic space.

As noted above, any general, all-encompassing parsimony metric would vastly oversimplify musical space, and it would not be difficult to construct or find a musical language which would require a definition of parsimony different from either Definition 8.1.4A or 8.1.4B. However, restricting our context to tonal space,[14] we can fine-tune our definition. While it is clear now that simply viewing our tonal objects in diatonic space is not an adequate means for defining tonal parsimony, we can bypass the problematic features outlined above with a carefully constructed definition that combines elements of diatonic and chromatic space. Let's begin with a definition that builds on Definition 8.1.3 by considering the direction of individual voices' VL-shifts.

Definition 8.1.7. Suppose X and Y are pc sets of equal cardinality, m, with $X = \{x_1, x_2, \ldots, x_n\}$ and $Y = \{y_1, y_2, \ldots, y_n\}$ where $x_1 = y_1$, $x_2 = y_2, \ldots, x_k = y_k$, and $x_i \neq y_j$ for any $i,j > k$. Furthermore, take σ to be the permutation of the x-indices $k+1$ to n that yields the VL-shift as specified in Definition 8.1.3. Then the **directional VL-shift** is given by

$$\left|\sum_{r=k+1}^{n}\left(y_r - x_{\sigma(r)}\right)\right|.$$

(If there is more than one possible choice for σ, choose the one which minimizes the expression).

Note that the evaluation of the expression within the summation is regular integer subtraction, not modular. Definition 8.1.7 gives us a way of distinguishing between, for example, the voice leading of the progression C major to F minor (one common tone, VL-shift = 2, directional VL-shift = 2) and of C major to E major (one common tone, VL-shift = 2, directional VL-shift = 0). Arguably, the contrary motion of the latter progression (delivering the zero value) produces a smoother voice-leading progression than the former.

Definition 8.1.8. Suppose X_1, X_2, Y_1, and Y_2 are pc sets embedded in both chromatic and diatonic space, and which belong to the same chromatic set class. Call the progression from X_1 to Y_1 "Pr_1", and the progression from X_2 to Y_2 "Pr_2". Pr_1 is said to have a **greater degree of tonal parsimony** than Pr_2 iff:

(a) Pr_1 has more common tones than Pr_2 in chromatic space, or
(b) Pr_1 and Pr_2 have the same number of common tones in chromatic space and Pr_1 is more parsimonious than Pr_2 in diatonic space, or
(c) Comparisons in (a) and (b) yield equal parsimony, and Pr_1 has a smaller VL-shift than Pr_2 in chromatic space, or
(d) Comparisons in (a), (b), and (c) yield equal parsimony, and Pr_1 has a smaller directional VL-shift than Pr_2 in chromatic space.

Definition 8.1.8 hierarchizes the various measures of parsimony to produce a parsimony metric applicable to tonal objects. The definition applies to tonal objects for which measures of relative CT- and VL-parsimony yield equivalent results in diatonic space (as is true for triads and seventh chords, by Theorem 8.1.6). The definition can be easily adapted by dividing comparison (b) into comparisons involving relative diatonic CT- and VL-parsimony measures. The definition is by no means meant to preclude other possible parsimony orderings, even those restricted to tonal objects. However, the metric it provides is consistent with both musical intuition and with the concept of parsimony as used in recent writings. The definition incorporates the features of parsimony discussed above: maximization of common tones, minimization of voice-leading intervals, consideration of both diatonic and chromatic space, and directional voice-leading measurement. The ordering of the features in the definition takes into account Cohn's preference for common-tone retention and mitigates the counterintuitive results of a purely diatonic measure of parsimony.

By way of example, figure 8.1.5 summarizes the parsimony values for progressions from the C-major triad to harmonic triads on all 12 pcs, with alternative enharmonic representations for some. The 32 selected progressions are then ordered by degree of tonal parsimony in figure 8.1.6, where a similar shape around the measure number indicates that the two adjacent progressions have equal degrees of tonal parsimony.

	# of Common-tones		VL-shift (chrom.)	Directnl. VL-shift
	Chrom.	Diat.		
C+	3	3	0	0
C♯+	0	3	3	3
D♭+	0	0	3	3
D+	0	0	6	6
D♯+	1	0	3	3
E♭+	1	2	3	3
E+	1	2	2	0
F+	1	1	3	3
F♯/G♭+	0	1	6	6
G+	1	1	3	3
G♯+	1	1	2	0
A♭+	1	2	2	0
A+	1	2	3	3
A♯+	0	2	6	6
B♭+	0	0	6	6
B+	0	0	3	3

	# of Common-tones		VL-shift (chrom.)	Directnl. VL-shift
	Chrom.	Diat.		
C-	2	3	1	1
C♯-	1	3	2	2
D♭-	1	0	2	2
D-0	0	0	5	5
D♯-	0	0	4	4
E♭-	0	2	4	4
E-	2	2	1	1
F-	1	1	2	2
F♯/G♭-	0	1	5	5
G-	1	1	4	4
G♯-	0	1	3	3
A♭-	0	2	3	1
A-	2	2	2	2
A♯-	0	2	5	5
B♭-	0	0	5	5
B-	0	0	4	4

Figure 8.1.5. Parsimony values for progressions to/from the C major triad.

A few comments on figure 8.1.5 and figure 8.1.6 are in order. By Theorem 8.1.6, there is no need to include a separate column for diatonic VL-shift in our table, since the resultant ordering would be identical to the ordering given in the diatonic common tone column. In figure 8.1.6, notice that the C major to C major progression is the most parsimonious, followed in order by C major to C minor, C major to E minor, and C major to A minor, representing respectively the P, L, and R operators on the harmonic triads. The next progression, C major to C♯ minor, corresponds to Lewin's SLIDE operator. This progression shows an important feature of the Definition 8.1.8 parsimony metric: enharmonically equivalent chords do not necessarily yield identical parsimony measures. The C major to D♭ minor progression does not appear until measure 15 of figure 8.1.6. This disparity is due to the peculiar nature of diatonic pitch-class space: although it usually serves to group chromatic pcs together (e.g., $\{G, G♯, G♭\}_7 = \{4\}_7$), it can also split a single chromatic pitch class into two diatonic classes (e.g., $\{C♯, D♭\}_{12} = \{1\}_{12}$, but $\{C♯, D♭\}_7 = \{0,$

Figure 8.1.6. Progression from C major ordered by parsimony. Measure numbers with the same enclosure shapes label equally parsimonious progressions.

1}$_7$). While C major to C♯ minor exhibits three diatonic common tones, C to D♭ minor contains none.

Should this parsimony differentiation between enharmonically equivalent progressions disturb us? Putting the current study in a broader context, the representation of collections of pitches by the twelve equivalence classes has been a basic assumption in neo-Riemannian writing; indeed, it could be argued that it is the key element that distinguishes the field from the nineteenth-century theory from which it is in part derived. Hyer's translation of Riemann's *Tonnetz* into a modular one opens triadic progressions to the application of the mathematics of group theory, on which the present study relies heavily.[15] The measure of tonal parsimony of Definition 8.1.8 melds the old with the "neo." While retaining enharmonic distinctions for certain comparisons, it imposes equivalence on octave-related pitches and, most significantly, assumes equal temperament.

It is important to keep in mind that the tonal parsimony metric given above makes no claim for the *acoustic* independence of two enharmonically related pcs, only for their *functional* independence. In order to judge the relative

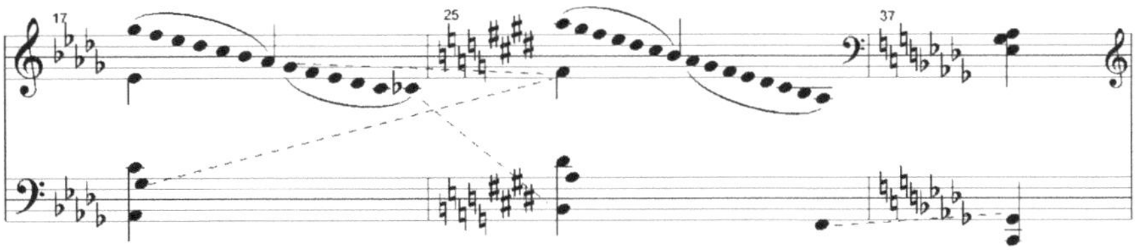

Figure 8.1.7. Schumann, *Fantasiestücke* Op. 12, No. 1.

parsimony of a given progression, it is therefore necessary first to label it correctly based on the *context* of that progression. Figure 8.1.7 provides a sketch of the middle section (mm. 17–38) of Schumann's *Fantasiestück* Op. 12, No. 1 ("Des Abends"), where the dotted lines represent enharmonic relationships. The first eight measures prolong an A♭-major chord, with a functionally ambiguous seventh, G♭, serving as a kind of pedal point. With the introduction of a C♭ in measure 24, these pitches are enharmonically reinterpreted as B and F♯, becoming the root and fifth of the B-major chord, which is prolonged (now with an A) until measure 37. Of course, this enharmonic shift is notational, not functional. Any uncertainty about the true nature of the shift disappears at the end of the section, where the A♭-triad returns with a strong G♭, leading into the repeat of the opening section of the piece in D♭ major. In retrospect, the middle section articulates a $V^{(7)}$–♭VII^7–V^7–I progression, and the B-major harmony is functionally C♭. Notice that this reinterpretation changes the number of diatonic common tones between ♭VII^7 and V^7 from 1 to 3 and restores the identity of this relatively common progression which is characterized by its tight voice leading.

Part II. Structural Representations of Diatonic Sequences

With a solid understanding of parsimony in the general setting, we can now turn our attention back to sequences. Returning the premises set up prior to our discussion of parsimony, we will begin by restricting our investigation to diatonic sequences and identifying two distinct types of progression in sequences: adjacent chords and adjacent sequential units.

Chord-to-chord parsimony needs little further explanation; it is simply a chord progression by which adjacent chords exhibit maximal CT- and VL-parsimony. Since we are working in the context of diatonic space, Theorem 8.1.6 shows us that these measurements yield identical relative parsimony orderings. Thus, while the *Leittonweschsel* and *Parallel* operators are more tonally parsimonious than the *Relative* in chromatic pitch space, under our current assumption of diatonic space, they are all equally parsimonious.

The definition of *unit-to-unit parsimony* requires more explanation. By definition, corresponding chords in adjacent sequential units must be transpositions of one another, and in this case, these transpositions are diatonic. In diatonic space, all triads are transpositions of one another. However, the particular type of transposition associated with corresponding chords in sequences also usually preserves the relative placement of pitches within the chord as seen in diatonic *pitch* space (allowing for possible octave displacement of an entire triad in the unit).[16] As a consequence, for example, a first inversion triad in one sequential unit will give rise to another first inversion triad in the corresponding place in the next sequential unit.[17] To distinguish this type of transposition from the more familiar modular transposition, the following definition is proposed.

Definition 8.2.1. The **ordered transposition operator** $\overline{T_n}$ acts as a standard transposition on ordered sets X (if $X = \{x_1, x_2, \ldots, x_k\}$, then $\overline{T_n}(X)$ is the ordered set $\{x_1 + n, x_2 + n, \ldots, x_k + n\}$ where each addition is computed modulo the cardinality of the space, if appropriate).

Under these circumstances, the notion of parsimony must be re-evaluated. Sequences provide some of the simplest examples for examining the contextuality of musical interpretation and perception. In examining the phenomenology of a sequence, we can temporarily put aside the complicated web of functional harmony and distill the progression to its repetitive voice-leading pattern. Through recognizing this pattern, we perceive a clear difference in the relationship between adjacent chords (whether those chords are part of a single sequential unit or span across two units) and the relationship between chords in corresponding positions in adjacent sequential units. Close temporal proximity and identical local context usually allows even the inexperienced listener to make a connection between the non-adjacent chords in this latter category.

How can we quantify this difference in the evaluation of the coherence of two chords in these two contexts? To formally represent the perceptual difference in distance measurement for elements of sequential units reflecting the temporal displacement of and the $\overline{T}$ relationship between corresponding elements of these units, the following definition is proposed.

Definition 8.2.2. Suppose X and Y are chords in corresponding positions in adjacent sequential units. The **unit-to-unit parsimony** between the units of X and Y is quantified by $|n|$ (see Definition 8.1.2), where n is the (diatonic) integer class such that $\overline{T_n}(X) = Y$.

Therefore, a sequence in which the sequential unit moves up (or down) by diatonic step exhibits greater unit-to-unit parsimony than does one that moves up (or down) by diatonic third.

Before applying our work on parsimony to specific diatonic sequences, let us investigate the structure of common types of such sequences using the tools of contextual transformation theory that I have developed elsewhere.[18] Formally, a key feature of the groups of contextual transformations acting on mod-12 pc sets is the partitioning of a set class—that of the harmonic triads in the most familiar cases—into two transposition classes, the orbits of the action of the transposition operators on the set class. In the diatonic setting, not only does the number of transposition operators change from twelve to seven, but the set class of the harmonic triads is limited to seven transpositionally related, inversionally symmetric diatonic pc sets, each representing a major and minor triad with a particular diatonic step as a root. However, the majority of harmonic sequences use repeating patterns of two chords. In such cases, a pair of chords whose tones belong to the same diatonic pc set but which appear in non-corresponding positions in their respective sequential units are contextually distinct entities. Thus, as a counterpart to Clough's departure from traditional neo-Riemannian treatment by allowing different set classes within a transformational representation of a diatonic sequence, we see that a single pc set can assume dual roles in such a representation.[19]

Figure 8.2.1 shows the piano part from an excerpt of the first movement of Mozart's Piano Concerto in A major, K. 488. Measures 259–60 contain an ascending diatonic sequence with a 5-6 intervallic voice-leading pattern in which the two-chord sequential unit is presented four times. Notice that the chord on the downbeat of measure 259 and the chord on the second beat of measure 260 are identified by the same harmonic label ("D-major triad") (and therefore by the same diatonic pc set label, "$\{1, 3, 5\}_7$"), but their respective roles in the sequence differ mostly as a consequence of their contrasting positions within the sequential units—the first D-major triad is in the first position in its unit, the second is in the second position. Thus, contextually they can be treated as separate chords and labeled as ordered pairs (D+, 1) and (D+, 2), where the first entry represents the chord type and the second represents the position within the sequential unit; the two chords are said to be members of two different *unit position classes.* This generalized dichotomization allows us to structurally model the progression of this sequence in a manner analogous to the diagrams representing repeated applications of contextual inversion operators in Kochavi 1998.

Such a diagram for Mozart's sequence is given in figure 8.2.2. The triads in the sequence are represented by the diatonic pitch-class integers of their roots subscripted by 0 or 1, distinguishing the triads in the first or second position within the sequential unit (the two unit position classes) respectively. The arrows between the triads of a given row represent the diatonic T_1 relationships among them, while those between the two rows represent the possible chordal progressions within the context of the sequence type, with bold arrows showing the actual succession used in the Mozart excerpt beginning on A major (5_1) and ending on A major (5_0). Note that these framing A-major triads are not part of

Figure 8.2.1. Mozart, Piano Concerto in A Major, K. 488, first movement.
(a) Measures 258–263.
(b) Intervallic voice leading.

the sequence (they lead into it and out of it), nor are they technically the same chord. Unlike with the familiar neo-Riemannian cycles (hexatonic, octatonic, and RL cycles), in this case returning to the pc set which began the progression does not imply that a full cycle (whether or not that cycle has proceeded through all of the triads in the space) has been traversed.

The operator defined by the arrows between the two transposition classes in a structure diagram associated with the usual mod-12 set classes is traditionally called a contextual inversion—"inversion" indicating the class of operators to which it belonged, and "contextual" to distinguish it from context-free labels of the traditional inversion operators. Although the members of the two unit position classes here are also related by inversion, it is more appropriate to model the row-connecting arrows with the more perceptually immediate transposition operators (as the triads are all inversionally symmetric in diatonic space). Therefore, we can model such a sequence with a contextual transposition operator.

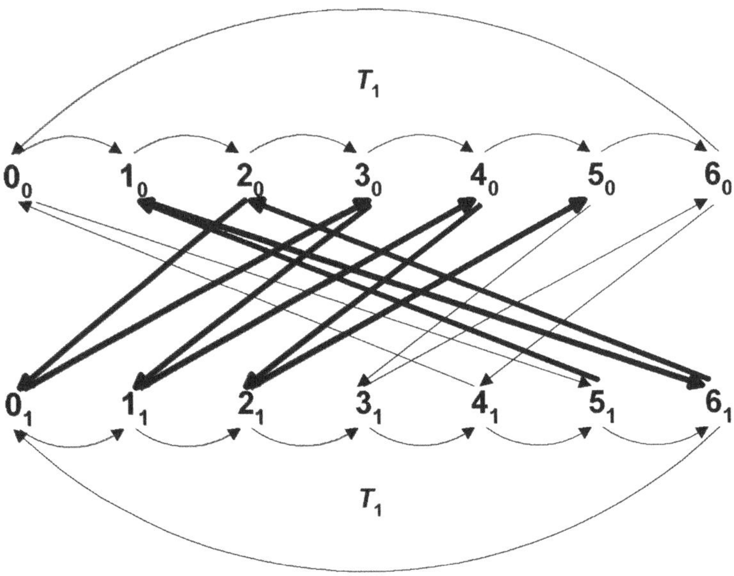

Figure 8.2.2. Structure diagram for diatonic sequence in Mozart, K. 488.

Definition 8.2.3. In a sequential progression of triads in diatonic space with two unit position classes, let the **sequence succession operator ("sso")** $F_{i,j}$ be the operator that maps a triad to the next triad in the sequence, where

$$F_{i,j} = \begin{cases} T_i \text{ on the first unit position class} \\ T_j \text{ on the second unit position class.} \end{cases}$$

Despite the fact that the $F_{i,j}$ operators are contextual transpositions, their mathematical behavior is similar in many ways to the contextual inversion operators ("cio's").[20] For example, applying an sso $F_{i,j}$ twice yields a (non-contextual) transposition operator that preserves unit position class. The main contrasts between the two arise from the differing cardinalities of the space on which the operators act. Since the cardinality of the unit position classes is seven, a prime number, the range of possible structures for the sso and the group that contains it is much narrower than for the cio and its group. An illustration of this is the restricted number of possible orders of $F_{i,j}$, as shown in the following lemma.

Lemma 8.2.4. If $i + j \equiv 0 \bmod 7$, then the sso $F_{i,j}$ has order 2. Otherwise, $F_{i,j}$ has order 14.

Proof. Note that $F_{i,j}{}^2 = T_{i+j}$. Therefore, if $i + j = 0$, $F_{i,j}$ has order 2, otherwise $F_{i,j}$ has order $2 \cdot 7 = 14$, since T_k has order 7 if $k \neq 0$.

In the Mozart example, the sso is $F_{5,3}$ which has order 14. Notice that in the current context, the sso's that are involutions have little musical interest, a marked contrast from their chromatic space counterparts; the most widely studied cio's (R, L, and P) are all involutions.

The strong constraint on $F_{i,j}$ shown in Lemma 8.2.4 vastly simplifies analysis of the possible group structures that it gives rise to when paired with the diatonic transposition operators; in fact, there is only one such structure.

Theorem 8.2.5. Let $G = \langle F_{i,j}, T_1 \rangle$, the group generated by the sso and the usual diatonic transpositions. Then $G \cong \mathbf{Z}_{14}$.

Proof. Note that

$$F_{i,j} \circ T_1 (X_0) = F_{i,j} (X_0 + 1) = X_1 + 1 + i,$$

$$T_1 \circ F_{i,j} (X_0) = T_1 (X_1 + i) = X_1 + 1 + i, \text{ and}$$

$$F_{i,j} \circ T_1 (X_1) = F_{i,j} (X_1 + 1) = X_0 + 1 + j,$$

$$T_1 \circ F_{i,j} (X_1) = T_1 (X_0 + j) = X_0 + 1 + j.$$

So G is abelian, and $g \in G$ must take the form $F_{i,j}{}^n \circ T_1{}^m$.

Suppose $i + j \not\equiv 0 \bmod 7$. Then by Lemma 8.2.4, $F_{i,j}$ has order 14, and there exists r such that $r(i + j) \equiv m \bmod 7$. Therefore, because $F_{i,j}{}^2 = T_{i+j}$, we have

$$T_1{}^m = T_m = T_{r(i+j)} = (T_{i+j})^r = F_{i,j}{}^{2r}.$$

So, $F_{i,j}{}^n \circ T_1{}^m = F_{i,j}{}^n \circ F_{i,j}{}^{2r} = F_{i,j}{}^{n+2r}$ for some r, and all $g \in G$ are powers of $F_{i,j}$. Therefore, $G \cong \mathbf{Z}_{14}$.

Now suppose $i + j \equiv 0 \bmod 7$. Then $F_{i,j}{}^2 = T_0$, so $n = 0$ or 1 and $0 \leq m \leq 6$. This produces 14 distinct elements. Because there is only one abelian group of order 14, $G \cong \mathbf{Z}_{14}$.

Part III. Parsimonious Diatonic Sequences

Now we are ready to apply our work on parsimony to diatonic sequences. A natural question to ask is, which diatonic sequence(s) maximize both unit-to-unit parsimony and chord-to-chord parsimony? Suppose each sequential unit is made up of two chords. To maximize unit-to-unit parsimony, each unit must be $\overline{T_1}$- or $\overline{T_6}$-related to the subsequent unit, and the choice of ordered transposition operator is consistent across the sequence. Take X_1 and X_2 to be corresponding triads in adjacent sequential units, and therefore separated by one diatonic step. For the time being, we will assume that the sequence is ascending, making X_2 one step

b	$Y \cap X_1$	$Y \cap X_2$	$\lvert Y \cap X_1 \rvert + \lvert Y \cap X_2 \rvert$
a	$\{a, a+2, a+4\}_7$	$\varnothing$	3
$a+1$	$\varnothing$	$\{a+1, a+3, a+5\}_7$	3
$a+2$	$\{a+2, a+4\}_7$	$\varnothing$	2
$a+3$	$\{a\}_7$	$\{a+3, a+5\}_7$	3
$a+4$	$\{a+4\}_7$	$\{a+1\}_7$	2
$a+5$	$\{a, a+2\}_7$	$\{a+5\}_7$	3
$a+6$	$\varnothing$	$\{a+3, a+5\}_7$	2

Figure 8.3.1. Summary of common tones for X_1–Y–X_2 progression.

higher than X_1. Furthermore, let Y be the second triad in the sequential unit of X_1, the intervening chord between X_1 and X_2.

What are the possible levels of chord-to-chord parsimony between Y and the two other triads? Let $X_1 = \{a, a+2, a+4\}_7$, implying that $X_2 = \{a+1, a+3, a+5\}_7$, and $Y = \{b, b+2, b+4\}_7$. Figure 8.3.1 gives the parsimony relations between Y and its neighbors based on the possible values of b. By Theorem 8.1.6, we know that to maximize parsimony, it is sufficient to maximize the number of common tones for each pair of chords. As shown in the rightmost column of figure 8.3.1, there are four progressions yielding a total of three common tones each, but two of these ($b = a$ and $b = a + 1$) simply produce chord repetitions, which, although potentially interesting, is not what we are looking for in the present context.[21] The other two possibilities for b that maximize parsimony are $a + 3$ and $a + 5$, and examples of how these two possibilities are realized in a progression from C major to D minor are presented in figure 8.3.2 where the upper path represents $b = a + 3$ and the lower path represents $b = a + 5$. The integers along the paths indicate the number of common tones between the two triads connected by the path. Notice the lower path (falling thirds, rising fourths–$F_{5,3}$) models the progression of the Mozart sequence from figure 8.2.1, diagramed in figure 8.2.2.

The other ascending diatonic sequence that maximizes parsimony, $b = a + 3$, reverses the order of the interval root progression, now rising fourths followed by falling thirds. An example of this sequence type can be found in the opening of the "Quoniam" section of the "Gloria" in Beethoven's *Mass in C*. Figure 8.3.3 gives a harmonic reduction of this passage, and figure 8.3.4 shows the structure diagram for the sequence type with the Beethoven progression in bold. Understanding the repetitive nature of a musical sequence, it is intuitively clear that the progressions derived from sso's of the form $F_{x,y}$ and $F_{y,x}$ should have the

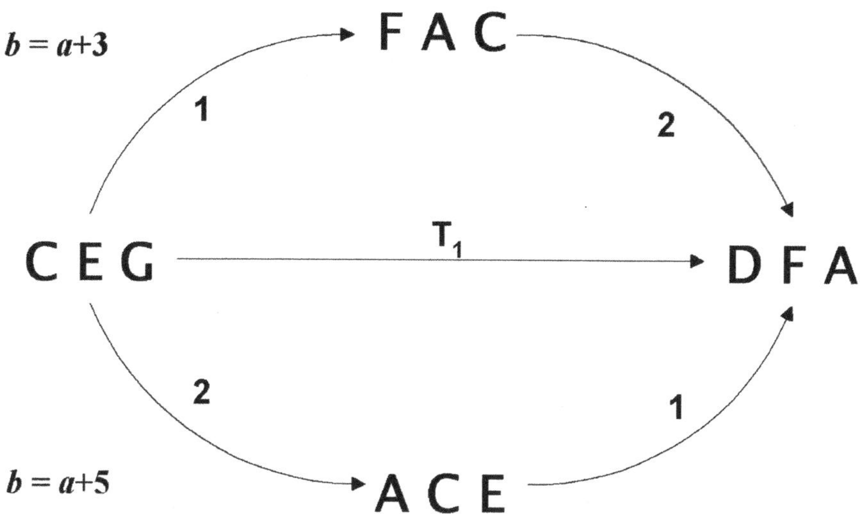

Figure 8.3.2. Two parsimonious paths resulting in an ascending diatonic step.

same degree of parsimony as well as the same overall shape (here, ascending by step). Obviously, these two sequences are closely related; in a sense, each contains the other; a redefinition of the unit position classes of one yields the other. However, they are musically quite distinct: $F_{5,3}$, with its characteristic 5-6 progressions, is more common than $F_{3,5}$, which is usually presented with root position triads, as in this example.[22]

Comparing figures 8.2.2 and 8.3.4, we can identify two mirror-image operators that convert one to the other. Inverting the transformational arrows about a horizontal line placed between the two rows of elements is akin to reversing the unit position class identities, as described. Therefore, such an inversion will always transform $F_{x,y}$ into $F_{y,x}$. Less immediate is the relationship via an inversion about a vertical line. Since all of the structure diagrams possess translational symmetry, the vertical line can be placed anywhere (as long as it is through an element or halfway between two elements). Note that the vertical line inversion can be represented algebraically by a modular inversion on the diatonic roots of the triads (which in our notation represent the entire diatonic triad), that is, X_0 becomes $(k - X)_0$ for some integer k mod the diatonic cardinality. When such an inversion is paired with a reversal of arrow direction, the sso is transformed as it was with the horizontal inversion. To show this, we begin by noting the relationship between inverse sso's.

Property 8.3.1. The sso's $F_{x,y}$ and $F_{-y,-x}$ are inverses of one another (where $-x$ and $-y$ are the additive inverses of x and y mod the diatonic cardinality).

Figure 8.3.3. $F_{3,5}$ sequence from Beethoven, *Mass in C*, "Gloria," mm. 214–21.

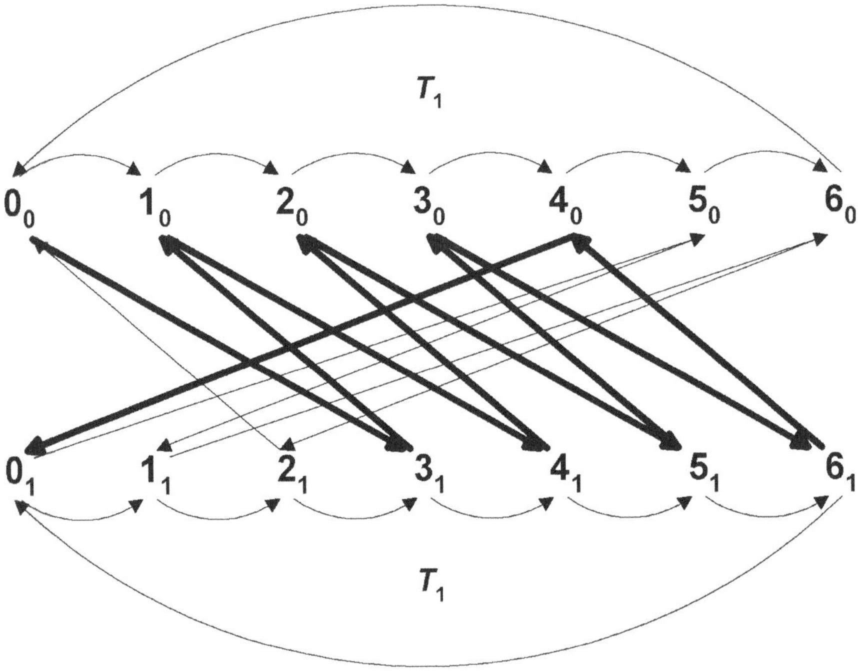

Figure 8.3.4. Structure diagram for Beethoven's $F_{3,5}$ sequence.

Proof. Remembering that every sso alters unit position class, we have

$$F_{-y,-x} \circ F_{x,y}\,(X_0) = F_{-y,-x}\,(X_1 + x) = X_0 + x - x = X_0, \text{ and}$$

$$F_{-y,-x} \circ F_{x,y}\,(X_1) = F_{-y,-x}\,(X_0 + y) = X_1 + y - y = X_1$$

Property 8.3.2. Suppose $F_{x,y}$ and $F_{x',y'}$ are sso's such that their structure diagrams are related by a mirror-image inversion about a vertical line coupled with a reversal of arrow direction. Then $F_{x',y'} = F_{y,x}$.

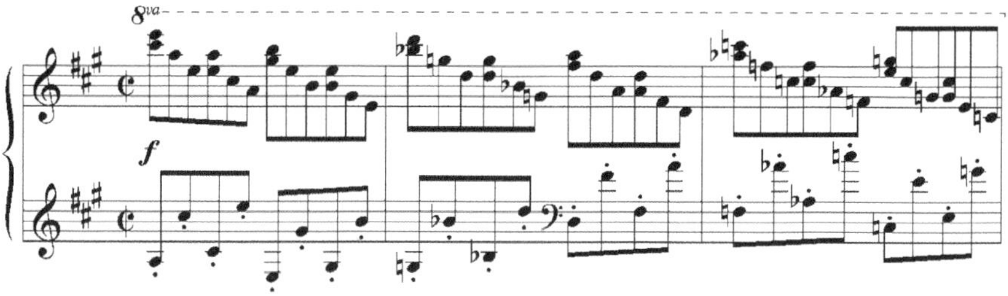

Figure 8.3.5. $F_{4,2}$ sequence from Liszt Piano Concerto No. 2 (piano part, mm. 25–27 of the *Allegro animato*).

Proof. First consider the structure diagram of $F_{v,w}$, the inversion of $F_{x,y}$ about a vertical line without a reversal of arrow direction. Then there exists an integer k modulo the diatonic cardinality where for a given chord in the first unit position class, X_0,

$$\begin{aligned} F_{v,w}(X_0) &= k - F_{x,y}(k - X)_0 \\ &= [k - (x + k - X)]_1 \\ &= (-x + X)_1, \end{aligned}$$

and similarly, $F_{v,w}(X_1) = (-y + X)_0$. Therefore, $v = -x$ and $w = -y$. Using Property 8.3.1, we have $F_{x',y'} = F^{-1}_{v,w} = F^{-1}_{-x,-y} = F_{y,x}$.

Our analysis began with the assumption that the two-chord-per-unit diatonic sequence ascended by step, but what are the maximally parsimonious sequences that descend by step? We have already essentially completed the work required to answer this question in our investigation of the ascending case. Reversing the directions of the arrows on the structure diagrams in figures 8.2.2 and 8.3.4 yields the two sequences we are looking for. The resulting sso's, $F_{4,2}$ and $F_{2,4}$, are easily derived from Property 8.3.1. Chromatically altered versions of $F_{4,2}$ with descending I/i–V progressions are not uncommon in the literature. The passage from Liszt's Piano Concerto No. 2 given in figure 8.3.5 shows a characteristic use of this progression.

The other common types of diatonic sequences that yield stepwise unit progression are those using ascending and descending fifths, and other than those sequences already discussed, these are the only sequences that maintain at least one common tone between every adjacent chord. The sso's corresponding to these sequences are $F_{4,4}$ for the ascending case and $F_{3,3}$ for the descending, inverses of one another. The structure diagram for $F_{3,3}$, the most common diatonic sequence in the literature, the descending fifths sequence, is given in figure 8.3.6. The horizontal and vertical symmetry of the structure diagram is due to Property 8.3.2: in this case, $F_{x,y} = F_{y,x}$.

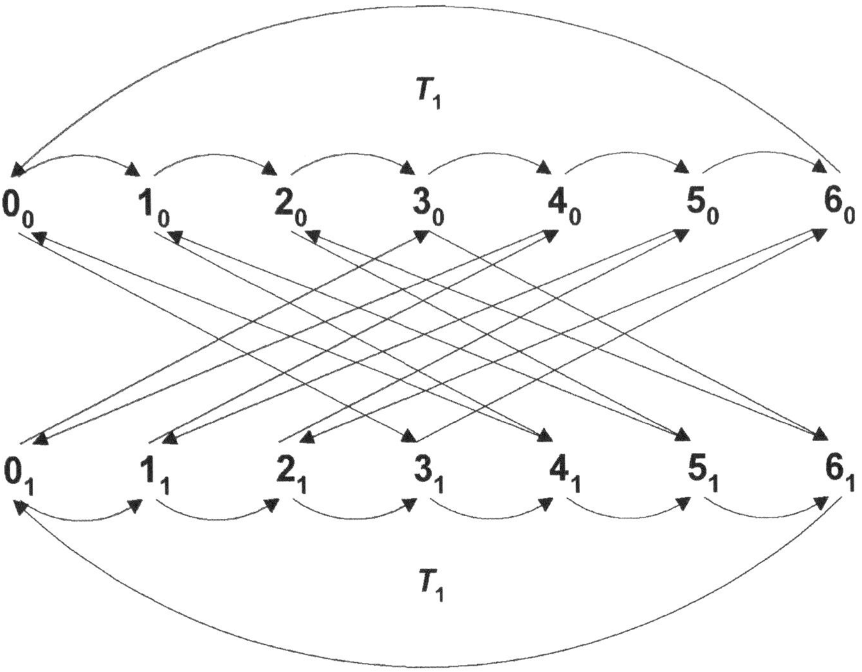

Figure 8.3.6. Structure diagram for $F_{3,3}$.

Another consequence of the fact that $F_{x,y} = F_{y,x}$ is that the repeated application of the sso beginning on the chord X yields exactly half of the elements of the two unit position classes before arriving at X in the opposite unit position class. In algebraic terms, for all integer chord roots X and non-zero integers x, $F_{x,x}{}^7(X_0) = X_1$ and $F_{x,x}{}^7(X_1) = X_0$. Mathematically, this property follows from the non-even primeness of our diatonic cardinality (7). Note that $\mathbf{Z}_{14} \cong \mathbf{Z}_7 \text{ X } \mathbf{Z}_2$, which can be generated by $(x, 1)$ for $x = 1, \ldots, 6$. Furthermore, any choice of x generates the mod 7 integer classes in the first position of the ordered pair, and therefore $(x, 1)^7 = (x, 0)$. Although this explanation is relatively simple to the mathematician, the musician untrained in mathematics understands the property even more immediately. Figure 8.3.7 presents the voice leading for a standard 6/3 – 5/3 $F_{3,3}$ sequence in C major. The sequence begins and ends on C-major chords, traversing all other diatonic triads in C major without repetition. The unit position classes of the two C-major chords are different: not only does the first C-major triad fall on a strong beat while the second falls on a weak beat, but the first is in first inversion and the second is in root position. Thus, in this, the most common of harmonic sequences, the listener almost must "momentarily lose sense of tonality" in order to allow for an adjustment in

Figure 8.3.7. $F_{3,3}$ in C major.

assessing which chord of each pair leads forward and which chord is itself led. In other words, a shift must be made as to the identity of the primary versus secondary unit position class. As is often true of key modulations, the moment of this identity shift usually cannot be pinned down; all that is certain is that when the sequence begins, the order of the unit position classes is conclusively established, but by the end that order has been reversed.

The discussion of parsimony and diatonic sequences in this chapter is meant to provide the structural backdrop for analysis. There is still much to be explored in relation to this topic—a generalization of sequence types, for example. However the work here supplies the tools and the structural perspective needed for further investigation and analysis.

Notes

1. Clough 2000. This paper sees its final form in Clough's contribution to the present volume.
2. Hook 2002a.
3. See Kochavi 2002, chapter 4.
4. A similar approach is taken in work by Adam Ricci 2002. Ricci makes a distinction between voice leading among chords "within the pattern" and "at the seam."
5. The mathematics that will be used in the present investigation does not stray beyond basic abstract algebra and elementary number theory. For the reader who needs a quick brush-up on sigma notation and the concept of permutations, please consult Goodaire and Parmenter 2005. For a review of the concepts of group theory that play a role in the second and third parts of this chapter, see Fraleigh 2002.
6. Cohn 1994, 1996, 1997.
7. Cohn 1997, 58.
8. Cohn 1994, 5–6. In Cohn 1996, these two properties fall under the single definition of a *maximally smooth* transition between adjacent chords.
9. Although this choice of pairing has obvious musical implications (which we will deal with in the paragraphs that follow), the mathematical implications are negligible. While it will not be proven here, removing the "common tone = common voice" constraint does not affect the total displacement value though additional pairings with the same value would emerge (see figure 8.1.3B).
10. Douthett and Steinbach 1998, 243.

11. The one-to-one correspondence between the four values of c_i and the four transposition indices both here and in part (a) is a direct consequence of the fact that the generic triads and seventh chords are generated by 2, which is relatively prime to 7, the cardinality of the space, and themselves have cardinalities equal to $\lfloor 7/2 \rfloor$ and $\lfloor 7/2 \rfloor + 1$ respectively. In general, such sets are deep. For a proof of this fact, see Lemma 1.2 in Clough, Engebretsen, and Kochavi 1999, 79–80.

12. This progression is produced by Lewin's SLIDE operator, which he defines as the operator that "preserves the third of a triad while changing its mode" (Lewin 1987, 178).

13. Technically, diatonic and chromatic set classes are incommensurable; that is, the diatonic set classes do not evenly partition the chromatic ones. For example, the pc sets {B, D, F} and {C♭, D, F} are in the same chromatic set class (indeed, they are identical chromatic pc sets), but belong to diatonic set classes $\{0, 2, 4\}_7$ and $\{0, 1, 4\}_7$, respectively.

14. While not wishing to underplay the complexity of the definition of "tonal space," I mean to provoke no controversy with its use here. If a definition is required, let a piece (or passage) be considered tonal if it is one that Schenker has analyzed (in good faith) or could have analyzed.

15. Hyer 1995. See figure 3, p. 119.

16. Certainly there are many exceptions to this general rule, and few would argue that minor alterations in note placement negate the effect of the sequence. See Fétis 1844, 26–30.

17. Ricci 2002 captures the distinction between chordal relationships within a unit and between units nicely in his description of a diatonic sequence as "a set of chords related by a repeating series of *pitch-class* transpositions, upon which a consistent *pitch* transposition acts. . . " (p. 4, italics added).

18. See Kochavi 1998 and 2002, chapter 2.

19. Clough 2000 and chapter 2 of the present volume.

20. There is a strong connection between sso's and Hook's 2002 uniform triadic transformations (UTT's). Clough uses Hook's formal system of UTT's to generate what are essentially sso's.

21. The notion of a "repeated" chord in a sequence is valuable in modeling the harmonic progression of it, whether the chord is literally repeated or not. Contextual re-interpretation of the harmonic function of a chord, a kind of localized pivot, is not uncommon in chromatic sequences.

22. This section of Beethoven's *Mass in C* actually integrates both types of maximally parsimonious sequences. A detailed analysis of the closing of the "Gloria" movement appears in Kochavi 2002, chapter 5.

Chapter Nine

Transformational Considerations in Schoenberg's Opus 23, Number 3

David Lewin

9.1 Introduction

Schoenberg's Opus 23, no. 3 has a clear subject, with a clear prime transpositional level: $B\flat_4$–D_4–E_4–B_3–$C\sharp_4$. The subject, with its thematic registral contour, opens the piece in the right hand. It recurs, transposed up 7 semitones, to open the next section of the piece (m. 6, right hand), and, an octave below its prime transpositional level, in the left hand of measure 6. It recurs, at its prime transpositional level, as a cantus firmus that opens the next section of the piece (mm. 9–10, *ruhig*). It recurs an octave below its opening level, to open the next section of the piece (mm. 12–13, *mf*). It recurs at its prime transpositional level, with B♭ and B♮ exchanged in register, to open the next section of the piece (end of m. 16, *ruhig*). It recurs transposed down 5 semitones, to open the next section of the piece (mm. 18–19, left hand). And finally, after some reprised material, it recurs to open the final sections of the piece (mm. 26–27, right hand), where it is mirrored against itself in an inverted form. The mirror texture continues to the end of the piece.[1]

The subject projects the pitch-class series B♭–D–E–B–C♯. We shall denote this series of pitch classes (not pitches-in-register) as S_0 (*S* for subject or series; 0 for its prime (pc-)transpositional level, as the pc series B♭–D–E–B–C♯). S_0 projects the unordered pc set {B♭,B,C♯,D,E}; we shall denote that unordered pentachord by P_0 (*P* for pentachord; 0 for its prime (pc-)transpositional level). The piece is saturated with transposed and inverted forms of P_0; some appear with the thematic *S*-ordering, while others do not.[2]

Traditionally, an analyst seeks a notation that will indicate each of the "transposed and inverted forms" of S_0, or P_0. Traditionally, one indicates by the twelve symbols S_n (respectively P_n) "S_0 (resp. P_0), transposed by the pc-interval *n*."

Traditionally, one labels as "s_0" and "p_0" some inverted form of S and the correspondingly inverted form of P. The various other inverted forms are then labeled "s_n" (resp. "p_n"), meaning "s_0 (resp. p_0), transposed by pc-interval n."

But which inverted form of S (resp. P) is to be labeled as s_0 (resp. p_0)? Traditionally, one might reply, "that inverted form most characteristically paired (and compared) in the music with S_0 (resp. P_0)." A problem arises for Op. 23, no. 3, though: it is not clear which inverted form of the subject/series/pentachord (if any) is "most characteristically paired" with the prime form.[3] The choice of notation here involves us immediately in considering what transformations (of the subject, the series S, and the pentachord P) are idiomatic for the piece. This article will explore the matter.

9.2 The Transformation I_7 (PC-Inversion-about-A-and-B♭)

I label the transformation "I_7" using the fixed-DO numerical system of Forte and others: if the pc C is labeled with the number 0, and the pcs C♯, D, ..., B♭, B are labeled with the numbers 1, 2, ..., 10, 11 mod 12, then I_7 transforms the pc labeled n into the pc labeled $(7 - n)$ mod12. Since $9 + 10 = 7$mod12, the transformation I_7 exchanges the pcs labeled 9 and 10, that is, the pcs A and B♭. I_7, which transforms pcs among themselves, then induces inversional transformations on forms of the subject, forms of the series S, and forms of the unordered pentachord P. Specifically, I_7 transforms the series S_0 = B♭–D–E–B–C♯ into the series A–F–E♭–A♭–G♭. For the moment, we shall refer to this series as "$I_7(S_0)$," resisting the temptation to label it as "s_0."

The temptation is strong because S_0 and P_0 are indeed paired and compared with $I_7(S_0)$ and $I_7(P_0)$ over the final sections of the piece, beginning a bit into measure 26 and continuing to the end. Furthermore, other clear forms of S and/or P are very audibly paired and compared with their I_7-inversions during these sections. For instance, in measures 28–29, S_7 as cantus firmus for the left hand (F–A–B–F♯–G♯) is audibly paired and compared with $I_7(S_7)$, the cantus firmus for the right hand (D–B♭–A♭–D♭–C♭). The extension of this phrase during measures 29–30 (*poco rit.*) projects an I_7 pc mirror between the hands. Then, starting at "tempo" in measure 30, the I_7-invariant dyad {C,G} sounds as a pedal while P_0 (upper register) is answered by $I_7(P_0)$ (lower register); the I_7-invariant {C,G} continues to sound as a pedal through measure 31, where $I_7(P_7)$ (upper register) is answered by P_7 (lower register); and so forth to the end of the piece.

9.3 Reservations about $I_7(S_0)$ as "s_0"

Though the transformation I_7 is acting powerfully during the last 10 measures of the piece (mm. 26–35), the force of the I_7 mirror does not become manifest until measure 26. Nor is the I_7-inverted form of S_0 (resp. P_0), i.e., the series A–F–E♭–G♯–F♯ (resp. the pentachord {E♭,F,F♯.G♯.A}), particularly prominent before measure 26.

True, $I_7(P_0)$ is, with some attention, audible in the melody that overlaps the phrase boundary at the end of measure 5 and the beginning of measure 6: the right hand there plays E♭–G♭–A–G♯–(A♭)–F, which is an ordering of $I_7(P_0)$. But the ordering is not the *S*-ordering, and the force of I_7 as pc transformation is not strong here.

Something closer to $I_7(S_0)$ might be asserted in the right hand melody of the phrase that begins at measure 12 (*tempo*), if one allows the idea of "diminution" or "ornamentation" into the listening context. We can then hear the melody A–(E)–F–E♭–A♭–(G)–F♯ as an "ornamented" version of $I_7(S_0)$. The notion of ornamentation is problematic in this context, however: what criteria determine an "unprepared chromatic neighbor" (E) and a "chromatic passing tone" (G) at just this moment, in a piece which is saturated by total chromaticism, along with many unornamented forms of *S* and *P*?

The phrase that begins at the end of measure 16 *(ruhig)* presents a succession of beamed *P*-forms, alternating between the hands, with close-to-*S* orderings. Of these forms, P_0 is the first and $I_7(P_0)$ the fourth. But is $I_7(P_0)$ "the inverted *P*-form most characteristically paired and compared with P_0" here? I think not; surely the inverted *P*-form most characteristically paired and compared with P_0 here is the third beamed *P*-form of the phrase, the form {D,B♭,G♯,C♯,B} in the right hand that directly precedes the $I_7(P_0)$-form under discussion. Calling that ordering of the third form "$K(S_0)$" for the time being, we can comfortably hear the first four forms of the phrase as P_0 [rh], $T_7(P_0)$ [lh], $K(P_0)$ [rh], and $T_7K(P_0)$ [lh]; of the four forms, the two in the left hand imitate the two from the right hand "at the fifth." A salient feature here, of what we shall soon call "K-inversion," is that $K(S_0)$, the series D–B♭–G♯–C♯–B, begins with the same dyad, {B♭,D}, as does S_0, the series B♭–D–E–B–C♯. $K(S_0)$ also ends with the same dyad, {B,C♯}, as does S_0. $K(P_0)$ thus has four common tones with P_0; the two pentachords are "maximally close" in total pc content (two distinct pentachords cannot have more than four common tones).

In similar fashion, $T_7K(P_0)$, the series A–F–E♭–A♭–G♭ audible within the fourth *P*-form here, begins with the same dyad, {F, A}, as does $T_7(S_0)$, the series F–A–B–F♯–G♯. $T_7K(S_0)$ also ends with the same dyad, {F♯,G♯}, as does $T_7(S_0)$. $T_7K(P_0)$ thus has four common tones with $T_7(P_0)$; the two pentachords are "maximally close" in total pc content.

In sum, while the fourth *P*-form (resp. *S*-form) here can abstractly be attained by I_7-inverting the first *P*-form (resp. *S*-form), the music suggests a different transformational pathway: the fourth *P*-form here is attained by K-inverting the second form, or by T_7-transposing the third form.[4] Figure 9.1 schematizes the set-up.

Depending on how we read the parallelogram, we can say that the upper horizontal K-relation is being vertically T_7-transformed to the lower horizontal

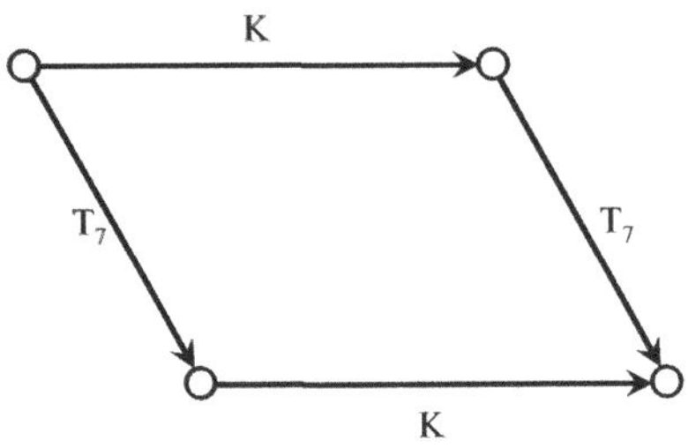

Figure 9.1. K and T_7 Network.

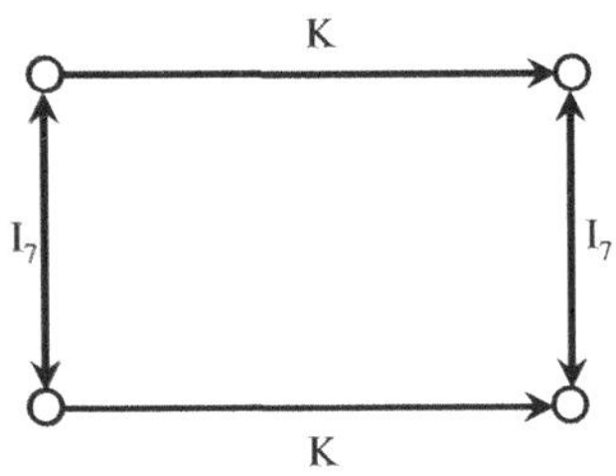

Figure 9.2. K and I_7 Graph of the *S*-Forms for mm. 26–29 in Schoenberg's Op. 23, no. 3.

K-relation, or we can equivalently say that the left-side vertical T_7-relation is being horizontally K-transformed to the right-side vertical T_7-relation. "Equivalently" is the key word here; it needs to be shored up with formal mathematics, which we shall defer for the present.

A similar sort of graph, figure 9.2, can be made for the cantus firmus forms of *S* in measures 26–29. The right hand cantus of 26–27 is in I_7-relation with the left hand cantus there; the right hand cantus of 28–29 is also in I_7-relation with the left hand cantus there; the right hand cantus of 28–29 is the K-inversion of the earlier right hand cantus (from mm. 26–27); the left hand cantus of 28–29 is the K-inversion of the earlier left hand cantus (from mm. 26–27).[5]

The graph of figure 9.2 also pertains to the four *P*-forms of measures 30–31 as they play out against the I_7-invariant pedal {C,G}. The right hand and left hand *P*-forms of measure 30 are I_7-related; so are the right hand and left hand *P*-forms of measure 31. The right hand *P*-form of 31 is the K-inversion of the right hand *P*-form from measure 30; the left hand *P*-form of measure 31 is the K-inversion of the left hand *P*-form from measure 30.

Depending on how we read the rectangle of figure 9.2, we can say that the upper horizontal K-relation is being vertically I_7-transformed to the lower horizontal K-relation, or we can equivalently say that the left-side vertical I_7-relation is being horizontally K-transformed to the right-side vertical I_7-relation. As with figure 9.1, "equivalently" will presently be shored up mathematically.

9.4 K-Inversion; Formalities

Our discussion of K-inversion so far can tempt us to consider $K(S_0)$ (resp. $K(P_0)$) as "that inverted form of S_0 (resp. P_0) most characteristically paired and compared with S_0 (resp. P_0)," at the expense of our earlier candidate $I_7(S_0)$ (resp. $I_7(P_0)$) for that position. Before continuing such thoughts, we should be more formal about just what "K-inversion" actually is, as a transformation of *S*-forms (or *P*-forms).

K-inversion, unlike I_7-inversion, is not induced by some transformation that permutes pitch classes. Rather, K-inversion is defined as operating only on forms of *S* (resp. forms of *P*), and not on individual pitch classes. K-inversion is what I have called a "contextual inversion."[6]

If *X* is some (any) transposed or inverted form of *S*, one can define K(*X*) as that inverted form of *X* which has the same opening {0,4} dyad as *X*, with the notes reversed. Thus K(E–C–B♭–E♭–D♭) is C–E–F♯–C♯–D♯. One observes that the last {0,2} dyad of *X* will be the last dyad of K(*X*), with the notes reversed. If *X* is now some (any) transposed or inverted form of *P*, one can define K(*X*) as that inverted form of *X* induced by the *S* orderings. Then K(*X*) will be that form of *X* which includes the same {0,4} dyad; K(*X*) will also be that form of *X* which includes the same {0,1,3,4} tetrachord. For example, K({B♭,**<u>C</u>**,**D♭**,**E♭**,**<u>E</u>**}) = {**<u>C</u>**,**D♭**, **E♭**,**<u>E</u>**,F♯}; the set of bold notes is the common {0,1,3,4} tetrachord, and the set of underlined notes is the common {0,4} dyad.

These definitions let us talk formally about K-inversion. To rigorously prove facts about the behavior of K-inversion, it is helpful to use numerical definitions. Readers who are averse to arithmetic can skim "Numerical Definition 1" following, and read only the statements of Observation 2.1 and Observation 2.2 beyond that, omitting the proofs of those observations, and returning to the main text following the proof of Observation 2.2.

NUMERICAL DEFINITION 1:

(a) If $\{n,\ n+1,\ n+3,\ n+4,\ n+6\}$ numerically represents, mod 12, a transposed form *X* of P_0, then K(*X*) is numerically represented by $\{n-2,\ n,\ n+1,\ n+3,\ n+4\}$. [Remark: the latter set is an inverted *P*-form which includes the same {0,1,3,4} tetrachord as the stipulated *X*.]

(b) If the set $\{m,\ m+2,\ m+3,\ m+5,\ m+6\}$ numerically represents a mod-12 inverted form *Y* of P_0, then K(*Y*) is numerically represented by the set $\{m+2,\ m+3,\ m+5,\ m+6,\ m+8\}$. [Remark: the latter set is a transposed *P*-form which includes the same {0,1,3,4} tetrachord as the stipulated *Y*.]

(c) K-related *S*-forms relate numerically in the corresponding fashion. If *X* is the series $\langle n,\ n+4,\ n+6,\ n+1,\ n+3\rangle$ (ordered set), a transposed form of

S_0, then K(X) is the series $\langle n+4, n, n-2, n+3, n+1\rangle$, an inverted form of S_0. [Remark: the two series share the same opening dyad, with notes reversed, and the same closing dyad, with notes reversed.] If Y is the series $\langle m, m-4, m-6, m-1, m-3\rangle$, an inverted form of S_0, then K(Y) is the series $\langle m-4, m, m+2, m-3, m-1\rangle$, a transposed form of S_0. [Remark: the two series share the same opening dyad, with notes reversed, and the same closing dyad, with notes reversed.]

***OBSERVATION 2.1*:**

The operation K commutes with all pc transpositions T_i, as those are construed to operate on forms of P (resp. forms of S). If one first K-inverts a P-form (resp. an S-form) and then T_i's that K-inverted form, the net result is the same as if one first T_i's the initial form, and then K-inverts the T_i'd form. Symbolically, $T_iK = KT_i$, as those transformations are construed to operate on P-forms (resp. S-forms).

Proof: Using formula (a) of Definition 1, we start with a sample prime form of P_0, namely $X = \{n, n+1, n+3, n+4, n+6\}$. Applying K-inversion according to the definition, we obtain the inverted P-form $K(X) = \{n-2, n, n+1, n+3, n+4\}$. Applying i-transposition to K(X), we obtain the inverted P-form

$$T_iK(X) = \{i+(n-2), i+n, i+(n+1), i+(n+3), i+(n+4)\}.$$

Starting again with $X = \{n, n+1, n+3, n+4, n+6\}$, we first transpose by i, obtaining the prime P-form $T_i(X) = \{i+n, i+(n+1), i+(n+3), i+(n+4), i+(n+6)\}$. K-inverting the latter form according to formula (a) of Definition 1, we obtain the inverted P-form

$$T_iK(X) = \{(i+n)-2, i+n, (i+n)+1, (i+n)+3, (i+n)+4\}.$$

Comparing the equation that concludes the preceding paragraph with the equation that concludes the paragraph before that, we see that $T_iK(X)$ in fact is the same P-form as $KT_i(X)$, for our sample prime P-form X.

Using the same methods, we can start with a sample inverted P-form $Y = \{m, m+2, m+3, m+5, m+6\}$ and demonstrate, using the formula of Definition 1 (b), that $T_iK(Y)$ is the same P-form as $KT_i(Y)$. So for any sample P-form W (whether prime or inverted), $T_iK(W) = KT_i(W)$. That is what it means to say that $T_iK = KT_i$ in the operational sense, as transformations of P-forms.

We can go through the same formalities, using formula (c) of Definition 1, to show that $T_iK = KT_i$ when the operations are construed as operations on S-forms.

***OBSERVATION 2.2*:**

The operation K commutes with all pc inversions I_i, as those are construed to operate on forms of P (resp. forms of S). If one first K-inverts a P-form (resp. an

S-form) and then I_i's that K-inverted form, the net result is the same as if one first I_i's the initial form, and then K-inverts the I_i'd form. Symbolically, $I_iK = KI_i$, as those transformations are construed to operate on *P*-forms (resp. *S*-forms).

Proof: Similar to the proof for Observation 2.1 above, using the arithmetic that tells us that the I_i-inversion of any number k mod 12 is the number $i - k$; thus if X is the sample prime *P*-form $\{n, n+1, n+3, n+4, n+6\}$, then $I_i(X)$ is the inverted *P*-form $I_i(X) = \{i-n, i-(n+1), i-(n+3), i-(n+4), i-(n+6)\}$, and so forth.

Observations 2.1 and 2.2 shore up formally the matters discussed earlier in connection with figures 9.1 and 9.2. As regards figure 9.1, we might start by imagining any sample *P*-form (resp. any sample *S*-form) in the upper left-hand node. If we K-transform it to fill the upper right-hand node, and then T_7-transform the result to fill the lower right-hand node, we obtain the T_7K transform in that lower-right node. Or, starting with the same sample *P*-form (resp. *S*-form) in the upper-left node, we can T_7-transform it to fill the lower-left node, and then K-transform the result to fill the lower-right node; we then find the KT_7-transform in the lower-right node. Since $KT_7 = T_7K$ in this context (Observation 2.1), we will find the same *P*-form (resp. *S*-form) in the lower-right node, no matter which transformational path we follow on the graph.

The same is true for figure 9.2, mutatis mutandis: no matter with what *P*-form (resp. *S*-form) we start in the upper-left node, we find I_7K(form) = KI_7(form) in the lower-right node, whether we follow the path K-then-I_7 or I_7-then-K.

The graphs of figures 9.1 and 9.2 are thus "well formed" in the sense of my *Generalized Musical Intervals and Transformations.*[7] They are, in fact, "product graphs" in the sense of that study.[8] In that sense, we can think of figure 9.2 as portraying a vertical I_7-motif, horizontally K-transformed from left to right; or, we can think of the example as portraying a horizontal K-motif, vertically I_7-transformed from top to bottom.

It is not so clear as it seemed that we can regard $I_7(S_0)$, rather than $K(S_0)$, as "that inverted form of S_0 most characteristically paired and compared with S_0," even in measures 26–35 (let alone earlier portions of the piece).

9.5 J-Inversion

Considering the series S_0 = B♭–D–E–B–C♯, we can contemplate yet another series, $J(S_0)$ = F–C♯–B–E–D, as a candidate for "that inverted form of S_0 most characteristically paired and compared with S_0." The retrograde of $J(S_0)$ is the series $RJ(S_0)$ = D–E–B–C♯–F; the first four notes of $RJ(S_0)$ are the last four notes of S_0, in the same order. That enables one (Schoenberg, or a listener) to link S_0 with $RJ(S_0)$, overlapping the four common tones. The phenomenon is manifest right at the beginning of the piece, in the first six notes (the final F of $RJ(S_0)$ comes in a bit prematurely, but it is in the proper register as regards the pitch-inverted subject). The J-relation continues in the lower registers of measures 2–3: the

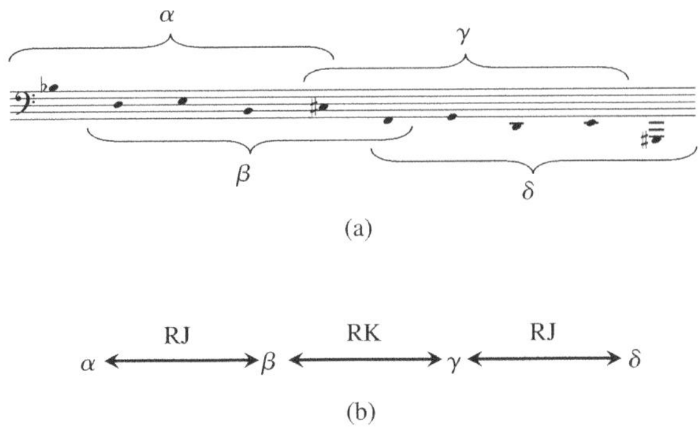

Figure 9.3. *S*-Forms and the J-Relation.

series $T_7(S_0)$ = F–A–B–F♯–G♯ is heard there in the left hand, while $J(T_7(P_0))$, the pentachord {A,C,B,F♯,(A),G♯}, is projected by the five notes of the piece so far which lie at C_3 or below.

The J-relation is also seen in the left-hand figuration of measure 14.2 (measure 14, beat 2) in figure 9.3a. Successive *S*-forms of the figuration are labeled α through δ on the figure. As figure 9.3b indicates, β is the J-retrograde of α, with a four-note overlap; δ is the analogous J-retrograde of γ. γ is the K-retrograde of β; the two forms are linked by their overlapping {0,4} dyad.[9]

In general, the J-inversion of a (prime or inverted) *S*-form *X* is defined as that inverted form of *X* which has the same final tetrachord as *X*, in reverse order. The J-inversion of a (prime or inverted) *P*-form *X* is that inverted form of *X* which shares with *X* the same {0,2,3,5} tetrachord. Thus J({G,A,B♭,C,D♭}) = {F♯, G,A,B♭,C}. If *X* is an *S*-form or *P*-form, then $J(X)$ will be that inverted version of *X* which shares with *X* a common {0,5} dyad.

J, like K, is a "contextual transformation" on the forms of *S* (resp. forms of *P*). If *X* is such a form, $J(X)$—like $K(X)$—has four common tones with *X*, and hence is "maximally close" to *X* in pc content. Aside from $J(X)$ and $K(X)$, no other inverted (or transposed) form of *X* has as many as four common tones with *X*; $J(X)$ and $K(X)$ are the unique two "maximally close" forms of *X*. The operation J resembles K in yet another respect, set forth in the following observation.

***OBSERVATION 3*:**
The operation J commutes with all transposition operations T_i, and all pc-inversion operations I_i when those are construed as operations on forms of *S* (resp. *P*): $T_iJ = JT_i$; $I_iJ = JI_i$.

The observation can be proved rigorously in the manner of Observations 2.1 and 2.2, once the operation of J-inversion is defined numerically, using formulas analogous to those for K-inversion in Definition 1.

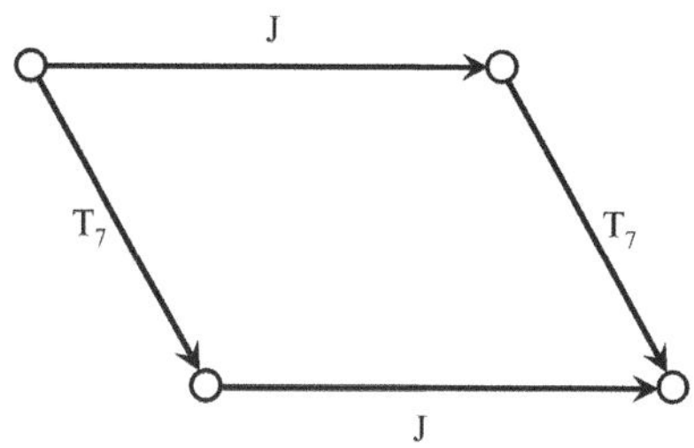

Figure 9.4. Transformation Graph involving J and T_7.

Figure 9.4 is a transformation graph involving J and T_7. It is of the same species as the graphs of figures 9.1 and 9.2: it is a well-formed product graph. We can imagine any sample *P*-form (resp. any sample *S*-form) in the upper left-hand node. If we J-transform it to fill the upper right-hand node, and then T_7-transform the result to fill the lower right-hand node, we obtain the T_7J-transform in that lower-right node. Or, starting with the same sample *P*-form (resp. *S*-form) in the upper-left node, we can T_7-transform it to fill the lower-left node, and then J-transform the result to fill the lower-right node; we then find the JT_7-transform in the lower-right node. Since JT_7 = T_7J in this context (Observation 3), we will find the same *P*-form (resp. *S*-form) in the lower-right node, no matter which transformational path we follow on the graph—"equivalently."

The graph of figure 9.4 pertains to the network of *P*-forms at the opening of the piece: the subject projects P_0 (upper-left node of figure 9.4); the F in the left hand of measure 2, along with the last four notes of the subject, projects J(P_0) (upper-right node of figure 9.4); $T_7(P_0)$ is projected in the bass register of measures 2–3 by a transposed form of the Subject there (lower-left node of figure 9.4); J$T_7(P_0)$ = T_7J(P_0) is projected in the bass register by the notes of measures 1–3 that lie at C3 or below (lower-right node of figure 9.4).

J shares yet another property with K. When a *P*-form is arranged in normal order, it projects a conjunct 5-note segment of an octatonic scale. When that *P*-form is merged with its J-inversion, the result is a conjunct 6-note segment of the scale, a segment which amounts to the scale minus some semitone dyad (for instance, the normally ordered *P*-form B♭–B–C♯–D–E merges with its J-inversion B–C♯–D–E–F to form the ordered hexachord B♭–B–C♯–D–E–F, which is an octatonic scale minus the semitone G–A♭). And when a normally ordered *P*-form merges with its K-inversion, the result is another conjunct 6-note segment of the scale, a segment which amounts to the scale minus some whole-tone dyad (for instance, the normally ordered *P*-form B♭–B–C♯–D–E merges with its K-inversion A♭–B♭–B–C♯–D to form the ordered hexachord A♭–B♭–B–C♯–D–E, which is an octatonic scale minus the whole-tone F–G dyad). The hexachords just discussed are the two possible conjunct 6-note subsegments of the octatonic scale. The octatonic character of the J-merging and K-merging is audible in the figuration of figure 9.3(a); the octatonic scale gives a rationale for the boundaries of the figuration there, since the entire scale is completely projected only with the final G♯ of the example.

9.6 L-Inversion

Let us turn our attention again to the final phrase of the music, which extends from measure 30, second beat (*tempo*) through the end of the piece. During the second beat of measure 30, the *P*-form that sounds together with the pedal {C,G} is P_0. We have called the answering, inverted P-form during the third beat "$I_7(P_0)$." The transformation I_7 is manifest during this music, as discussed earlier, but other forces are here as well. One of the strongest is the concentration of total chromaticism: the two *P*-forms together project the total chromatic of pitch classes, along with the {C,G} dyad; the total chromaticism is "concentrated" because no pitch classes in this aggregate are repeated, save for those of the pedal dyad {C,G}. The musical discourse, which has been more loosely "totally chromatic" over most of the piece, here becomes quite tightly totally chromatic.[10]

Does it not make sense, then, to describe the *P*-form of beat three as "that inverted form of P_0 (beat two) which has no common tones with P_0"? Such a description is in the spirit of the other contextual inversions we have been considering (I_7 is not a contextual inversion operation on *P*-forms; it derives from an inversion operation on individual pitch classes).

Specifically, we can define a contextual inversion called "L-inversion" on *P*-forms (resp. *S*-forms). If *X* is such a form, L(*X*) is that unique inverted form of *X* which shares no common tones with *X*. The *P*-form on the third beat of measure 30 then manifests $L(P_0)$, as it L-inverts the form P_0 of the second beat.

I am taking some care with language here. I do not say that the third-beat P-form "is" $L(P_0)$—"as opposed to" $I_7(P_0)$. The beat-three form "is" both $I_7(P_0)$ and $L(P_0)$. Better yet, it manifests both $I_7(P_0)$ and $L(P_0)$, manifesting both the working of transformation I_7 and the working of transformation L during the last two beats of measure 30. Using mathematical symbols, we can write that $I_7(P_0) = L(P_0)$. The two indicated *P*-forms are the same; transformations I_7 and L have the same effect when applied to P_0. But the two transformations do not have the same effect on *any* sample *P*-form. For instance, the two *P*-forms in measure 31 are I_7-related, continuing to project the mirror texture. But they are not L-related; they have common tones. In measures 7–8, the left-hand *S*-form D–B♭–A♭–D♭–B is L-related to the right-hand *S*-form E♭–G–A–E–F♯; the two forms have no common tones. However, the two forms here are not I_7-related, though they are I_5-related.[11]

The L transformation can be heard in measures 18–19, during the passage that begins "*dolce*." The left hand presents that prime form of the series earlier called "$T_7(S_0)$," F–A–B–F♯–G♯. The notes of the series are brought out of the left-hand texture by metric and registral accents. The right hand, meanwhile, projects E–C–B♭–E♭–D♭ by dynamic and registral accents (presumably there should be an accent symbol, missing in the Hansen edition, over the high B♭). E–C–B♭–E♭–D♭ and F–A–B–F♯–G♯, sharing no common tones, are L-transforms of each other.[12]

As a contextual inversion, L behaves in a similar fashion to K and J. It commutes with every T_i and with every I_i (this can be proved in the fashion of our earlier definitions and observations).

9.7 Transformational Group Structure So Far Generated

We have admitted the familiar group of all pc transpositions T_i and all pc inversions I_i (I_7, in particular, along with all the T_i, will generate this group; in fact, I_7 with T_7 will generate this group).

We shall now determine what further transformations are generated, once we admit J, K, and L into the picture. Each of J, K, and L is of order two. That is, if we start with a given *S*-form (resp. *P*-form), the J-transform of its J-transform is the given form itself. Symbolically, $J^2 = JJ = T_0$. Likewise $K^2 = T_0$, and $L^2 = T_0$.

We shall now calculate the K-transform of the J-transform. Starting with the generic prime *P*-form $X = \{n, n+1, n+3, n+4, n+6\}$, we first take its J-transform, obtaining the *P*-form $J(X) = \{n+1, n+3, n+4, n+6, n+7\}$. Then we take the K-transform of the latter set, obtaining the *P*-form $KJ(X) = \{n+3, n+4, n+6, n+7, n+9\}$. This is *X*, transposed by 3. On the other hand, if we start with a generic inverted *P*-form, $Y = \{n, n+2, n+3, n+5, n+6\}$, the application of J will yield the form $J(Y) = \{n-1, n, n+2, n+3, n+5\}$, and a subsequent application of K will yield the form $KJ(Y) = \{n-3, n-1, n, n+2, n+3\}$. This is *Y*, transposed by 9.

In sum, KJ, as it operates on *P*-forms (resp. *S*-forms), is the transformation Q_3, where Q_3 transforms prime forms of *P* (resp. *S*) into their 3-transpositions, and inverted forms of *P* (resp. *S*) into their 9-transpositions.

The phenomenon is manifest in figure 9.3. There, the *P*-form α is a prime form, the *P*-form β is $J(\alpha)$, and the *P*-form γ is $KJ(\alpha)$. We see that γ is the (pitch-class) 3-transpose of α. The *P*-form δ is an inverted form, the *P*-form α is $J(\delta)$, and the *P*-form β is $KJ(\delta)$. We see that β is the (pitch-class) 9-transpose of δ.

Now $(KJ)(JK) = K(JJ)K = KK = T_0$. So JK is that transformation inverse to KJ. JK is then the operation Q_9, where Q_9 transforms prime forms of *P* (resp. *S*) into their 9-transpositions, and inverted forms of *P* (resp. *S*) into their 3-transpositions.

Similar reasoning shows us that LJ is the operation Q_8, and JL is the operation Q_4 (Q_8 transposes prime forms by 8 and inverted forms by 4; Q_4 transposes prime forms by 4 and inverted forms by 8).

Since Q_9 and Q_4 are in our group, Q_9Q_4 ("Q_9-of-Q_4") will be in the group. That is, Q_1 will be in the group, where Q_1 transposes prime forms by 1 and

T/I Group:	Contextual Group:
Pc Transpositions and Inversions Acting on Forms of *S* (resp. *P*)	Contextual Inversions such as J, $K = Q_3J$, $L = Q_8J$, Acting on Forms of *S* (resp. *P*)
$T_iT_j = T_{i+j}$	$Q_mQ_n = Q_{m+n}$
$T_i(T_jI) = T_{i+j}I$	$Q_m(Q_nJ) = Q_{m+n}J$
$(T_iI)T_j = T_{i-j}I$	$(Q_mJ)Q_n = Q_{m-n}J$
$(T_iI)(T_jI) = T_{i-j}$	$(Q_mJ)(Q_nJ) = Q_{m-n}$

All members of the T/I commute with all of the contextual group, to form "mixed" transformations like $(T_iI)Q_m = Q_m(T_iI)$, $T_i(Q_mJ) = (Q_mJ)T_i$, etc.

Figure 9.5. Formulas for the T/I Group and the Contextual Group.

inverted forms by 11. Then every Q_i will be in the group, as i ranges over the numbers mod 12, since Q_i is generated by applying Q_1 i times.

Since $KJ = Q_3$, $K = (KJ)J = Q_3J$. Since $LJ = Q_8$, $L = (LJ)J = Q_8J$. It follows: J, along with the 12 Q_i transformations, generates a group of 24 members that includes both K and L. The group members are the 12 transformations of form Q_i and the 12 transformations of form Q_iJ. Analogous considerations show that this group is also the group of the 12 Q_i and the 12 Q_iK; it is also the group of the 12 Q_i and the 12 Q_iL. We shall call this the contextual group of transformations on forms of *P* (resp. forms of *S*).

Since J, K, and L all commute with every T_i and every I_i, so do the various products of J, K, and L. Since $Q_1 = Q_9Q_4 = (JK)(JL)$, Q_1 commutes with every T_i and with every I_i. Since Q_n is Q_1 applied n times, every Q_n commutes with every T_i and with every I_i. Since J also commutes with the T's and I's, so does Q_nJ. Thus: every member of the contextual group commutes with every member of the T/I group, considering the latter as a group of transformations on forms of *P* (resp. forms of *S*).

Figure 9.5 summarizes the group structure of our transformations with synoptic formulas. All the arithmetic in these formulas is mod 12. "I" here denotes any fixed inversion operation (such as the earlier "I_7"), when that operation is construed as operating on *S*-forms or *P*-forms.

The T/I group is "simply transitive" on the 24 forms of *S* (resp. *P*): if form 1 and form 2 are any such forms, then there is a unique (NB) T/I operation which transforms form 1 to form 2. The contextual group is also simply transitive on those 24 forms.

The mixed group is the group product of the T/I group with the contextual group. It comprises 288 distinct operations of form TIOP-preceding-CXTOP (= CXTOP-preceding-TIOP) on the 24 forms. There are 24 × 24 = 576 distinct such "names," but each distinct operation has two names. That is because Q_6, which

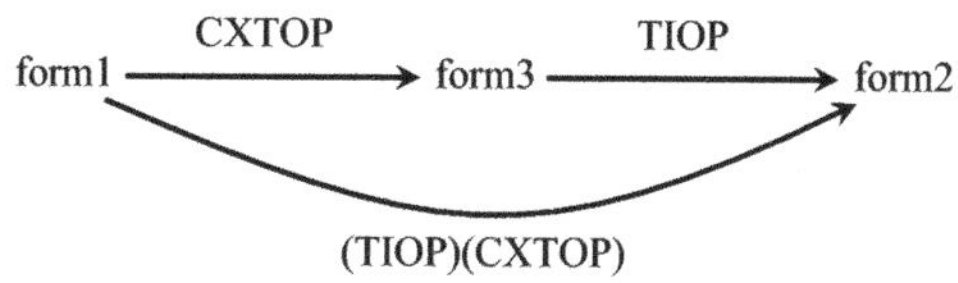

Figure 9.6. *S*-Forms and Mixed T/I and Contextual Operations.

transposes prime forms "up" a pc tritone and inverted forms "down" a pc tritone, is the same operation as T_6 (i.e., it has the same effect on every sample form). So T_6Q_6 is another name for the identity operation. Then, if NAME is the name of an operation, T_6Q_6NAME is another (different) name for the same operation.

If form 1 and form 2 are any *S*-forms (or *P*-forms), then there exist 12 distinct mixed operations which transform form 1 to form 2. To see this, choose any of 24 possible intermediate forms and call it "form 3." Let CXTOP be the unique contextual operation mapping form 1 to form 3, and let TIOP be the unique contextual operation mapping form 3 to form 2 (see figure 9.6).

The mixed operation (TIOP)(CXTOP) then maps form 1 to form 2. If form 3′ is some other intermediate form, then CXTOP′ will be different from CXTOP, and TIOP′ will be different from TIOP (TIOP′ maps form 3′ to form 2; TIOP mapped form 3′ differently, to form 2. The inverse of CXTOP′ maps form 3′ to form 1; the inverse of CXTOP mapped form 3 differently, to form 2). So, for each choice of form 3 (including form 3 = form 1 and form 3 = form 2) there will be a distinct mixed-operation-name mapping form 1 to form 2. Thus there will be 24 distinct mixed-operation names that map form 1 to form 2. If form 3′ is the tritone-transpose form 3, CXTOP′ will be Q_6(CXTOP), and TIOP′ will be T_6(TIOP). The mixed composition of those operations, (TIOP′)(CXTOP′), will be T_6Q_6((TIOP)(CXTOP)), which is another name for the operation (TIOP)(CXTOP). So, while there are 24 distinct names for mixed operations mapping form 1 to form 2, there will be only 12 distinct operations that bear such names—each operation bearing two of the names.

9.8 Analytic Implications and Applications

As a result of the mathematical structure just discussed, when we consider *S*-forms (*P*-forms) form 1 and form 2, we cannot speak of—or even properly conceptualize—such a thing as "the" idiomatic mixed-operational relation between the forms. We shall have to decide, depending on the phenomenology of the musical context, just which of the 12 possible such relations we wish to assert.

We have already noted an example of this in the music from measure 26 to the end, where S_0 and P_0 are being compared with certain inverted forms of *S* and *P*. In the context of the mirror-texturing we saw the relation as an I_7-

α = {A, F, E♭, A♭, G♭} = *P*-form in rh across the barline, mm5–6.

β = F – A – B – F♯ – G♯ = *S*-form in rh, m6.

γ = B♭ – D – E – B – C♯ = *S*-form in lh, m6.

δ = D – B♭ – A♭ – D♭ – B = *S*-form in lh, mm7–8.

ε = E♭ – G – A – E – F♯ = *S*-form in lower rh, mm7–8.

ϕ = D – F♯ – G♯ – D♯ – F = *S*-form in upper rh, mm7–8.

(a)

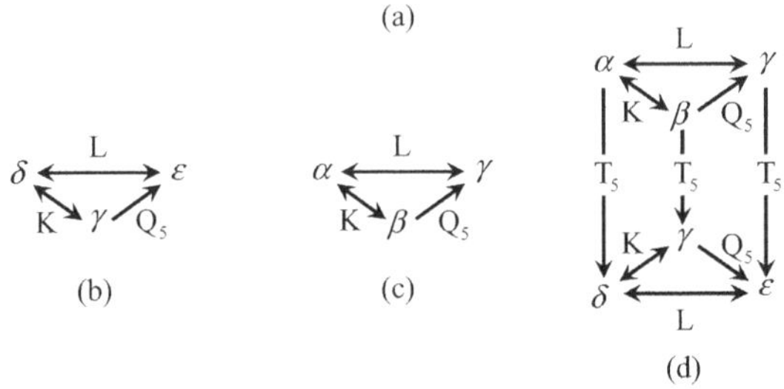

Figure 9.7. Network Models for mm. 5–8 in Op. 23, no. 3.

relation; in the context of the total chromaticism (quasi-combinatoriality), we saw the relation as an L-relation. We would do violence to our perceptions if we were to deny either of those relations.

For a more extended example, let us consider the forms tabulated in figure 9.7a. We have already discussed some relations involving γ and δ. A K-relation can be heard between γ (beginning B♭–D, ending B–C♯) and δ (beginning D–B♭, ending D♭–B). δ and ε have no common tones—all the more audible if ε is played louder than ϕ—so that an L-relation can be heard between δ and ε. The transformational network of figure 9.7b can then be asserted in connection with γ, δ, and ε.

On this network the arrow from γ to ε must be labeled "Q_5," not "T_5" or anything else. That is because $Q_5 = LK$. To be well-formed, the (graph of the) network must be labeled so that the same net transformation is asserted when we go from γ directly to ε as when we follow the composite pathway (from γ via δ to ε).

The network of figure 9.7c is constructed in analogous fashion. *P*-form α has no common tones with *S*-form γ, so we can hear α-to-γ as an L-relation. α embeds the dyad {A,F}, while the inverted β-form begins F–A, so we can hear α-to-β as a K-relation. Then the direct arrow from β to γ must be labeled "Q_5," since β via α to γ gives the composite transformational pathway K-then-L, and $LK = Q_5$.

The triangular networks of figures 9.7b and 9.7c are isographic—that is, they have isomorphic graphs. In this case, they have the same graph. Figure 9.7d conflates 7b and 7c into a larger network that (inter alia) emphasizes the isography. Figure 9.7d was constructed as follows: when we analogize 7c with 7b, we see that α is analogized with δ, γ with ε, and β with γ. Consulting the score (or figure 9.7a), we see that δ is the 5-transpose of α, that ε is the 5-transpose of γ, and that γ is the 5-transpose of β. Searching for one transformation that will equivalently perform all of the 5-transposes, we find that such a transformation must be T_5. (Q_5 would transform β to γ and γ to ε, but it would not transform the inverted *P*-form α to the inverted form δ. Q_5 transposes inverted forms "down" 5 semitones, not "up.")

Inspecting the network of figure 9.7d, we observe that it is a product network, since T_5 commutes with K, with L, and with Q_5. We can say that the motif of the upper triangle is vertically transformed by T_5 into the lower triangle. Or, equivalently, we can say that a vertical T_5-motif is arranged in triangular fashion by the K/L/Q_5 complex of transformations.

On figure 9.7d, form γ—the "tonic form" of the piece—appears in two locations. In the visually higher location, γ "is" the Q_5 transform of β; in the lower location, γ "is" the T_5-transform of β. This illustrates a matter discussed at the opening of the present section: considering form 1 (e.g., β) and form 2 (e.g., γ), we have to decide, depending on the musical context, which of the 12 possible relations we wish to assert.

In this connection we should stress that γ is not a location on the graph of 7d; it is the contents of two different locations in the network. The transformations that label arrows here indicate relations among locations, relations which must be consistent with the contents of those locations, but which already exist on the graph of locations and arrows, without contents.[13]

As for form ϕ of figure 9.7a, I see no reason not to analyze it as $T_{11}(\varepsilon)$. Formally, it could be analyzed as $Q_{11}(\varepsilon)$, or as $T_iQ_n(\varepsilon)$, where $n = 11 - i$, or as $T_iI_7Q_nJ(\varepsilon)$, where $n = 5 - i$. But I do not hear any musical phenomena that suggest any of these other relations, except possibly Q_{11}, which will be discussed later. T_{11} seems well suited to portray the "syncopated forms" going on between ε and ϕ in the right hand (mm. 7–8).

The idea of "syncopated forms" continues in the right hand during the last two beats of measure 8; the forms now involve three voices, not two. Figure 9.8a lists forms ε and ϕ again, along with the three new forms ρ, σ, and τ over the end of measure 8. We have decided, as portrayed in figure 9.8b, that the "symphony" of the ε/ϕ "organum" will be denoted by T_{11}. Figure 9.8c shows the analogous T_{11} organal symphony from ρ to τ, now articulated by the third organal voice σ, into T_5 (ρ to σ) and T_6 (σ to τ).

The T_{11} symphony from ε to ϕ, portrayed in figure 9.8b, can also be articulated into a T_5-plus-T_6 (syncopated) symphony, as suggested by the left half of figure 9.8d. Here ρ articulates the ε/ϕ T_{11}-organum into T_5 (ε to ρ) and T_6 (ρ to ϕ). The T_5 relation from ε to ρ is audible in that the bottommost and topmost voices of the right hand during the last two beats of measure 8 can easily be heard to transpose up by 5 semitones the lower and upper voices of the right hand during

ε = E♭ – G – A – E – F♯ = *S*-form in lower rh, mm7–8.

ϕ = D – F♯ – G♯ – D♯ – F = *S*-form in upper rh, mm7–8.

ρ = A♭ – C – D – A – B = *S*-form in lower rh, m8, beats 2–3.

σ = D♭ – F – G – D – E = *S*-form in middle rh, m8, beats 2–3.

τ = G – B – C♯ – G♯ – A♯ = *S*-form in upper rh, m8, beats 2–3.

(a)

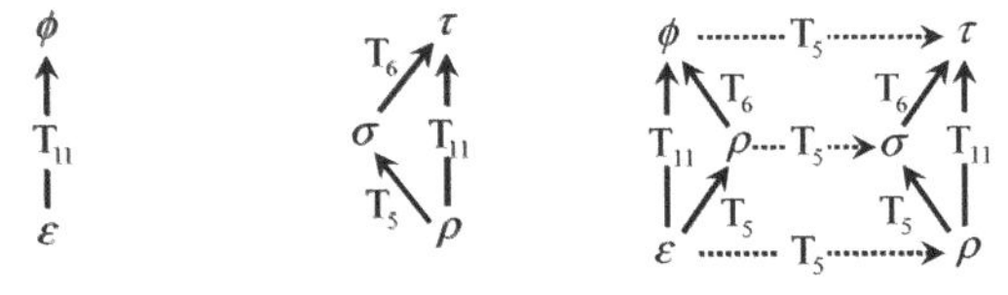

(b) (c) (d)

Figure 9.8. Syncopated Organum and Related Networks for mm. 7–8 in Op. 23, no. 3.

the preceding beats. ε is the lower right hand voice of measures 7.2–8.1; ρ is the bottom-most right hand voice of measure 8.2–8.3.

Figure 9.8d is a product network, since all the T_i operations commute among themselves. One could say that the triangular motif on the left of the network (solid arrows) is T_5'd (dotted arrows) into the triangular motif on the right (solid arrows). Or, one could say that the horizontal dotted arrows show a T_5-motif which is organized vertically into the triangular motif of the syncopated organum. Since all the *S*-forms involved here are prime, and none inverted, the whole network will remain a valid product network if Q_5, Q_6, and Q_{11} are asserted throughout instead of T_5, T_6, and T_{11} (the Q's commute among themselves). Or, since Q's commute with T's, the network will remain a valid product network if all solid T_i arrows are replaced respectively by solid Q_i arrows, while the dotted T_5 arrows remain dotted T_5 arrows. Likewise, the network will remain a valid product network if the dotted T_5 arrows are replaced by dotted Q_5 arrows, while the solid T_i arrows retain their T_i labels.

In the latter case, the horizontal dotted arrow from ρ to σ would be labeled "Q_5," while the solid arrow toward the lower right of the network, from ρ to σ, would be labeled "T_5." As discussed before, such behavior of our networks is not a problem (if they are well-formed). The symbol ρ designates the contents of each of two distinct locations on the graph; the arrow labels are associated with arrow motion between distinct locations on the graph. The contents of the various locations have to behave "plausibly," according to the formal criteria cited in note 13, as they would in the case under discussion here.[14]

Figure 9.8d shows a product network with a T_5-motif in one of its dimensions. Figure 9.7d showed another product network with a T_5-motif in one of

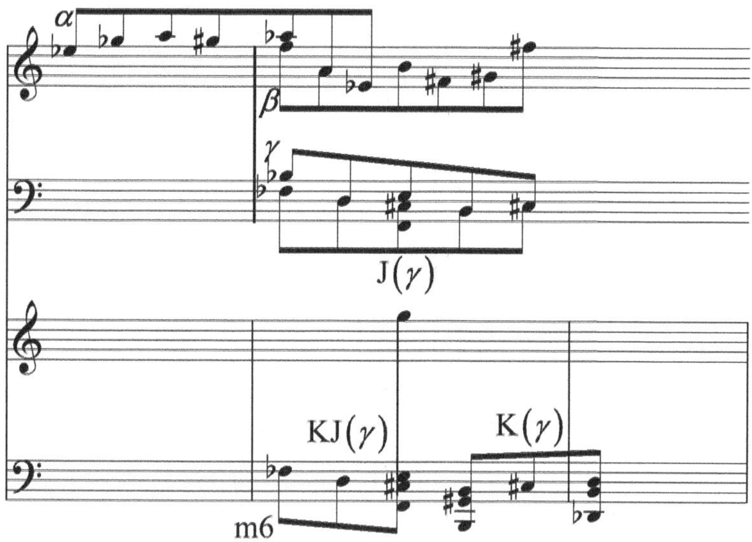

Figure 9.9. *P*-Segmentations for m. 6 in Op. 23, no. 3.

its dimensions. Figure 9.4 showed a product network with a T_7-motif in one of its dimensions. The idea is clearly a constructive feature of the composition's "middle ground." In that regard it expands upon the original subject of measures 1–2.1 and the "answer" in the bass of measures 2–3, which answers the subject "at T_7."

Figure 9.9 shows a plausible segmentation of measure 6, including the accompaniment, into *P*-forms. Other *P*-forms could be asserted, but the ones laid out in the example stay transformationally "close" to α, β, and γ.[15] γ, as noted earlier, has the *P*-form associated with the subject and with series S_0. On figure 9.9, the low F of $J(\gamma) = J(S_0)$ comes in "prematurely," as regards the possibility of retrograde inversion. That mimics the behavior of $J(S_0)$ in measure 2.1: the low F there also came in "prematurely," and in measure 6 the F is even more premature. The serial structure of $J(S_0)$ is not so clear in measure 6. But the registral resemblance between the premature low F of measure 2.1 and the more premature low F of measure 6 renders quite audible the assertion of $J(\gamma)$ on the second staff of figure 9.9. The melding of $J(\gamma)$ with $KJ(\gamma)$, on the lower two staves of figure 9.9, is by now a familiar transformational motif. Indeed, the example shows how α is melded with $K(\alpha) = \beta$; then $J(\gamma)$ is melded with $KJ(\gamma)$, and γ itself is melded with $K(\gamma)$, the latter appearing on the bottom staff of figure 9.9 across the bar line into measure 7. It does not seem quite possible to work out a product network here: K and J do not commute. But K-transformation does persist over the forms of figure 9.9, and that renders the segmentation plausible (and

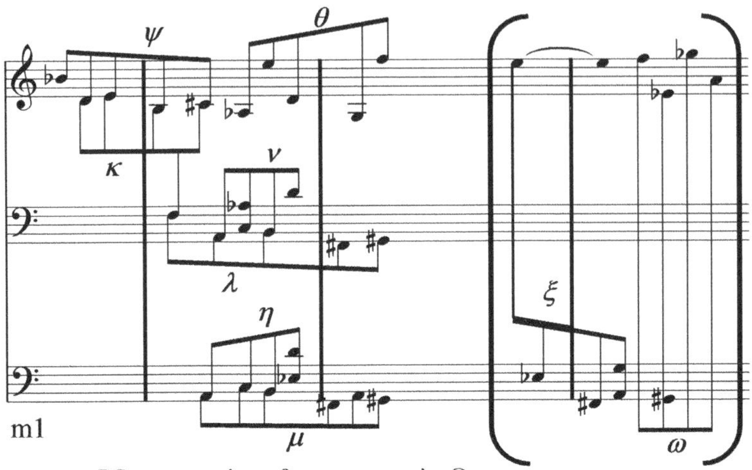

Figure 9.10. *P*-Segmentations for mm. 1–3 in Op. 23, no. 3.

thematic). As noted earlier, other *P*-forms could be asserted in this music, but the forms cited in figure 9.9 reference every note in measure 6.

Throughout the piece, in the manner of figure 9.9, our thematic transformations can be helpful in determining *P*-segmentation for music where *S*-forms are not clearly audible. Figure 9.10, which represents the music from the beginning through the first beat of measure 3, further illustrates that point.

In figure 9.10, the forms $\kappa = \mathrm{J}(\psi)$, $\lambda = \mathrm{T}_7(\psi)$, and $\mu = \mathrm{J}(\lambda) = \mathrm{T}_7(\kappa)$. Hence, either $\mathrm{T}_7\mathrm{J}$ or JT_7 sends the "tonic" series ψ to μ ($\mu = \mathrm{JT}_7(\psi) = \mathrm{T}_7\mathrm{J}(\psi)$). The discussion of figure 9.4 covered these matters, as they fell into a product network. That discussion pointed out that *P*-form μ is articulated by register: it comprises those pitch classes whose representative pitches lie at C_3 or below in the music. The registral articulation of μ suggests to my ear another registral articulation: the *P*-form labeled θ on figure 9.10 comprises those pitch classes whose representative pitches lie at G3 or above in the music of the pertinent time-span.[16]

The forms labeled ν and η, on figure 9.10, address the two little "oom-pah" figures in the accompaniment. Listening to those figures by themselves, one hears a six-note conjunct octatonic subset—a subset associated in this piece with the melding of a *P*-form and its J-transform. ν and η are here the pertinent *P*-form and its J-transform. The J-melding motif is consistent with transformational behavior so far: ψ and κ are J-melded, so are λ and η; the J-melding of η and ν is consistent with the motif.

A transformational rationale for θ takes some concentration, none being obvious; soon, I shall attempt to make one plausible. Meanwhile, I find the registral articulation of θ quite convincing. Helpful in this connection is the fact that

θ is ordered serially as an inverted form (A♭–E–D–G–F) of the *S*-form ψ, not just the *P*-form ψ. To be sure, the registral contour is distorted from that of the thematic subject, but the serial order of pitch classes can be heard with a little concentration.

I have laid out figure 9.10 in staves that emphasize the following relations: ψ and θ meld to form a complete octatonic scale (the one that embeds κ). λ and ν meld to form an octatonic scale missing E♭ (the missing E♭ could be borrowed from η). η and μ meld to form an octatonic scale missing F (the missing F could be borrowed from λ). The octatonic character of each staff, in figure 9.10, expands on pertinent octatonic aspects of *S* and *P*, especially in connection with J and K transformation. Discussion of figure 9.3 earlier considered such matters.

George Perle, analyzing the first 4½ measures, articulates forms ψ, κ, λ, and ν, but not μ, η, or θ.[17] In addition, he articulates forms ξ and ω, which appear to the right of figure 9.10 in parentheses. ω seems significant as ψ; it would be quite consistent with my analysis on the left side of figure 9.10, an analysis which does not extend so far in the piece. Perle's ξ and ω evidence a consistent "harmonic" hearing of the pertinent time-span, at some variance with my "contrapuntal" articulations of μ and θ (my octatonic-hexachordal hearing of η-cum-ν-in-J-relation is closer to Perle's). ξ and ω are bonded together by a contextual inversion that preserves their common tritone (and their common diminished triad); perhaps such a contextual inversion can be found elsewhere in the piece as well, in which case it could be given a letter label of its own (the operation as such is $Q_{11}J$).

Jan Maegaard, in his dissertation of 1972, does mark off form θ with a beam, probably because θ manifests the series, but also puts the θ-beam in parentheses, probably because θ's registral layout does not manifest the contour of the subject itself. He does not discuss or indicate the registral feature of θ's being at or above G3. Maegaard does not mark off with beams the forms κ, μ, ν, or η of figure 9.10, nor does he mark off Perle's ξ (his analytic sketch does not go as far into the piece as to reach Perle's "ω"). Maegaard cites Schoenberg's now-famous quote about measure 2 of the piece.[18]

Ethan Haimo, in his study of 1990, marks form θ from figure 9.10; he also marks forms ξ and ω. He does not articulate μ, ν, or η from that example.[19]

Figure 9.11 attempts to organize the *P*-forms of figure 9.10 into a coherent and thematic transformational network. Figure 9.11a lists the *P*-forms in their *S*-orderings. Figure 9.11b organizes the forms transformation ally, along with a bracketed form "π," which is not as yet manifest in our analytic discussions of measures 1–3. The relevance of π to those discussions will emerge presently. Figure 9.11b is organized transformationally in a familiar way. The T_7-motif is portrayed with vertical arrows; that draws to our attention the fact that η is T_7 of θ, a fact which may well seem fortuitous at this point, but which will soon take on greater significance. The horizontal lines of figure 9.11b project another

ψ = B♭ – D – E – B – C　　ν = A♭ – C – D – A – B

κ = F – C♯ – B – E – D　　η = E♭ – B – A – D – C

λ = F – A – B – F♯ – G♯　　θ = G♯ – E – D – G – F

μ = C – G♯ – F♯ – B – A　　π = C♯ – F – G – D – E

(a)

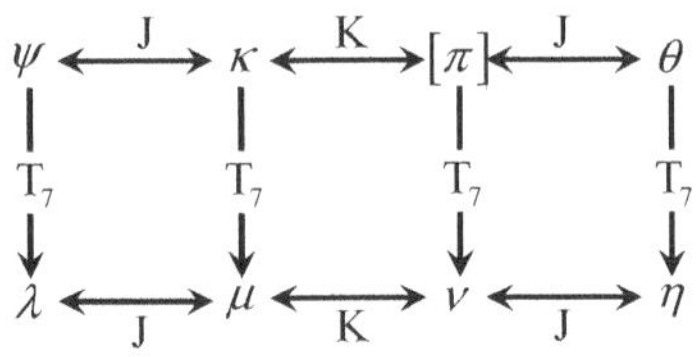

(b)

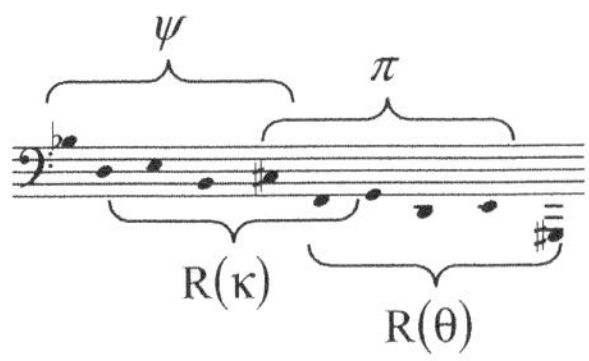

(c)

(d)

Figure 9.11 Transformational Network for figure 9.10.

transformational motif, a chain J–K–J of contextual inversions. Since J and K commute with T_7, the figure 9.11b manifests a product network of a by-now familiar species: the J–K–J chain organizes horizontally the vertical T_7-motifs; equivalently, T_7 organizes vertically the horizontal J–K–J motifs.

The "*Hilfgestalt*" π is implied, as it were, by the organization of the other forms into the product network. π enables the J–K–J motif to manifest itself horizontally along the top rank of figure 9.11b, and the T_7 motif to manifest itself vertically above form ν. Though there is no π-form of *P* in the music of measures 1–3, I find the network of figure 9.11b convincing.

Particularly convincing in this connection is the later music of figure 9.11c (m. 14.2), already studied in figure 9.3. Here, as portrayed in figure 9.11d, the J–K–J chain, which implicates the succession ψ–κ–π–θ, becomes manifest and explicit in the musical foreground. So does form π itself, here manifest as $KJ(\psi) = Q_3(\psi)$. When figure 9.11d is compared with the top rank of figure 9.11b, the musical gesture of figure 9.11c seems a good deal less arbitrary or willful: not simply an amusing trick, it realizes a transformational potential already inherent in the opening measures of the piece—a "promissory note," in the felicitous words of Edward T. Cone.[20]

9.9 Possible GIS-Structures on the 24 Forms of *P* (RESP. *S*)

Our large mixed-operation group, as its operations affect *S*-forms (resp. *P*-forms), includes quite a variety of subgroups that are simply transitive on those forms.[21] As already observed, the T/I group is simply transitive, and so is the Q/J ("contextual") group. The group comprising the twelve operations T_i and the twelve operations T_iJ is simply transitive (and commutative); so is the group comprising the twelve operations T_i and the twelve operations T_iK; as is the group comprising the twelve operations T_i and the twelve operations T_iL.[22] Various other subgroups of the large group are simply transitive on *S*-forms (resp. *P*-forms).

Of concern to us in this connection is the strong mathematical relation between simply transitive groups of operations and GIS-structuring. The mathematics is set forth in Generalized Musical Intervals and Transformations (*GMIT*); the interested reader can explore the index there under "simply transitive group." The matter can be summarized as follows: in any GIS, the group of GIS-transpositions is simply transitive on the objects of the GIS; so is the group of interval-preserving operations. If the GIS is commutative, the two groups are the same. If the GIS is not commutative, then the two groups are the "commuting groups" of each other: an operation preserves GIS-intervals if, and only if, it commutes with all GIS-transpositions; an operation is a GIS-transposition if, and only if, it commutes with all interval-preserving operations. If STRANS is a simply transitive group of operations on a certain family of objects, and STRANS′ is the commuting group for STRANS, then there exists a GIS-structure whose formal GIS-transpositions are the members of STRANS, and whose formal interval-preserving operations are the members of STRANS′; when the groups are non-commutative, there also exists a "dual" GIS-structure, whose formal GIS-transpositions are the members of STRANS′, and whose formal interval-preserving operations are the members of STRANS.

GMIT devotes some attention to a group of operations called "PETEY," defined on the objects of a general non-commutative GIS.[23] PETEY comprises all products of operations, one from the group of the GIS-transpositions, and the other from the group of the interval-preserving operations. If we consider the 24 *P*-forms (resp. *S*-forms) for Op. 23, no. 3 to form a GIS whose formal GIS-transpositions are the 24 operations in the simply transitive T/I group, then PETEY will comprise the 288 operations of our "mixed-operation group." If we consider the 24 *P*-forms (resp. *S*-forms) to form a GIS whose formal GIS-transpositions are the 24 operations in the simply transitive Q/J ("contextual") group, then PETEY will again comprise the 288 operations of our "mixed-operation group."

The paragraph above shows that our mixed-operation group has a certain abstract theoretical rationale, when we consider our 24 *P*-forms (resp. *S*-forms) to be formal GIS-objects being acted upon by the 24 members of the contextual group as formal "GIS intervals" spanning pairs of those objects, or when we consider our 24 *P*-forms (resp. *S*-forms) to be formal GIS-objects being acted upon by the 24 members of the T/I group as formal "GIS intervals" spanning pairs of those objects.

To further appreciate the generic character of our PETEY group, we can compare it to the analogous PETEY group for harmonic (major and minor) triads, rather than the *S/P* pentachord forms. In this connection, we now consider the group of pitch class T's and I's as those operating on triads. Its commuting group contains the Riemannian contextual inversions P, R, and L, where P maps any triad into its parallel (inverting a triad about its constituent fifth dyad), R maps any triad into its relative (inverting a triad about its constituent major third dyad), and L maps any triad into its Riemannian *Leittonwechselklang* (inverting a triad about its constituent minor third dyad).

One verifies that RP = Q_3, where the operation Q_3 maps any major triad into its 3-tranpose minor and any minor triad into its 9-transpose major (e.g., RP(C) = R(c) = E♭; RP(c) = R(C) = a). In like fashion, one verifies that LP = Q_8, where the operation Q_8 maps any major triad into its 8-tranpose minor and any minor triad into its 4-tranpose major (e.g., LP(C) = L(c) = *A♭*; LP(c) = L(C) = *e*). The commuting group for the T/I group in this context (i.e., operating on triads) thus contains both Q_3 and Q_8. So, it contains Q_{11} (= Q_3Q_8). Thus, it contains Q_1, the inversion of Q_{11}. It follows that it contains Q_n for any *n* mod 12, since $Q_n = Q_1$ applied *n* times. Then, since it also contains the parallel operation, P, it also contains the twelve operations Q_nP. Both R and L are of the form Q_nP: since RP = Q_3, R = RPP = Q_3P; since LP = Q_8, L = LPP = Q_8P.

It can be proved that the "contextual group" of operations Q_n and Q_nP is, in fact, the (entire) commuting group for the T/I group, as the group of operations on triads. The mixed group (of T_n*s*, T_nI*s*, Q_n*s*, and Q_nP*s*) is then a PETEY group for the GIS whose objects are the 24 triads, and whose formal GIS-transpositions are the 24 Q_n/Q_nP operations on triads. Readers familiar with Riemann's theories will recognize the contextual group here as the group of "*Schritte*" (Q_nP) and "*Wechsel*" (Q_n), recently elaborated and explored in the work of Henry Klumpenhouwer.[24]

Notes

1. In returning again and again to the subject at its prime transpositional level, to begin fresh sections of this highly contrapuntal piece, Schoenberg's procedure reminds me strongly of Bach's, in those fugues of the WTC that involve extended contrapuntal manipulations of their subjects. I'm thinking of the C♯-minor and D♯-minor fugues in Book I, and above all of the B♭-minor fugue in Book II. Bach, when using a new contrapuntal combination of the subject in these fugues, manifests a strong tendency at odds with the tonal procedure of the textbook "school fugue": to return to his subject in its original key, when he shows a new trick it can perform (one might imagine the tonic key as a closet, to which Bach returns when he wants to show a new toy). Other aspects of Op. 23, no. 3 remind me particularly of the B♭-minor fugue in Book II. In both pieces, the sense of the note B♭ as incipit is strong, whenever the subject recurs. Schoenberg's piece has a number of prominent transposed entries for its subject "at the fifth" (above, or fourth below); his serial sketches

show that he continued to use such diatonic terminology through his life. One prominent inverted form of Schoenberg's subject is also "at the fifth"—beginning on F. The denouement of the Bach fugue highlights a mirror texture involving its subject, as do the final sections of Schoenberg's piece.

2. The unordered pitch-class pentachord projected by the opening of the B♭-minor fugue from Book II, {A,B♭,C,D♭,E♭}, is a transposed form of P_0 = {B♭,B,C♯,D,E}. I would not make much of this, were it not for the other resemblances outlined in note 1.

3. George Perle 1962, 43–45, analyzing Op. 23, takes as what I call "s_0" the inverted form of S_0, which shares the same incipit, the note B♭. This convention for labeling inverted forms of a series is also found in the theoretical works of Milton Babbitt; it facilitates the construction of Babbitt's serial matrices. It has historical roots in the practice that would refer to F–C♯–B–E–D as "inversion at the fifth," hearing the incipit F of that series as "a fifth above" B♭. Perle's convention labels F–C♯–B–E–D analogously, as a version of my "s_7," hearing the incipit F as "7 semitones above" B♭. As a general theoretical practice, the convention is problematic so far as it implicitly asserts a "zero" tonicity about the incipit note for the prime form of a generic series (or fugue subject, etc.). In analyzing Op. 23, no. 3, the convention is problematic so far as its "zero" inverted form, B♭–F♯–E–A–G, is not prominent in the piece, whereas a number of other inverted forms are prominent, and are prominently paired and compared with the referential prime form s_0.

4. In his dissertation, the topic of (alternate) transformational pathways is central to Edward Gollin 2000. The interested reader will find much material in that work that touches on methodological issues germane to the present analysis.

5. I_7 organizes much more than the cantus forms in measures 26–27. In a brilliant analysis, John Roeder 1989, 247–48, shows how not only those forms but also the accompanimental notes in these measures manifest a compositional array involving ten simultaneously unfolding forms of s_0, organized into five I_7-related pairs of forms.

6. The reader can consult the index in Lewin 1993 for discussion of "contextual operations or transformations," along with a certain number of relevant citations. More recent work includes articles by John Clough 1998 and Jonathan Kochavi 1998.

7. See Lewin 1987, 195 (Definition 9.2.1 (D)).

8. Ibid. The interested reader should consult "Product of graphs" in the index (p. 257).

9. The interested reader can find a study of "retrograde-inversional chaining" involving forms of generalized series in *GMIT,* section 8.2 (pp.180–88). The chaining technique of figure 9.3(a), from Schoenberg's piece, can be compared in particular with Example 8.2.5 in GMIT, 183–84, an example which studies analogous figuration from Bach's first Two-Part Invention.

10. While it takes only two-thirds of a measure to project the total chromatic in measure 30, beats 2 and 3, and only the pedal {C,G} is repeated therein, it takes the opening phrase over two measures to project the total chromatic (up to the G in m. 3), and three pitch classes (B, D, and A) have been repeated by the time that G sounds. After that G, from the third sixteenth of measure 3 on, one must wait until the fourth sixteenth-note of measure 4 for the B♭ there to complete the next aggregate; the wait goes on through 21 notes, in the course of which relatively many pitch classes are repeated (E♭, D, F♯, E♭ again, G♯ again, C♯ = D♭, D again, A, F, and D yet a third time, before we hear the B♭). Withholding the B♭ for so long is an effective

means of accenting the entry of the K-inverted subject, during the first beat of measure 4. But the total chromaticism as such is quite diffuse here.

11. The *S*-form in the left hand here is marked *mezzo forte*, standing out in relief from the strutting accompanimental material in the left hand, marked *pianissimo*. When the *S*-forms in the right hand enter, during the second beat, there is a dynamic mark of *forte*, which appears just below the treble-clef staff in the Hansen edition. I wonder if this *forte* is not meant to apply only to the form of *S* that begins on E♭, but not to the higher form that begins on D. The higher *S*-form might then be continuing from earlier, at a *piano* dynamic. If that were the case, then the two L-related forms here, D–B♭–A♭–D♭–B and E♭–G–A–E–F♯, would be brought into sharp dynamic relief from their surroundings, as *Hauptstimme* and *Nebenstimme*.

12. Other forms of the series (subject) are audible in this passage; Milton Babbitt 1974 discusses them. But none of the other series forms is so forcefully accented as are the two just discussed.

13. The formalities of the situation are carefully set forth in *GMIT*, chapter 9, pp.193–219.

14. In this connection it is wise to keep in mind that a symbol like "ε" which appears to occupy a Euclidean "point" on a network, is actually standing for a phenomenon that takes a considerable amount of Newtonian (clock) time to "fill," as we listen to the music. During that time span, the phenomenon can easily enter into a number of different musical relationships. And even if the phenomenon were to be projected by an "instantaneous" chord (say) in the music, it could still occupy many different phenomenological locations in our perception(s), as the music proceeds. The interested reader will find a thorough discussion of these matters in Lewin 1986. N.B. there, especially the discussion (on p. 360) of the "unique notehead at the barline of m. 12" in Schubert's "*Morgengruss*," a notehead whose geometric setting on the Euclidean plane of the score contributes enormously to "the fallacious idea that there is one unique [phenomenological] object called 'the B♭ of measure 12,' an object which impinges upon us at one unique phenomenological time, the time in which the B♭ 'is.'" The Euclidean "points" with which we represent formal nodes of our networks court the same sorts of phenomenological dangers, if carelessly apprehended.

15. The Gollin 2000 dissertation (cited in note 4) has much relevant material on this and related matters.

16. I am influenced by the fact that Schoenberg was a cellist. The articulations of μ and θ, as "at or below viola C (C_3)" and "at or above violin G (G_3)," would be less salient for a non-string player.

17. The analysis appears in Perle 1962, 44.

18. See Maegaard 1972. Maegaard's analysis appears on p. 67 of Vol. III. He cites the quotation from Schoenberg's sketch—I believe that Maegaard was the first to discover and publish it—on p. 96 Vol. I: *"Das ist nachträglich!! konstruiert und stellt sich heraus als der 2te Takt: die Form, die ich mit vieler Mühe hier gesucht habe, ist mir im 2. Takt sofort richtig eingefallen."* (This [Schoenberg's analysis] was constructed subsequently!! [after the compositional process], and is manifest from the second measure on. The form, for which I painstakingly searched here, [had] already occurred to me currently as early as measure 2.)

19. See Haimo 1990, 93–94.

20. See Cone 1982, 233–41.

21. To recapitulate an earlier definition, a group of operations STGRP is "simply transitive" on the 24 forms when, given any form 1 and any form 2 (possibly form 2 = form 1), there exists a unique operation OP in STGRP, such that OP(form 1) = form 2.

22. I developed and used an analogous sort of group for analyzing Stockhausen's *Klavierstück III* in *Musical Form and Transformation* (New Haven: Yale University Press, 1993), 16–67.

23. GMIT, p. 57.

24. The most recent exposition can be found in Henry Klumpenhouwer 2002, 456–76.

Chapter Ten

Transformational Etudes

Basic Principles and Applications of Interval String Theory

Stephen Soderberg

The following study is based on a general definition of a *string* of any objects such as that of Eric Weisstein: "A string of length n on an alphabet l of m objects is an arrangement of n not necessarily distinct symbols from l. There are m^n such distinct strings."[1] We will begin by defining the alphabet as a set of intervals between pitches in order to introduce six string transformations. We will then begin to redefine the alphabet in a series of studies taking us from a traditional pitch-centric viewpoint toward basic digital manipulations of electro-acoustic source material.

10.1 Interval Strings

A *generalized interval string* α is an ordered n-tuple of integers which sum to $\hat{m} = km$:

$$\alpha = \langle \alpha_i \rangle_{i=1}^{n}, \text{ such that } \sum \alpha_i = \hat{m} = km \quad (k \geq 1, m > 1). \tag{1}$$

Here n is the length of the string, $\hat{m}$ is the chromatic span of the string, m is its modulus, and k is an integer. When $k = 1$ (i.e., $\hat{m} = m$), the string is said to be "compact," or in its "minimal configuration." We choose $m > 1$ to avoid the trivial case (mod 1) in the present context.

Interval strings are circular in that element α_1 is considered the successor of α_n. Thus an interval string can be represented as a polygon inscribed in a circle such that its n points coincide with n of the $\hat{m}$ equally spaced sites around

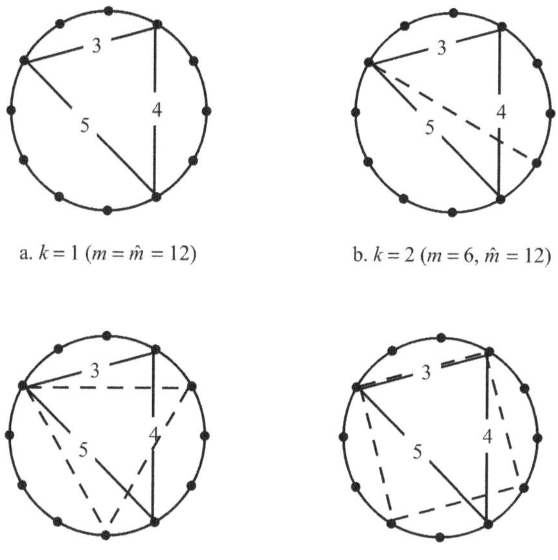

a. $k = 1$ ($m = \hat{m} = 12$)

b. $k = 2$ ($m = 6$, $\hat{m} = 12$)

c. $k = 3$ ($m = 4$, $\hat{m} = 12$)

d. $k = 4$ ($m = 3$, $\hat{m} = 12$)

Figure 10.1.

the circle's circumference. Figure 10.1a shows the interval string $\langle 3, 4, 5 \rangle$ represented as a triangle inscribed in a 12-circle. Note that the sites on the circle's circumference are not (yet) labeled, so the rotational orientation of the triangle is undefined in strictly intervallic terms.[2] Note also that the 12 equal intervals between contiguous sites around the circumference represent the chromatic span, which should not be confused with the modulus. The modulus in this case might be 12 (figure 10.1a), 6 (figure 10.1b), 4 (figure 10.1c), 3 (figure 10.1d), 2 or, trivially, 1 (corresponding to the values 1,2,3,4,6, or 12 for k). This will be further clarified when pitch structures are introduced in the next section.

Repetitions of elements are allowed, and therefore any given interval string may be treated as a multiset. Thus, $\langle 2, 2, 3, 4 \rangle$ and $\langle 1, 1, 2, 1, 2, 6, 2 \rangle$ are valid interval strings. While zero and negative values for string elements are valid and have interesting music-theoretic interpretations,[3] to simplify the presentation we will assume that, in general, $\alpha_i > 0$ for all i. Any zero element resulting from a string manipulation will be deleted from the string as redundant.

To clarify the length, span, and modulus of a string under discussion, we will use the notation $\alpha|_{n,m,k}$ and will refer loosely to "an n-string over $\hat{m}$" or more specifically to "an $[n, m, k]$-string." For example, $\alpha|_{5,12,3}$ is a 5-string over 36, or a $[5, 12, 3]$-string.

A *segment*[4] $\varsigma_j(\alpha)$ is a string of n' consecutive elements from α that sum to s. The subscript j indicates that the segment begins on the j-th element of α and will generally be dropped unless needed for clarity. A segment of $\alpha\big|_{n,m,k}$ will be referenced as $\varsigma_j\big|_{n',s}$, where n' is the length of the segment $(1 \le n' \le n)$ and s is the chromatic span of the segment $(s \le \hat{m})$. Specific segments will be enclosed within ceiling brackets ($\lceil\ \rceil$).

For example, $\lceil 1,2,2,4 \rceil = \varsigma_2\big|_{4,9}$, $\lceil 2 \rceil = \varsigma_3\big|_{1,2} = \varsigma_4\big|_{1,2}$, $\lceil 2,4,3 \rceil = \varsigma_4\big|_{3,9}$, and $\lceil 1,1,2,2,4,3,1,8 \rceil = \varsigma_1\big|_{8,22}$ are segments of $\langle 1,1,2,2,4,3,1,8 \rangle$. Strings such as $\langle 1,8,1,1 \rangle$ and $\langle 2,4,3,1,8,1,1,2 \rangle$ are technically valid segments of $\langle 1,1,2,2,4,3,1,8 \rangle$; however we will generally prefer to handle segments that "turn the corner" as segments of rotations (see below under string transformations).

10.2 Pitch Structures

As implied in the above definition of an interval string, we will be dealing mostly with pitch structures embedded in a referential pitch space as defined by Robert Morris[5] and only occasionally with pitch-class sets. Many notations have been developed to deal with octave equivalence and the modular nature of pitch spaces. We will adopt one that indicates octave location with a numeric superscript. Thus, if C is the referent pitch, C^1 is one octave above the referent C, C^2 is two octaves above, etc. We will fix 0 as the numeric referent in any pitch space and set $C = 0$ when dealing with mod-12 structures. Thus, using the latter as a model, we have the following correspondences:

0	1	2	⋯	11	12	13	14	⋯	23	24	25	26	⋯	35	36	⋯
0	1	2	⋯	11	0^1	1^1	2^1	⋯	11^1	0^2	1^2	2^2	⋯	11^2	0^3	⋯
C	$C\#$	D	⋯	B	C^1	$C\#^1$	D^1	⋯	B^1	C^2	$C\#^2$	D^2	⋯	B^2	C^3	⋯

We define a generalized pitch structure in relation to an interval string in the following way. If p is any integer and α is an n-string over $\hat{m}$, a *pitch structure* is the ordered pair (p,α)[6] defined as the set

$$(p,\alpha) = \{p_i = p_{i-1} + \alpha_{i-1}\}_{i=1}^{n} \quad (p_1 = p) \tag{2}$$

If we have occasion to write $(p,\alpha)_m$ it will indicate we are taking the pitch structure mod m, creating a *pitch-class set*. The distinction is subtle but important. Some examples will clarify.

Begin with $\langle 2,2,3,4 \rangle$, a 4-string over 11. Since 11 is prime, k must be 1 and $m = 11$. If we then choose $p = 0$, then $(0,\langle 2,2,3,4 \rangle) = \{0,2,4,7\} = (0,\langle 2,2,3,4 \rangle)_{11}$. Alternatively (remembering we are using 0 as the referent in any pitch space), if we choose $p = 17 (= 6^1)$, then we have the pitch structure

$(17,\langle 2,2,3,4\rangle)=\{17,19,21,24\}=\{6^1,8^1,10^1,2^2\}\equiv(17,\langle 2,2,3,4\rangle)_{11}=\{6,8,10,2\}$. In general we will choose $p=0$ for examples.

Now consider $\langle 1,1,2,1,2,6,2\rangle$, a 7-string over 15. 15 is composite, so k can take the value of its divisors 1, 3, or 5 (we will always ignore the trivial case $k=m$). If $k=1$, then $m=\hat{m}=15$. Choosing $p=0$, $(0,\langle 1,1,2,1,2,6,2\rangle)=\{0,1,2,4,5,7,13\}\equiv(0,\langle 1,1,2,1,2,6,2\rangle)_{15}$. On the other hand, if $m=5$, $k=3$, $\hat{m}=15$, then $(0,\langle 1,1,2,1,2,6,2\rangle)=\{0,1,2,4,5,7,13\}$ as before; but now $\{0,1,2,4,5,7,13\}=\{0,1,2,4,0^1,2^1,3^2\}\equiv(0,\langle 1,1,2,1,2,6,2\rangle)_5=\{0,1,2,4,0,2,3\}=\{0,1,2,3,4\}$.

For $m=12$ and $k=1$, $\langle 3,4,5\rangle$ is a 3-string over 12. $\langle 7,8,9\rangle$ is a 3-string over 24 for which we will stipulate $m=12$ and $k=2$. Taking $p=0$ as usual, we can form the two structures $(0,\langle 3,4,5\rangle)=\{0,3,7\}$ and $(0,\langle 7,8,9\rangle)=\{0,7,15\}=\{0,7,3^1\}$. Translating these into traditional letter notation for pitches, the former is the triad *C–E♭–G*, and the latter is *C–G–E♭*^{1}. This is in agreement with the way we traditionally speak about a chord in "close position" (what we are calling "compact") and the "same" chord in "open position."

At times we will want to go in the reverse direction and find an interval string for a given pitch structure. To do this we must already know either the chromatic span $\hat{m}$ or the minimal configuration modulus m and, in the latter case, by convention assume the lowest possible value of k. Given the pitch set $S=\{p_1,p_2,p_3,\ldots,p_{n-1},p_n\}$ which has been ordered such that $p_i\le p_{i+1}$ we then define the most compact interval string of S,

$$STR(S)=\langle p_2-p_1,p_3-p_2,\ldots,p_n-p_{n-1},m-p_n+p_1\rangle \qquad (3)$$

If we are in $m=7$ and wish to determine the interval string for the pitch set $\{2,8,11,4,3\}$, we first reorder it to read $\{2,3,4,8,11\}$. Then $STR(S)=\langle 3-2,4-3,8-4,11-8,k\cdot 7-11+2\rangle=\langle 1,1,4,3,k\cdot 7-9\rangle$. Choosing $k=1$ would yield a negative value for the last term (which we wish to avoid in the present context), but any value of k greater than or equal to 2 would make a valid string. For the "most compact" string, we choose the least of these, $k=2$, which makes the last term 5. So $STR(S)=\langle 1,1,4,3,5\rangle$. Checking this result by summing the elements, we get $\hat{m}=14$. If we want to consider S as a pitch-class set, then we first "mod out" the set: $\{2,8,11,4,3\}\bmod 7\equiv\{2,1,4,4,3\}=\{1,2,3,4\}=S'$; then $STR(S')=\langle 1,1,1,4\rangle$. This example was chosen to confront what appears at first to be a discrepancy. It would seem that $\langle 1,1,4,3,5\rangle$ and $\langle 1,1,1,4\rangle$ ought to be forms of the same string; in other words, there should be some interval transformation taking us directly from one string to the other without first devising a pitch representation to discover any unison/octave "doublings." This problem will be resolved below in the discussion of the configuration transformation which will place both of these strings in the same configuration class.

In summary, the presence of k in (1) is what generalizes the usual definition of an interval string. If $k=1$, all the elements of the interval string are contained within one octave (a minimally configured string); if $k = 2$, within two octaves; and so on. When $k = 1$, the generalized interval string is identical to the one the author defined in 1998[7] in the sense that taking a collection of pitches (spanning any number of octaves) as a pitch-class set is analogous to (but not the same as) placing all pitches within the span of one octave (i.e., minimally configuring the string). When $k = 2$, the interval string restricts the corresponding collection of pitches to two octaves, but without changing their pitch-class values.

Formally recognizing octave placement of pitches can be useful in more ways than reintroducing "open positions" of pitch structures as in the triad example above. It is also helpful in discovering characteristics which may be hidden when pitches are consistently changed to pitch classes. For example, the 12-chromatic's string can be (doubly) generated by concatenation of the string (segment) $\langle 4,3 \rangle$ twelve times. This generation can be short-circuited to get the usual diatonic's string expressed as a stack of interlocking triad string segments, $\delta = \langle 4,3,4,3,4,3,3 \rangle$, a $[7,12,2]$-string with the related pitch structure $(0,\delta) = \{0,4,7,11,2',6',9'\}$. The same procedure applies in the 24-chromatic where a 13-note hyperdiatonic scale can have the $[13,24,5]$ form, $\delta^+ = \langle 10,9,9,9,10,9,9,9,10,9,9,9,9 \rangle$. In the 12-chromatic case, the three primary chords are immediately recognized by their characteristic root-position segments, $\lceil 4,3 \rceil$, $\lceil 3,4 \rceil$ and $\lceil 3,3 \rceil$. Likewise, the 13-note hyperdiatonic has five primary pentachords characterized by the segments of δ^+: $\lceil 10,9,9,9 \rceil$, $\lceil 9,9,9,10 \rceil$, $\lceil 9,9,10,9 \rceil$, $\lceil 9,10,9,9 \rceil$, and $\lceil 9,9,9,9 \rceil$. If δ^+ is expressed in its minimal configuration, $\langle 2,2,2,2,2,1,2,2,2,2,2,2,1 \rangle$, the resulting related pentachords' second-order maximally even strings $\langle 4,6,3,6,5 \rangle$, $\langle 3,6,4,5,6 \rangle$, $\langle 4,5,4,5,6 \rangle$, $\langle 4,5,4,6,5 \rangle$, and $\langle 3,6,3,6,6 \rangle$, are likewise in minimal configuration, but not immediately recognizable as related to the corresponding segments in "generator configuration."

10.3 Transformations on Interval Strings

It should now be clear that we are not following the usual course of treating "pitch" as the "primitive" concept, then defining "interval" derivatively as the distance between two pitches. Instead, we have been proceeding precisely backwards by giving primitive status to the interval as an *abstract* element, and only then introducing pitch as a way of "fixing" the position (rotational orientation) of a pattern of intervals.

This approach raises an interesting question. How far is it possible to go with interval strings alone as abstract (unfixed, pitch-independent) music-theoretic objects? The question is similar to asking how much geometry can be done without embedding its objects in a coordinate system; or, put another way, what sorts of transformations of an object are possible with no reference system other than the object itself?

While pitch structures as defined above will be used in examples to help illustrate and understand some of the concepts, what follows is a list of transformations performed solely on interval strings with no necessary reference to pitch.

10.3.1 *Rotation*

Rotation (ROT) is the permutation that rotates the first element to the last element of the string (or segment).

$$\alpha|_{n,m,k} = \langle \alpha_1, \alpha_2, \ldots, \alpha_n \rangle \xrightarrow{ROT} \langle \alpha_2, \ldots, \alpha_n, \alpha_1 \rangle = ROT(\alpha) = \beta|_{n,m,k} \quad (4)$$

Note:

- n, m, k remain unchanged
- $ROT \circ ROT = ROT^2$; $ROT^n(\alpha) = \alpha$.

ROT produces a "rereading" of the string that yields what, in the tonal literature, is usually referred to as a "chord inversion." Choosing $\alpha|_{3,12,1} = \langle 4, 3, 5 \rangle$, $(7, \alpha) = (7, \langle 4, 3, 5 \rangle) = \{7, 11, 14\} = \{7, 11, 2^1\} = S$. (Notice we are not reducing the pitch structure mod 12, but the span is still over one octave.) Then $\beta|_{3,12,1} = ROT(\alpha) = \langle 3, 5, 4 \rangle$ and $(7, \beta) = (7, \langle 3, 5, 4 \rangle) = \{7, 10, 15\} = \{7, 10, 3^1\} = T$. Thus a G-major triad in root position has become an E♭-major triad in first inversion, and ROT can be seen as a "quasi-figured-bass notation" transformation.

10.3.2 *Retrograde*

This transformation is the permutation that reads the original string (or segment) in reverse order:

$$\alpha|_{n,m,k} = \langle \alpha_1, \alpha_2, \ldots, \alpha_n \rangle \xrightarrow{RET} \langle \alpha_n, \ldots, \alpha_2, \alpha_1 \rangle = RET(\alpha) = \beta|_{n,m,k} \quad (5)$$

Note:

- n, m, k remain unchanged.
- RET is its own inverse: $RET \circ RET(\alpha) = \alpha$.

RET corresponds to the chromatic (atonal) inversion of a pitch structure. For example, let $\alpha|_{3,12,1} = \langle 4, 3, 5 \rangle$ and $\beta|_{3,12,1} = RET(\alpha) = \langle 5, 3, 4 \rangle$. Then $(7, \alpha)_{12} = (7, \langle 4, 3, 5 \rangle)_{12} = \{7, 11, 2\} = S$ and $(7, \beta)_{12} = (7, \langle 5, 3, 4 \rangle)_{12} = \{7, 0, 3\} = T$. The elements of S and T are related by I_2.

10.3.3 *Sum*

The one-to-one transformation SUM is applied to clearly defined segments within a string. Let $\varsigma_{j_1} = \left[\alpha_{j_1}, \ldots, \alpha_{j_v} \right]$ be some segment of a string α. Then

$$\alpha|_{n,m,k} = \langle \ldots, \alpha_{j_1-1}, \varsigma_{j_1}, \alpha_{j_v+1}, \ldots \rangle \xrightarrow{SUM(\varsigma)} \langle \ldots, \alpha_{j_1-1}, \varsigma'_{j_1}, \alpha_{j_v+1}, \ldots \rangle = \beta|_{n-v+1,m,k} \quad (6)$$

where $\varsigma'_{j_1} = \alpha_{j_1} + \ldots + \alpha_{j_v}$.

Note:

- m, k remain unchanged, but $n \to n - v + 1$
- *SUM* has no clear inverse, but is closely related to *SPLIT*.

If we are given the segment $\varsigma_1 = \lceil 1, 3, 2 \rceil$ in the interval string $\alpha|_{5,15,1} = \langle 1, 3, 2, 4, 5 \rangle$, then $SUM(\varsigma_1) = \lceil 6 \rceil$; so *SUM* maps α to $\beta|_{3,15,1} = \langle 6, 4, 5 \rangle$. *SUM* corresponds to the deletion of notes within a pitch structure. $(0, \alpha) = (0, \langle 1, 3, 2, 4, 5 \rangle) = \{0, 1, 4, 6, 10\} = S$ in the example's mod 15 system. After applying *SUM* we have $(0, \beta) = (0, \langle 6, 4, 5 \rangle) = \{0, 6, 10\} = T$ (1 and 4 have been deleted). $T \subset S$, and, by analogy, we may write $\beta \subset \alpha$, i.e., β is a *substring* of α.

10.3.4 *Split*

This transformation expands an interval string by generating a segment from one of its elements.

$$\alpha|_{n,m,k} = \langle \ldots, \alpha_{j-1}, \alpha_j, \alpha_{j+1}, \ldots \rangle \xrightarrow{SPLIT(\alpha_j)} \langle \ldots, \alpha_{j-1}, \lceil \pi(\alpha_j) \rceil, \alpha_{j+1}, \ldots \rangle$$
$$= \beta|_{n+\#\pi-1,m,k} \quad (7)$$

where $\pi(\alpha_j)$ is one of the permutations of the numerical partitions[8] of α_j, and $\#\pi$ is the number of elements in that partition.

Note:

- m, k remain unchanged, but $n \to n + \#\pi - 1$
- *SPLIT* is closely related to *SUM*.

SPLIT is a multiple-valued (one-to-many) transformation. Hence, if $\alpha = \langle 1, 2, 2, 5 \rangle$, $SPLIT(5)$ (on α's last element) will produce the set of strings $\mathrm{B} = \{\langle 1, 2, 2, 5 \rangle, \langle 1, 2, 2, 4, 1 \rangle, \langle 1, 2, 2, 1, 4 \rangle, \langle 1, 2, 2, 3, 2 \rangle, \langle 1, 2, 2, 2, 3 \rangle, \langle 1, 2, 2, 3, 1, 1 \rangle, \langle 1, 2, 2, 1, 3, 1 \rangle, \langle 1, 2, 2, 1, 1, 3 \rangle, \langle 1, 2, 2, 2, 2, 1 \rangle, \langle 1, 2, 2, 2, 1, 2 \rangle, \langle 1, 2, 2, 1, 2, 2 \rangle, \langle 1, 2, 2, 2, 1, 1, 1 \rangle, \langle 1, 2, 2, 1, 2, 1, 1 \rangle, \langle 1, 2, 2, 1, 1, 2, 1 \rangle, \langle 1, 2, 2, 1, 1, 1, 2 \rangle, \langle 1, 2, 2, 1, 1, 1, 1, 1 \rangle\}$. Note that $(p, \alpha) \subseteq (p, \beta)$ for all $\beta \in \mathrm{B}$; e.g., $(0, \langle 1, 2, 2, 5 \rangle) = \{0, 1, 3, 5\} \subseteq \{0, 1, 3, 5, 8, 9\} = (0, \langle 1, 2, 2, 3, 1, 1 \rangle)$. Thus α is a substring of β for all $\beta \in \mathrm{B}$.

10.3.5 *Scalar Multiplication*

$$\alpha|_{n,m,k} = \langle \alpha_1, \alpha_2, \ldots, \alpha_n \rangle \xrightarrow{M_q} \langle q\alpha_1, q\alpha_2, \ldots, q\alpha_n \rangle = q \cdot \alpha = M_q(\alpha) = \beta|_{n,m,kq} \quad (8)$$

Note:

– n, m remain unchanged, but $k \to k \cdot q$

– $M_q \circ M_{1/q} = M_{1/q} \circ M_q = M_1$.

Scalar multiplication of strings, where each element is multiplied by the positive number q, produces a result similar to a homothetic transformation in geometry where the entire plane stretches or shrinks by q, retaining the "shape" and relative orientation of any figure but making it larger or smaller by q. If $q > 1$, the initial string expands along with the entire chromatic span; if $0 < q < 1$, it contracts; if $q = 1$, M_q is an identity and returns the original string.

Scalar multiplication leads to several well-known music-theoretic applications, some in the tonal and others in the atonal sphere. If we begin with $\alpha|_{7,12,1} = \langle 1,1,1,1,1,1,6 \rangle$, a 7-string subset of the 12-chromatic, and choose $q = 7$, we arrive at a familiar pitch/pitch-class structure. First, $M_7(\alpha) = \langle 7,7,7,7,7,7,42 \rangle = \beta|_{7,12,7} = \beta|_{7,84,1}$, a 7-string over $12 \cdot 7 = 84$; and we can return to α by applying the multiplicative inverse of $p = 7$, $p' = 1/7$: $M_{1/7}(\beta) = \langle 1,1,1,1,1,1,6 \rangle = \gamma|_{7,84,1/7} = \beta|_{7,12,1} = \alpha|_{7,12,1}$. Moving this into pitch space and selecting $p = 0$, $(0,\alpha) = \{0,1,2,3,4,5,6\}$ and $(0,\beta) = \{0,7,2^1,9^1,4^2,11^2,6^3\}$, more familiar as the projection of six perfect fifths above *C*, $\{C, G, D^1, A^1, E^2, B^2, F\#^3\}$. And, of course, $(0,\beta)_{12} = \{0,7,2,9,4,11,6\}$ which, translated into letter notation and rearranged, is the usual *G*-major scale $\{G, A, B, C, D, E, F\#\}$. The following transformation, *CONF*, can also be used to reduce a multiplied string to its least configuration without reference to related pitch structures.

10.3.6 *Configure*

Any segment whose length is 3 or more may be reconfigured with *CONF*, thereby reconfiguring its string.

$$\alpha|_{n,m,k} = \langle \ldots, \alpha_{j_1-1}, \varsigma, \alpha_{j_{v+1}+1}, \ldots \rangle \xrightarrow{CONF(\varsigma)} \langle \ldots, \alpha_{j_1-1}, \varsigma', \alpha_{j_v+1}, \ldots \rangle = \alpha'|_{n,m,k'} \quad (9)$$

where

$$\begin{aligned}
\varsigma &= \left[\alpha_{j_1}, \alpha_{j_2}, \ldots, \alpha_{j_{v-1}}, \alpha_{j_v}, \alpha_{j_{v+1}}\right] \\
\varsigma' &= \left[\alpha'_{j_1}, \alpha''_{j_1}, \alpha_{j_2}, \ldots, \alpha_{j_{v-1}}, \alpha'_{j_v}\right] = CONF(\varsigma) \\
\alpha'_{j_1} &= \left(\alpha_{j_1} + \alpha_{j_2} + \ldots + \alpha_{j_{v-1}} + \alpha_{j_v}\right) \pm q_1 m && s.t.\ 0 \le \alpha'_{j_1} \le m \\
\alpha''_{j_1} &= \left(\alpha_{j_1} - \alpha'_{j_1}\right) \pm q_2 m && s.t.\ 0 \le \alpha''_{j_1} \le m \\
\alpha'_{j_v} &= \left(\alpha_{j_v} + \alpha_{j_{v+1}}\right) \pm q_3 m && s.t.\ 0 \le \alpha'_{j_v} \le m
\end{aligned}$$

Note:

– m remains unchanged; n remains unchanged unless α'_{j_1}, α''_{j_1}, or $\alpha'_{j_v} = 0$, in which cases n is reduced by 1; $k \to k' = k - (q_1 - q_2 + q_3)$.

– α_{j_1} *SPLIT*s into α'_{j_1} and α''_{j_1}, while α_{j_v} and $\alpha_{j_{v+1}}$ *SUM* to α'_{j_v}.

CONF expresses what happens to interval strings as notes are shifted by octaves in corresponding pitch structures. Iterated reconfigurations of segments may reduce any string to its minimal configuration. All strings that can be reduced by *CONF* to some rotation of the same minimally configured string belong to the same *configuration class.*

If we begin with the 6-string over 12, $\alpha = \langle 1,1,1,2,3,4 \rangle$, and multiply by 5, we arrive at the 6-string over 60, $5\alpha = \langle 5,5,5,10,15,20 \rangle = \beta$. We can immediately reduce this by subtracting multiples of 12 from elements greater than 12 and begin reconfiguring from the string $\langle 5,5,5,10,3,8 \rangle$. One of the many possibilities for reducing this string to its minimal configuration using *CONF* is shown is:

$$\begin{array}{cc} \langle \lfloor 5,5,5,10 \rfloor, 3, 8 \rangle & k=3 \\ \updownarrow & \\ \langle \lceil 3,2,5,3,3 \rceil, 8 \rangle & k=2 \\ \updownarrow & \\ \langle 1,2,\lceil 2,5,6,8 \rceil \rangle & k=2 \\ \updownarrow & \\ \langle 1,2,1,1,5,2 \rangle & k=1 \end{array}$$

We can now resolve the confusion that was created in the *STR* example above. It will be recalled that one string representation of the pitch set $\{2,8,11,4,3\}$ is the $[5,7,2]$-string $\langle 1,1,4,3,5 \rangle$, while the string of the corresponding pitch-*class* set, $\{1,2,3,4\}$, is the $[4,7,1]$-string $\langle 1,1,1,4 \rangle$. *CONF* now gives us a way to map between these two interval strings directly: $\langle 1,1,\lceil 4,3,5 \rceil \rangle \xrightarrow{CONF} \langle 1,1,\lceil 0,4,1 \rceil \rangle \xrightarrow{(drop\ zero)} \langle 1,1,4,1 \rangle \xrightarrow{ROT^3} \langle 1,1,1,4 \rangle$. Thus $\langle 1,1,4,3,5 \rangle$ and $\langle 1,1,1,4 \rangle$ belong to the same configuration class.

10.4 Inclusion Relations

Strings may be concatenated to form new strings. If $\alpha = \langle \alpha_1, \ldots, \alpha_a \rangle$ is an *a*-string over *d* and $\beta = \langle \beta_1, \ldots, \beta_b \rangle$ is a *b*-string over *e*, then $\alpha \oplus \beta$ is $\langle \alpha_1, \ldots, \alpha_a, \beta_1, \ldots, \beta_b \rangle$, an $(a+b)$-string over $(d+e)$. Thus, by the definition of a segment, any string may be viewed as a concatenation of segments which effectively partition the string. In other words, any $[n,m,k]$-string α can be partitioned into $t(\leq n)$ segments:

$$\alpha = \varsigma_1 \big|_{a,e} \oplus \varsigma_2 \big|_{b,f} \oplus \ldots \oplus \varsigma_t \big|_{d,h} \qquad (10)$$

such that $a+b+\ldots+d=n$ and $e+f+\ldots+h=\hat{m}$.

Thus the string $\langle 1,1,2,2,4,3,1,8\rangle$ can be partitioned in various ways such as $\langle 1,1\rangle\oplus\langle 2,2\rangle\oplus\langle 4,3,1,8\rangle$, a 3-partition; $\langle 1\rangle\oplus\langle 1,2,2\rangle\oplus\langle 4,3\rangle\oplus\langle 1,8\rangle$, a 4-partition; and $\langle 1\rangle\oplus\langle 1\rangle\oplus\langle 2\rangle\oplus\langle 2\rangle\oplus\langle 4\rangle\oplus\langle 3\rangle\oplus\langle 1\rangle\oplus\langle 8\rangle$, the (unique) 8-partition.

If $\bigoplus_{i=1}^{t} \varsigma_i\big|_{n_i',s_i} = \varsigma_1\big|_{n_1',s_1} \oplus \varsigma_2\big|_{n_2',s_2} \oplus \cdots \oplus \varsigma_t\big|_{n_t',s_t}$ is some *t*-partition of $\alpha\big|_{n,m,k}$, we define α's *partition string*

$$\gamma = \langle n_i' \rangle_{i=1}^{t} = \langle n_1', n_2', \ldots, n_t' \rangle = \langle \#\varsigma_1, \#\varsigma_2, \ldots, \#\varsigma_t \rangle \tag{11}$$

(where # indicates "cardinality of") and its partition-related *substring*

$$\phi = \langle s_i \rangle_{i=1}^{t} = \langle s_1, s_2, \ldots, s_t \rangle = \langle SUM(\varsigma_1), SUM(\varsigma_2), \ldots, SUM(\varsigma_t) \rangle \tag{12}$$

Thus each element of the partition string takes the value of n' for the corresponding segment, and each element of the substring is the sum of the elements in the corresponding segment.

For the 4-partition $\langle 1\rangle\oplus\langle 1,2,2\rangle\oplus\langle 4,3\rangle\oplus\langle 1,8\rangle$, the partition string is $\langle 1,3,2,2\rangle$, and the related substring is $\langle 1,(1+2+2),(4+3),(1+8)\rangle = \langle 1,5,7,9\rangle$. Laying this example out fully:

$$\begin{array}{llcccc l}
\gamma & = \langle n_i' \rangle_{i=1}^{4} = & \langle 1, & 3, & 2, & 2\rangle & \text{(cardinality of each segment)} \\
 & & \updownarrow & \updownarrow & \updownarrow & \updownarrow & \\
\bigoplus_{i=1}^{4} \varsigma_i\big|_{n_i',s_i} & = & \langle 1\rangle \oplus & \langle 1,2,2\rangle \oplus & \langle 4,3\rangle \oplus & \langle 1,8\rangle & (\alpha \text{ partitioned into segments}) \\
 & & \updownarrow & \updownarrow & \updownarrow & \updownarrow & \\
\phi & = \langle s_i \rangle_{i=1}^{4} = & \langle 1, & 5, & 7, & 9\rangle & \text{(sum of elements in each segment)}
\end{array}$$

In the author's "White Note Fantasy,"[9] α was interpreted as a "scale" string, and the "generic" (partition) string γ was "warped" as it passed through α to map a "resultant" string (substring) ϕ. Put another way, γ partitioned α, whose segments were then summed to generate ϕ,

$$WARP(\gamma, \alpha) = \phi \tag{13}$$

and the transformation took γ to ϕ.[10] So in the example, the action of *WARP* can be followed thus: the second element of γ (i.e., 3) "collects" or "defines" segment $\varsigma_2 = \langle 1,2,2\rangle$, the elements of which are then summed to obtain ϕ's second element (i.e., 5). Similar paths can be described for the first, third and fourth elements of γ.

We will now shift our viewpoint, using the same string types previously used in *WARP* but exploiting the string inclusion relationship between α and ϕ via γ. $\phi \subseteq_\gamma \alpha$ will indicate ϕ is a substring of α indicated by γ's partition of α. Using

43
52
61
322
331
421
511
2221
3211
4111
22111
31111
211111
1111111

Figure 10.2. Partitions of 7.

43 34
52 25
61 16
322 232 223
331 313 133
421 412 241 214 142 124
511 151 115
2221 2212 2122 1222
3211 3121 3112 2311 2131 2113 1321 1312
1231 1213 1132 1123
4111 1411 1141 1114
22111 21211 21121 21112 12211 12121 12112 11221
11212 11122
31111 13111 11311 11131 11113
211111 121111 112111 111211 111121 111112
1111111

Figure 10.3. Permutations of the partitions of 7.

this approach, we will first develop an algorithm for generating all the interval strings in any given chromatic universe.

10.4.1 *Algorithm for generating the substrings of* $\alpha|_{n,n,1}$

Working by example, consider the interval string $\alpha|_{7,7,1} = \langle 1, 1, 1, 1, 1, 1, 1 \rangle$ which is the only 7-string over 7.

1. List the numerical partitions of 7 in lexicographic order (figure 10.2).[11]
2. List all distinct multiset permutations of each numerical partition in reverse lex order (figure 10.3).

34
25
16
223
133
124 214
115
1222
1123 2113 1213
1114
11122 11212
11113
111112
1111111

Figure 10.4. "Transposition class" interval strings mod 7.

34
25
16
223
133
124
115
1222
1123 1213
1114
11122 11212
11113
111112
1111111

Figure 10.5. "Set class" interval strings mod 7.

3. Delete all but one permutation for strings related by rotation (figure 10.4); this will be a list of all interval strings corresponding to all "Transposition classes" mod 7. (We have chosen to retain the strings corresponding to what has come to be known as "normal order.") We will identify this list as $\mathbf{L}_1$.
4. Delete all but one permutation for strings related by retrograde (figure 10.5). In this case, there are only two, 214 and 2113. This is now a list of all interval strings corresponding to all "set classes" mod 7. We will identify this list as $\mathbf{L}_2$.

Generating $\mathbf{L}_2$ provides all the information necessary to generate and compare interval-class vectors (without the need to generate pitch-class lists), and, as we shall now see, $\mathbf{L}_1$ provides a basis for fully analyzing the substring content of any interval string.

10.4.2 *Algorithm for generating the substrings of* $\alpha|_{n,m,k}$

Continuing with the example, we now generate all 7-strings over $\hat{m}$ for any choice of $\hat{m} \geq 7$; that is, we are looking for a way to list $\alpha|_{7,m,k}$ for any choice of m and k.

1. Write an abstract 7-string as *ABCDEFG*
2. Take the first entry from $\mathbf{L}_1$, $\langle 3, 4 \rangle$
3. Use $\langle 3, 4 \rangle$ as a partition string to partition the 7-string as $(ABC)(DEFG)$
4. Sum within each segment to obtain the substring $\langle (A+B+C)(D+E+F+G) \rangle$
5. Rotate the 7-string to *BCDEFGA* and repeat steps 2 through 5 for all rotations
6. Return to step 2 and replace $\langle 3, 4 \rangle$ with the second entry from $\mathbf{L}_1$, $\langle 2, 5 \rangle$; repeat steps 2 through 6 for all entries in $\mathbf{L}_1$

If we choose $\alpha = \langle 1,1,1,1,1,1,1 \rangle$ as the 7-string, the algorithm will return $\mathbf{L}_1$ with each of its entries repeated 6 times. The choice $\alpha = \langle 2,2,1,2,2,2,1 \rangle$ will generate all the well-known substrings in the diatonic 7-string (i.e., the harmonic content in the usual diatonic set). Using strings of 1's is a good way to set up this algorithm using a spreadsheet application with functions created according to the algorithm. When the 1's in the α n-string are written over, the rest of the values in the table (the substrings) change accordingly. Once a spreadsheet has been created for one n-string, it can be used for any n-string. For example, the table of strings for the 12-note chromatic can be used to investigate any 12-strings and their associated substrings such as the catalog of 12-strings in the quartertone system.

To see the need to use $\mathbf{L}_1$, which does not recognize retrograde equivalence, rather than $\mathbf{L}_1$ as the source for generic strings, consider the asymmetric 7-string $\alpha = \langle 1,2,3,4,5,6,7 \rangle$ for which $\hat{m} = 28$. Figure 10.6 is a comparison of the results for choosing a non-symmetric partition string and its retrograde. α^v is the v-th rotation of α; the third column is the result of $WARP(\langle 1,1,2,3 \rangle, \alpha)$ and the fourth column is the result of $WARP(\langle 3,2,1,1 \rangle, \alpha)$. Unlike the case for a symmetric string such as $\langle 2,2,1,2,2,2,1 \rangle$ for which the two right columns could be paired as retrograde-related, the contents for $\langle 1,2,3,4,5,6,7 \rangle$ appear to be related only by their retrograde-related partition strings.

10.5 String Modulation

In tonal theory, the presence of a note or chord X that is "foreign" with respect to the prevailing harmonic structure can be handled in one of two ways, broadly speaking. Either X is "nonharmonic," or X signals a change in the local pitch

ν	α^{ν}	$\phi_{\nu} \subseteqq_{\langle 1,1,2,3\rangle} \alpha_{\nu}$	$\phi_{\nu} \subseteqq_{\langle 3,2,1,1\rangle} \alpha_{\nu}$
0	$\langle 1,2,3,4,5,6,7\rangle$	$\langle 1,2,7,18\rangle$	$\langle 6,9,6,7\rangle$
1	$\langle 2,3,4,5,6,7,1\rangle$	$\langle 2,3,9,14\rangle$	$\langle 9,11,7,1\rangle$
2	$\langle 3,4,5,6,7,1,2\rangle$	$\langle 3,4,11,10\rangle$	$\langle 12,13,1,2\rangle$
3	$\langle 4,5,6,7,1,2,3\rangle$	$\langle 4,5,13,6\rangle$	$\langle 15,8,2,3\rangle$
4	$\langle 5,6,7,1,2,3,4\rangle$	$\langle 5,6,8,9\rangle$	$\langle 18,3,3,4\rangle$
5	$\langle 6,7,1,2,3,4,5\rangle$	$\langle 6,7,3,12\rangle$	$\langle 14,5,4,5\rangle$
6	$\langle 7,1,2,3,4,5,6\rangle$	$\langle 7,1,5,15\rangle$	$\langle 10,7,5,6\rangle$

Figure 10.6.

structure often called a "modulation." In the former case, the result can sometimes lead to exotic juxtapositions or, in the extreme case, a break with tonality itself. In the case of modulation, X ushers in, or is discovered as belonging to, a new harmonic structure; but the new structure is usually "new" only in the sense of a change wrought by pitch transposition. The structure is reoriented with respect to pitch, but maintains its harmonic shape.

In terms of the interval string notation we have been exploring, traditional modulation is typically of the form $(p,\delta)\rightarrow(p\pm x,\delta')$ where δ is some tonal string (usually one of the diatonic modes), x is the index of transposition, and δ' is some rotation of δ. The point is that actions on the interval string (other than rotation) are traditionally rare in order to preserve an underlying harmonic unity that will support stacks of thirds and traditional voice leading (from which Schenkerian and neo-Riemannian principles are derived).

In this context, composers at tonality's edge, such as Debussy, Stravinsky, and Bartók, present some real analytical problems. One study that takes us directly to the modulation problem is Allen Forte's "Harmonic Syntax and Voice Leading in Stravinsky's Early Music."[12] Forte provides a list of forty pitch-class sets that can be associated in significant ways with Stravinsky's music during the period 1908–14. Forte labels many of these pc sets as "octatonic" (twenty-one of the forty), "diatonic" (fourteen) and "whole tone" (three). But Forte does not relate eleven pc sets in this list to any of these scales. Three of these eleven are abstractly contained in what Forte calls the "master diatonic set" (interval string $\langle 1,1,1,2,2,1,2,2\rangle$), but the remaining eight can not be contained in any of the three underlying scales normally associated with the pre-atonal Stravinsky.

The representative problem is this: Are these eight "maverick" pc sets nonharmonic structures that might arise through some non-tonal procedure (or even accidentally), or do they arise in some sort of quasi-tonal modulation to/through underlying scale structures that have not yet been identified with Stravinsky? The former choice implies that the presence of one of these mavericks would have to be approached contextually and ad hoc. The latter implies that there may be deeper formal connections between the mavericks (and possibly other

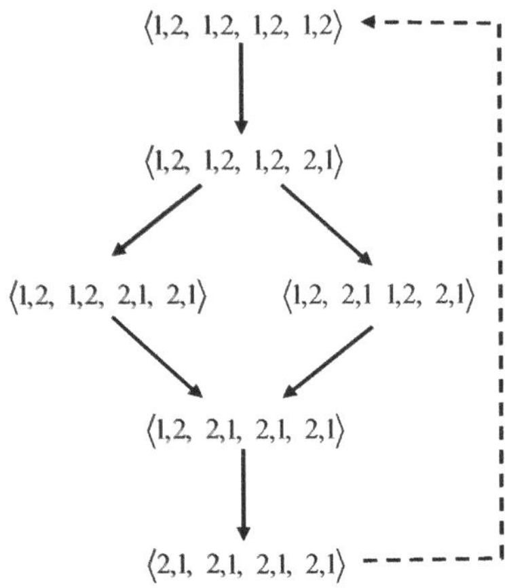

Figure 10.7.

similar structures); this is the least explored possibility, but interval string theory allows us to make some conjectures in this direction.

Despite Stravinsky's well-known claim, "I compose with intervals," nothing that will now be set forth is meant to suggest that Stravinsky consciously or even subconsciously composed with interval strings. The justification for our Stravinsky-inspired conjectures rests not in the drive for correct, insightful analysis, but in expanding compositional possibilities. Even questionable analyses can occasionally lead to new ideas in compositional theory.

We begin by noting the fourfold rotational symmetry of the octatonic interval string, $o = \langle 1,2,1,2,1,2,1,2 \rangle$. This symmetry invites a 4-partition into equal segments, $\langle 1,2 \rangle \oplus \langle 1,2 \rangle \oplus \langle 1,2 \rangle \oplus \langle 1,2 \rangle$, which further invites operating on the segments separately. Each segment has only two elements. Rotation and retrograde produce identical results in this case, so by random choice we will refer to these two transformations collectively as a rotation. Successive applications of *ROT* to these segments reveals that there are only five resulting 8-strings (counting the original) that are distinct (i.e., not rotation-related to one another), two of which are asymmetric and retrograde related. In figure 10.7 each arrow signifies the rotation of one 2-segment except for the dashed line, which represents the rotation of the entire 8-string.

Segmentation into four identical 2-strings is not the only partition of o available. If we rotate a $\langle 2,1 \rangle$ segment following the rotation of a $\langle 1,2 \rangle$ segment, we can reach any of the ten possible 8-strings containing four 1's and four 2's. For example, $\langle 1, \lceil 2,1 \rceil, 2,2,1,2,1 \rangle \xrightarrow{ROT} \langle 1, \lceil 1,2 \rceil, 2,2,1,2,1 \rangle$ reaches an 8-string not available with the initial partition. Thus, by partitioning the octatonic string in various ways and applying *ROT*, we can "modulate" to a different 8-string with a different harmonic structure, i.e., one with a different set of substrings.

Forte Number	Prime Form	Interval String	
4-5	0126	1146	
4-7	0145	1317	
4-8	0156	1416	
4-16	0157	1425	
4-19	0148	1344	("Classic Atonal Set")
5-Z18	01457	13125	
5-22	01478	13314	
6-Z28	013569	122133	
6-Z29	013689	123213	
6-34	013579	122223	(Scriabin's "Mystic Chord")
7-22	0125689	1131213	

Figure 10.8.

o-Derived String	Stravinsky's Maverick Strings
12211221*	1146, 1416, 1425, 1344, 13314, 122133, 122223
12122121*	1317, 1416, 1425, 1344, 13125, 13314, 122133, 123213
12121221*	1317, 1416, 1425, 1344, 13125, 13314, 122133, 123213, 122223
12212121*	1146, 1317, 1416, 1425, 1344, 123213, 13125, 13314, 122133, 123213
12121212*	
11112222	1146, 1425, 1344, 122223
1121222	1146, 1317, 1416, 1425, 1344, 13125, 122223
22212111	1146, 1317, 1416, 1425, 1344, 122223
11122122	1146, 1416, 1425
11211222	1146, 1317, 1416, 1425, 1344, 13125, 122223

Figure 10.9.

We now return to the eleven "maverick" Stravinsky pitch-class sets identified by Forte, and list them for reference in figure 10.8 along with their corresponding interval strings. Figure 10.9 then lists all the 8-strings containing four 1's and four 2's and, to the right of each 8-string, the mavericks that can be found among their substrings (the 8-strings listed in figure 10.8 are given with asterisks). While all of the mavericks can be found in one or another of all of these 8-strings with the exception of the pure octatonic, the 7-string $\langle 1, 1, 3, 1, 2, 1, 3 \rangle$ cannot be located as a substring of an o-related 8-string. This suggests we should try other transformations in search of a larger string that would include all the mavericks.

SUM applied to any 2-segment of *o* will collapse it into the 1-segment $\lceil 3 \rceil$; and *SPLIT* (in this case, a single-valued transformation) will *expand* any segment's "2" element resulting in the 3-segment $\lceil 1,1,1 \rceil$. Thus $\langle 1, \lceil 2 \rceil, 1, 2, 1, 2, 1, 2 \rangle \xrightarrow{SPLIT} \langle 1, \lceil 1, 1 \rceil, 1, 2, 1, 2, 1, 2 \rangle$.[13] This is the string we are seeking. All of the maverick strings can be located in the 9-string $\langle 1, 1, 1, 1, 2, 1, 2, 1, 2 \rangle$.

The introduction of *SUM* and *SPLIT* at this point suggests further string modulations that can include diatonic and whole-tone strings. This in turn invites the construction of closed or open chains of string modulations that might serve as the basis for compositional plans. Figure 10.10 illustrates this possibility. The left side indicates string modulations using *ROT*, *SUM* and *SPLIT* which take us from the octatonic to the diatonic to the whole-tone and back to the octatonic. Between each of these more familiar areas are string regions where we might find any of the maverick strings. The right side indicates the corresponding pitch actions: *ROT* corresponds to the neo-Riemannian transformation of adding or subtracting 1 from a pitch; *SUM* and *SPLIT* correspond to the deletion and addition of a note, respectively. Obviously, there are many such paths; and, if we do away with the desire to traverse ground that will take us to and through the octatonic, diatonic and whole-tone regions (i.e., allow any transformation-connected strings), there are an infinity of such paths.

10.6 Tapes, Samples, and Index Shuffling

Shortly after the tape recorder made its appearance in the twentieth century, composers began to think of the magnetic tape itself as a manipulable sound source. At the most basic, "mechanical" level, there are only two things one can do musically with a tape: vary the playback speed or cut it into segments which are reordered and spliced.

Figure 10.11a shows a section of a hypothetical tape divided into five unequal segments. Each segment is marked with one or two arrows to help us keep track visually of how various transformations will reorder the segments. That the arrows are all oriented left-to-right is not overly significant as a representation of any segment's content. We are merely indicating content orientation for each segment of the tape; when any transformation moves a segment it retains the orientation of that segment's content, but physically flipping any segment is not ruled out in practice (tapes may be played backward but the background "temporal chromatic" (clock time) may not flow in reverse).

In accordance with our previous discussion, we can treat the section of tape shown in figure 10.11a as an interval string. Thus, $\alpha = \langle \alpha_1, \alpha_2, \alpha_3, \alpha_4, \alpha_5 \rangle = \langle 1, 2, 1, 1, 3 \rangle$ is a 5-string over 8. As such, all of the string formations described above may be

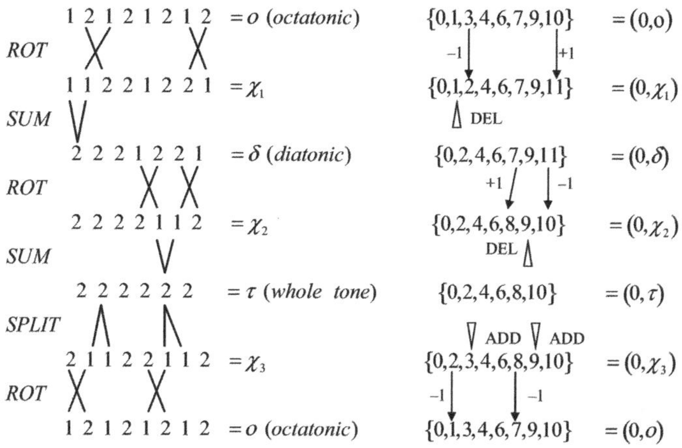

Figure 10.10.

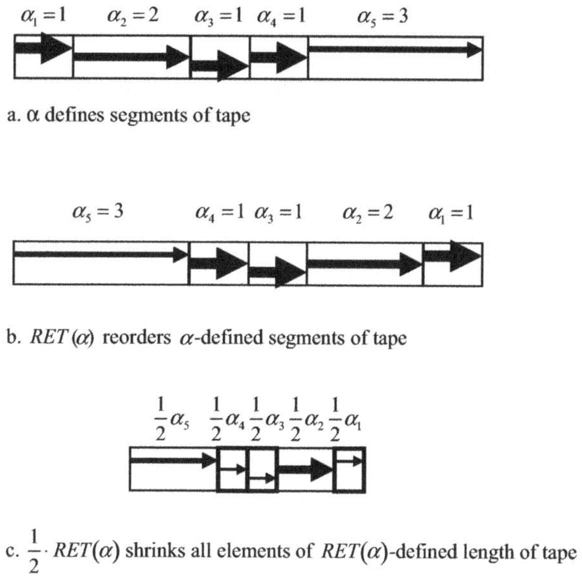

Figure 10.11.

applied to it. For illustration, figure 10.11b shows the action of *RET* on the tape. Of the six transformations presented in section 3, *RET* and *ROT* are index transformations, while *SUM*, *SPLIT*, *M*, and *CONF* are content transformations. A content transformation such as *CONF* may at first seem inapplicable here, but the reader should keep in mind that we are no longer shuffling pitches, but ordered segments of "tape content."[14] Thus we now may have occasion to allow zero values for string elements since two or more segments may be initiated simultaneously. Figure 10.11c shows the effect of another transformation, $M_{1/2}$, on $RET(\alpha)$; the shrinking of the tape is analogous to increasing the playback speed.

We began with a generalized (but still pitch-centric) definition of an interval string which recognized octave displacement of pitches. We will now conclude by redefining the interval string abstractly as a *"sample" string*:

$$\alpha = \langle \varsigma_i \rangle_{i=1}^{n} . \tag{14}$$

This is simply an ordered *n*-tuple of segments (or samples), each of which contains "information" which may be left undefined. This information is usually thought to include pitches, durations, intensities, etc., but this is slightly misleading since we are not necessarily describing the coordinate values making up an infinitesimally small singular aural event such as what happens in analog-digital conversion. Each segment (sample) may contain many such singular events—a measure and a half from a Beethoven piano sonata, 1/9 second of a purple finch's song, three hours of the sounds of rush hour traffic, a message sent in Morse code. A segment's length—important in content transformations—is now irrelevent with respect to the composite ordering transformations we will now define.

For the purposes of presenting the final three string transformations, we will assume that n is an even integer, though in practice it need not be. Figure 10.12a is a representation of a string α with $n = 10$ segments (samples).

We first make a connection to a transformation many readers will recognize. *ALPHA* is generally thought of as a twelve-tone operation on pitch classes that rotates only the even (or odd) integers; e.g., using permutation notation, $(1)(3)(5)(7)(9)(B)(02468A)$. This transformation also makes sense as a permutation of string segments. Begin by forming two new strings from α, $\chi_{odd} = \langle \varsigma_1, \varsigma_3, \varsigma_5, \varsigma_7, \varsigma_9 \rangle$ and $\chi_{even} = \langle \varsigma_2, \varsigma_4, \varsigma_6, \varsigma_8, \varsigma_{10} \rangle$. Next define a transformation $\Diamond$, "interleaf" or "shuffle," such that $\chi_{odd} \Diamond \chi_{even} = \langle \varsigma_i \rangle_{i=1}^{10} = \alpha$ (see figure 10.12b). Then one version of *ALPHA* is

$$ALPHA(\alpha) = \chi_{odd} \Diamond ROT(\chi_{even}) \tag{15}$$

as illustrated in figure 10.12c. We say "one version" because there are three other ways to perform these composite transformations: $ROT(\chi_{even}) \Diamond \chi_{odd}$, $\chi_{even} \Diamond ROT(\chi_{odd})$, and $ROT(\chi_{odd}) \Diamond \chi_{even}$.

The final two transformations were developed by Roger Reynolds[15] at the suggestion of David Wessel and used extensively throughout Reynolds' compositional works as "editorial algorithms." The first of this pair Reynolds calls "SPLITZ." In our current notation, this is defined

$$SPLITZ(\alpha) = \chi_{odd} \Diamond RET(\chi_{even}) \tag{16}$$

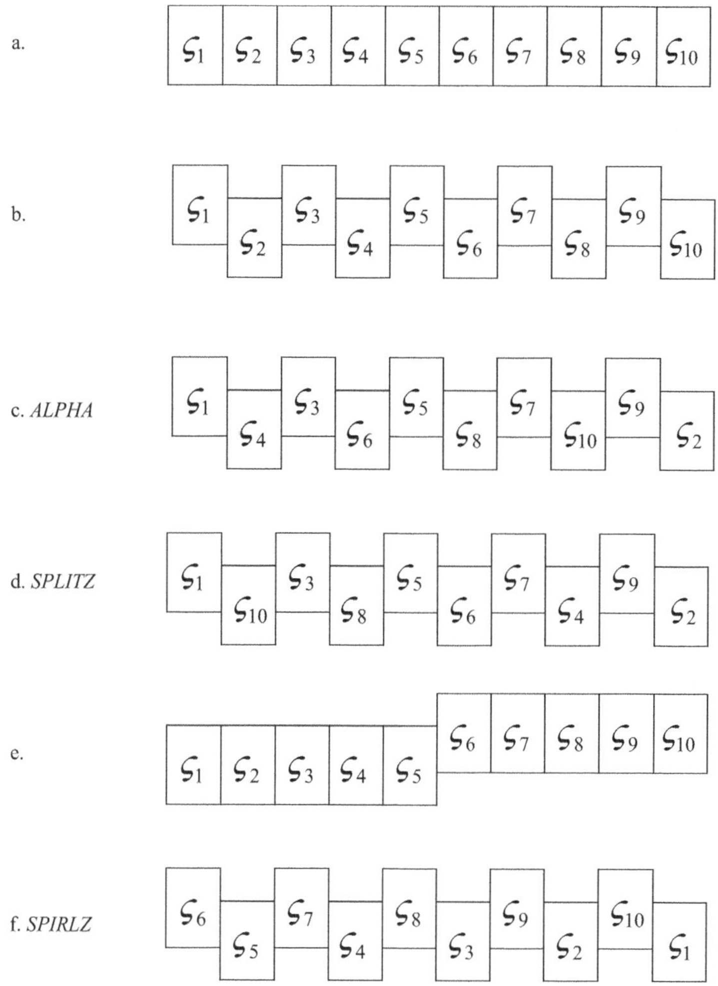

Figure 10.12.

and illustrated in figure 10.12d. As with *ALPHA*, *SPLITZ* is actually any one of the four possible interleaf combinations of χ_{odd} and χ_{even} with *RET* applied to one or the other (but not both).

"SPIRLZ" is the second transformation in this pair of editorial algorithms. First, we must create another two strings from α, $\phi_{left} = \left\langle \varsigma_1, \varsigma_2, \varsigma_3, \varsigma_4, \varsigma_5 \right\rangle$ and $\phi_{right} = \left\langle \varsigma_6, \varsigma_7, \varsigma_8, \varsigma_9, \varsigma_{10} \right\rangle$ (see figure 10.12e). Then, in our notation, one of the four versions of *SPIRLZ* is

$$SPIRLZ(\alpha) = \phi_{right} \lozenge RET(\phi_{left})$$

as represented in figure 10.12f.

To see how *SPLITZ* and *SPIRLZ* are "complementary" transformations (meaning, in this case, composite transformations which form inverse-related pairs), we bring back an idea initially presented above. We have been assuming that the shuffle transformation works in one and only one way, i.e., as we have defined $\lozenge$. In fact, there are many ways to shuffle two or more strings, one of which is concatenation, $\oplus$. This makes possible the construction of compositionally useful inverse-related pairs such as

$$\begin{aligned}&\left[RET(\chi_{odd}) \oplus \chi_{even}\right] \circ \left[RET(\phi_{left}) \lozenge \phi_{right}\right] \\ &= \left[RET(\phi_{left}) \lozenge \phi_{right}\right] \circ \left[RET(\chi_{odd}) \oplus \chi_{even}\right] = \alpha \end{aligned} \tag{18}$$

Devising a shorthand makes the pattern in the four inverse-related procedures easier to see:

$$\left.\begin{matrix} \left(\overleftarrow{o \oplus e}\right) * \left(\overleftarrow{l} \lozenge r\right) \\ \left(o \oplus \overleftarrow{e}\right) * \left(l \lozenge \overleftarrow{r}\right) \\ \left(\overleftarrow{e \oplus o}\right) * \left(r \lozenge \overleftarrow{l}\right) \\ \left(e \oplus \overleftarrow{o}\right) * \left(\overleftarrow{r} \lozenge l\right) \end{matrix}\right\} = \alpha \tag{19}$$

where α is any (even) sample, $o = \chi_{odd}$, $e = \chi_{even}$, $l = \phi_{left}$, $r = \phi_{right}$, $\overleftarrow{x} = RET(x)$, and $x * y = (x \circ y)\ or\ (y \circ x)$. The top equation in (19) is thus a "stripped-down" version of (18).

We have been purposely biasing our interpretation of interval string theory toward the segmentation and manipulation of "source material" as a generalized or abstract category, be it material captured on a magnetic tape so it can be physically manipulated, or digitized via a sampling process for processing by computer. But one interpretation of "source material" may, of course, still be pitch or interval.

Working again by example, we focus on equation (18). An abstract sample string is really no more than an indexed *n*-tuple which, choosing $n = 8$, we will write as $\alpha = \langle 1,2,3,4,5,6,7,8 \rangle$. Thus, $\chi_{odd} = \langle 1,3,5,7 \rangle$, $\chi_{even} = \langle 2,4,6,8 \rangle$, $\phi_{left} = \langle 1,2,3,4 \rangle$, and $\phi_{right} = \langle 5,6,7,8 \rangle$. Equation (18) gives us two round trips from α to α. On the first, we go to $RET(\phi_{left}) \lozenge \phi_{right} = \langle 4,5,3,6,2,7,1,8 \rangle = \pi_1$ before $RET(\chi_{odd}) \oplus \chi_{even}$ (applied to π_1) returns us to α. On the second, we go to $RET(\chi_{odd}) \oplus \chi_{even} = \langle 7,5,3,1,2,4,6,8 \rangle = \pi_2$ before $RET(\phi_{left}) \lozenge \phi_{right}$

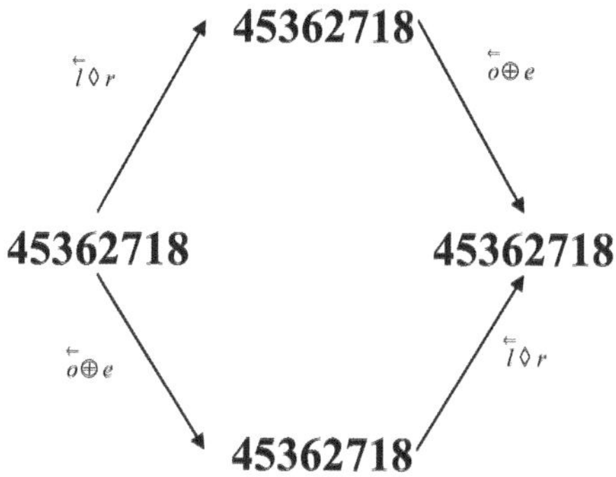

Figure 10.13.

	1	2	3	4	5	6	7	8
ints:	1	2	1	2	1	2	1	2
pcs:	0	1	3	7	10	9	6	4

Figure 10.14.

(applied to π_2) returns us to α (see figure 10.13). Next we make two index assignments (i.e., we "put stuff in the index buckets"), one for an interval-string interpretation, the other for a serial pitch-class structure interpretation (figure 10.14). Substituting, we find first that the interval strings represented by interval assignments to π_1 and π_2 are $\langle 2,1,1,2,2,1,1,2\rangle$ and $\langle 1,1,1,1,2,2,2,2\rangle$, placing us back in the world of octatonic "string modulations."

The ordered pitch-class sets represented by pitch-class assignments to π_1 and π_2 are $S=[7,10,3,9,1,6,0,4]$ and $T=[6,10,3,0,1,7,9,4]$, respectively. Note that the original assignment, $X=[0,1,3,7,10,9,6,4]$ was designed as a partition of the octatonic set such that on the left is the all-interval set class 4-Z29 based on interval string $\langle 1,2,4,5\rangle$, and on the right is its Z-related "twin," 4-Z15 based on $\langle 2,3,1,6\rangle$. In the transformed set S these subsets are chromatically inverted (resulting strings are $\langle 4,2,1,5\rangle$ and $\langle 1,3,2,6\rangle$) as $S=I_{10}X$. But in T the result is much different due to the built-in design of X's odd-even partition. T's left subset (*RET* of X's odd partition) is $[6,10,3,0]$, traditionally called a half-diminished seventh chord whose root is C; the right subset (X's even partition) is $[1,7,9,4]$, a dominant seventh chord with root A. Depending on which (10-tone serial) path we choose in this particular example, we either travel $Atonal(1) \leftrightarrow Atonal(2)$ or $Atonal \leftrightarrow Tonal$.

Notes

1. Weisstein 2005.

2. The consequence of this is that there is no interval-string transformation that is analogous to the music-theoretic transposition of pitches.

3. For example, a zero element in a string might indicate a unison or a simultaneity; mixed positive and negative values are useful in serial applications and in describing and comparing contours; and negating all the elements in a given string would indicate a generalized "dual" string (cf. Lewin's 1982, 37–39, use of this term).

4. What is called a "segment" here was labeled a "substring" in Soderberg 1998, sec.1.8. Paul Nauert's reduction of transpositionally symmetric interval strings (Nauert 2003, 190–91) is a special case that treats the segment as a generator.

5. Morris 1987, 33–58.

6. This notation is a generalization of David Lewin's "(p, sign)" Klang notation (Lewin, 1987, 176), where "p" is a pitch class serving as the root of a diatonic triad, and "sign" takes the value + or – as the triad based on p is major or minor.

7. Soderberg 1998, section 1.7.

8. A "numerical partition of n" is any string of positive integers summing to n. The numerical partitions of n up to $n = 22$ as well as permutations of multisets are available on the *Combinatorial Object Server* website at http://www.theory.csc.uvic.ca/~cos (outputs of higher values may be available upon request).

9. Soderberg 1998.

10. This was a generalization of the insight of Clough and Douthett 1991, 170, which noted that diatonic triads and seventh chords are "second-order maximally even" structures.

11. To simplify the presentation, $n = 7$ is being used here. 7 (as a partition of itself) is being omitted as trivial. For $n = 12$, this algorithm will generate the list of interval strings corresponding to the list of set classes mod 12. The partitions of 7 are also used in Clough, Cuciurean, and Douthett 1997, 81.

12. Forte 1986, 95–129

13. This 9-string also has an alternative derivation as the string complement of the diminished triad T = {0,3,6}. T's set-theoretic complement is T' = {1,2,4,5,7,8,9,10,11} whose string is <1,2,1,2,1,1,1,1,2>.

14. When the abstraction is carried out fully, string, segment or element "content" need not even be recognizably "musical."

15. Reynolds 2002.

Works Cited

Agmon, Eytan. 1989. "A Mathematical Model of the Diatonic System." *Journal of Music Theory* 33: 1–25.

———. 1991. "Linear Transformations Between Cyclically Generated Chords." *Musikometrika* 3: 15–40.

———. 1996. "Coherent Tone Systems: A Study in the Theory of Diatonicism." *Journal of Music Theory* 40: 39–59.

———. 2003. "Numbers and the Western Tone-System: Beyond Psychoacoustics." Paper presented at a meeting of the American Mathematical Society, Baton Rouge, LA, March 14–16.

Ahn, So-Yung. 2003. "Harmonic Circles and Voice Leading in Asymmetrical Trichords." PhD diss., State University of New York at Buffalo.

Antokoletz, Elliott. 1984. *The Music of Béla Bartók: A Study of Tonality and Progression in Twentieth-Century Music.* Berkeley and Los Angeles: University of California Press.

———. 2000. "Organic Expansion and Classical Structure in Bartók's Sonata for Two Pianos and Percussion." In *Bartók Perspectives,* edited by Elliott Antokoletz, Victoria Fischer, and Benjamin Suchoff, 77–94. New York: Oxford University Press.

Babbitt, Milton. 1960. "Twelve-Tone Invariants as Compositional Determinants." *Musical Quarterly* 46: 246–59.

———. 1965. "The Structure and Function of Music Theory." *College Music Symposium* 5: 49–60. Reprinted in *Perspectives on American Composers,* edited by Benjamin Boretz and Edward T. Cone, 10–21. New York: Norton, 1972.

———. 1973–74. "Since Schoenberg." *Perspectives of New Music* 12: 3–28.

Balzano, Gerald. 1980. "The Group-Theoretic Description of 12-Fold and Microtonal Pitch Systems." *Computer Music Journal.* 4, no. 4: 66–84.

Brinkman, Alexander R. 1986. "A Binomial Representation of Pitch for Computer Processing of Musical Data." *Music Theory Spectrum* 8: 44–57.

Browne, Richmond. 1981. "Tonal Implications of the Diatonic Set." *In Theory Only* 5, no. 6–7: 3–21.

Brun, Viggo. 1919. "La série 1/5 + 1/7 + 1/11 + 1/13 + 1/17 + 1/19 + 1/29 + 1/31 + 1/41 + 1/43 + 1/59 + 1/61 . . . , où les dénominateurs sont 'nombres premiers jumeaux' est convergente ou finie." *Bulletin des Sciences Mathématiques* 43: 124–28.

———. 1961. "Musikk og Euklidske Algoritmer." *Norsk Matematisk Tidskrift* 9: 29–36.

Callender, Clifton. 1998. "Voice-leading Parsimony in the Music of Alexander Scriabin." *Journal of Music Theory* 42: 219–33.

Carey, Norman. 2002. "On Coherence and Sameness, and the Evaluation of Scale Candidacy Claims." *Journal of Music Theory* 46: 1–56.

———. 2007. "Coherence and Sameness in Well-formed and Pairwise Well-formed Scales." *Journal of Mathematics and Music* 1: 79–98.

Carey, Norman, and David Clampitt. 1989. "Aspects of Well-Formed Scales." *Music Theory Spectrum* 11: 187–206.

———. 1996a. "Regions: A Theory of Tonal Spaces in Early Medieval Treatises." *Journal of Music Theory* 40: 113–47.

———. 1996b. "Self-Similar Pitch Structures, Their Duals, and Rhythmic Analogues." *Perspectives of New Music* 34, no. 2: 62–87.

Chapman, Alan. 1981. "Some Intervallic Aspects of Pitch-Class Set Relations." *Journal of Music Theory* 25: 275–90.

Childs, Adrian. 1998. "Moving Beyond Neo-Riemannian Triads: Exploring a Transformational Model for Seventh Chords." *Journal of Music Theory* 42: 181–93.

Clampitt, David. 1995. "Some Refinements on the Three Gap Theorem, with Applications to Music." *Muzica* 6, no. 2: 12–22.

———. 1997. "Pairwise Well-Formed Scales: Structural and Transformational Properties." PhD diss., State University of New York at Buffalo.

———. 1998. "Alternative Interpretations of Some Measures from *Parsifal*." *Journal of Music Theory* 42: 321–34.

———. 1999. "Ramsey Theory, Unary Transformations, and Webern's Op. 5, No. 4." *Intégral* 13: 63–93.

Clough, John. 1979. "Aspects of Diatonic Sets." *Journal of Music Theory* 23: 45–61.

———. 1979–1980. "Diatonic Interval Sets and Transformational Structures." *Perspectives of New Music* 18: 461–82.

———. 1994. "Diatonic Interval Cycles and Hierarchical Structure." *Perspectives of New Music* 32, no. 1: 228–53.

———. 1996. "Aspects of Sequence in Music of the First Viennese School." Keynote address, *Austria, 996–1996: Music in a Changing Society*, Ottawa.

———. 1998. "A Rudimentary Geometric Model for Contextual Transposition and Inversion." *Journal of Music Theory* 42: 297–306.

———. 2000. "Uniform Flip-Flop Circles and Their Groups." Paper presented at the annual meeting of the Society for Music Theory, Toronto.

Clough, John, John Cuciurean, and Jack Douthett. 1997. "Hyperscales and the Generalized Tetrachord." *Journal of Music Theory* 41: 67–100.

Clough, John, and Jack Douthett. 1991. "Maximally Even Sets." *Journal of Music Theory* 35: 93–173.

Clough, John, Jack Douthett, N. Ramanathan, and Lewis Rowell. 1993. "Early Indian Heptatonic Scales and Recent Diatonic Theory." *Music Theory Spectrum* 15: 36–58.

Clough, John, Nora Engebretsen, and Jonathan Kochavi. 1999. "Scales, Sets, and Interval Cycles: A Taxonomy." *Music Theory Spectrum* 21: 74–104.

Clough, John, and Gerald Myerson. 1985. "Variety and Multiplicity in Diatonic Systems." *Journal of Music Theory* 29: 249–70.

———. 1986. "Musical Scales and the Generalized Circle of Fifths." *American Mathematical Monthly* 93: 695–701.

Cohn, Richard. 1991a. "Bartók's Octatonic Strategies: A Motivic Approach." *Journal of the American Musicological Society* 44: 262–300.

———. 1991b. "Properties and Generability of Transpositionally Invariant Sets." *Journal of Music Theory* 35: 1–32.

———. 1994. "Generalized Cycles of Fifths, Some Late-Nineteenth Century Applications, and Some Extensions to Microtonal and Beat-Class Spaces." Keynote address, annual meeting of Music Theory Midwest, Bloomington, IN.

———. 1996. "Maximally Smooth Cycles, Hexatonic Systems, and the Analysis of Late-Romantic Triadic Progressions." *Music Analysis* 15: 9–40.

———. 1997. "Neo-Riemannian Operations, Parsimonious Trichords, and Their *Tonnetz* Representations." *Journal of Music Theory* 41: 1–66.

———. 1998a. "Introduction to Neo-Riemannian Theory: A Survey and A Historical Perspective." *Journal of Music Theory* 42: 167–80.

———. 1998b. "Music Theory's New Pedagogability." *Music Theory Online* 4, no. 2.

———. 1998c. "Square Dances with Cubes." *Journal of Music Theory* 42: 283–96.

———. 2000. "Weitzmann's Regions, My Cycles, and Douthett's Dancing Cubes." *Music Theory Spectrum* 22: 89–103.

Cone, Edward T. 1982. "Schubert's Promissory Note: An Exercise in Musical Hermeneutics." *19th-Century Music* 5: 233–41.

Cowell, Henry. 1930. *New Musical Resources*. New York: A. A. Knopf.

Dahlhaus, Carl. 1990. *Studies on the Origin of Harmonic Tonality*. Translated by Robert Gjerdingen. Princeton, NJ: Princeton University Press.

Douthett, Jack. 1993a. "The Twin Primes Problem and Musical Scales and Chords." Unpublished manuscript.

———. 1993b. "MP, CV, and Twin Primes." Unpublished manuscript.

Douthett, Jack, and Richard Krantz. 1996. "Energy Extremes and Spin Configurations for the One-Dimensional Antiferromagnetic Ising Model with Arbitrary-Range Interaction." *Journal of Mathematical Physics* 37: 3334–53.

———. 2007. "Maximally Even Sets and Configurations: Common Threads in Mathematics, Physics, and Music." *Journal of Combinatorial Optimization* 14:385–410.

Douthett, Jack, and Peter Steinbach. 1998. "Parsimonious Graphs: A Study in Parsimony, Contextual Transformations, and Modes of Limited Transposition." *Journal of Music Theory* 42: 241–63.

Dummit, David S., and Richard M. Foote. 2003. *Abstract Algebra*, 3rd ed. New York: Wiley.

Fétis, François-Joseph. 1844. *Traité complet de la théorie et de la pratique de l'harmonie contenant la doctrine de la science et de l'art*. Paris: Maurice Schlesinger.

Forte, Allen. 1973. *The Structure of Atonal Music*. New Haven: Yale University Press.

———. 1986. "Harmonic Syntax and Voice Leading in Stravinsky's Early Music." In *Confronting Stravinsky: Man, Musician, and Modernist*, edited by Jann Pasler, 95–129. Berkeley and Los Angeles: University of California Press.

Fraleigh, John B. 2002. *A First Course in Abstract Algebra*. Boston, MA: Addison-Wesley Publishing Company.

Gamer, Carlton. 1967. "Some Combinatorial Resources of Equal-Tempered Systems." *Journal of Music Theory* 11: 32–59.

Gauldin, Robert. 1983. "The Cycle-7 Complex: Relations of Diatonic Set Theory to the Evolution of Ancient Tonal Systems." *Music Theory Spectrum* 5: 39–55.

Gollin, Edward. 1998. "Some Aspects of Three-Dimensional *Tonnetze*." *Journal of Music Theory* 42: 195–206.

———. 2000. "Representations of Space and Conceptions of Distance in Transformational Music Theories." PhD diss., Harvard University.

Goodaire, Edgar G. and Michael M. Parmenter. 2005. *Discrete Mathematics with Graph Theory*. Upper Saddle River, NJ: Prentice Hall.

Haimo, Ethan. 1990. *Schoenberg's Serial Odyssey: The Evolution of his Twelve-Tone Method, 1914–1928*. Oxford: Clarendon Press.

Hauptmann, Moritz. 1991. *The Nature of Harmony and Metre*, translated and edited by W. E. Heathcote. New York: Da Capo Press.

Helmholtz, Hermann von. 1954. *On the Sensations of Tone as a Physiological Basis for the Theory of Music*, 2nd English ed., translated and revised by Alexander J. Ellis. New York: Dover Publication.

Hilborn, Robert C. 1994. *Chaos and Nonlinear Dynamics: An Introduction for Scientists and Engineers*. New York: Oxford University Press.

Hook, Julian. 2002a. "Uniform Triadic Transformations." PhD diss., Indiana University.

———. 2002b. "Uniform Triadic Transformations." *Journal of Music Theory* 46: 57–126.

———. 2007. "Cross-Type Transformations and the Path Consistency Condition." *Music Theory Spectrum* 29: 1–39.

Hooker, Lynn. 2001. "Modernism Meets Nationalism: Béla Bartók and the Musical Life of Pre-World War I Budapest." PhD diss., University of Chicago.

Hyer, Brian. 1995. "Reimag(in)ing Riemann." *Journal of Music Theory* 39: 101–38.

Imig, Renate. 1970. *Systeme der Funktionsbezeichnung in den Harmonielehren seit Hugo Riemann*. Düsseldorf: Verlag Gesellschaft zur Förderung der systematischen Musikwissenschaft.

Johnson, Timothy A. 2003a. *Foundations of Diatonic Theory: A Mathematically Based Approach to Music Fundamentals*. Emeryville, CA: Key College Publishing.

———. 2003b. *Instructor Resources: Foundations of Diatonic Theory*. Emeryville, CA: Key College Publishing.

Karpati, Janos. 1975. *Bartók's String Quartets*, translated by Fred Macnicol. Budapest: Corvina Press.

———. 1994. *Bartók's Chamber Music*. Stuyvesant, NY: Pendragon Press.

———. 1995. "Perfect and Mistuned Structures in Bartók's Music." *Studia Musicologica* 36: 365–80.

Klumpenhouwer, Henry. 1994. "Some Remarks on the Use of Riemann Transformations." *Music Theory Online* 0, no. 9.

———. 2002. "Dualist tonal space and transformation in nineteenth-century musical thought." In *The Cambridge History of Western Music Theory*, edited by Thomas Christensen, 456–76. Cambridge: Cambridge University Press.

Kochavi, Jonathan H. 1997. "Contextually-Defined Inversion Operators: A General Mathematical Setting." Paper presented at the annual Meeting of the Society for Music Theory, Phoenix.

———. 1998. "Some Structural Features of Contextually-Defined Inversion Operators." *Journal of Music Theory* 42: 307–20.

———. 2002. "Contextually Defined Musical Transformations." PhD diss., State University of New York at Buffalo.

Kopp, David. 1995. "A Comprehensive Theory of Chromatic Mediant Relations in Mid-Nineteenth-Century Music." PhD diss., Brandeis University.

———. 2002. *Chromatic Transformations in Nineteenth-Century Music.* Cambridge: Cambridge University Press.

Krantz, Richard, Jack Douthett, and John Clough. 2000. "Maximally Even Sets: A Discovery in Mathematical Music Theory is Found to Apply in Physics." In *Bridges: Mathematical Connections in Art, Music, and Science, Conference Proceedings,* edited by Reza Sarhangi, 193–200. Winfield, Kansas: Bridges Conference.

Krantz, Richard J., Jack Douthett, and Steven D. Doty. 1998. "Maximally Even Sets and the Devil's-Staircase Phase Diagram for the One-Dimensional Ising Antiferromagnet with Arbitrary-Range Interaction." *Journal of Mathematical Physics* 39: 4675–82.

Lendvai, Ernö. 1971. *Béla Bartók: An Analysis of His Music.* London: Kahn and Averill.

Leong, Daphne. 1999. "Metric Conflict in the First Movement of Bartók's *Sonata for Two Pianos and Percussion.*" *Theory and Practice* 24: 57–90.

Lerdahl, Fred. 2001. *Tonal Pitch Space.* New York: Oxford University Press.

Lerdahl, Fred, and Ray Jackendoff. 1983. *A Generative Theory of Tonal Music.* Cambridge, MA: MIT Press.

Lester, Joel. 1992. *Compositional Theory in the Eighteenth Century.* Cambridge, MA: Harvard University Press.

Lewin, David. 1982. "A Formal Theory of Generalized Tonal Functions." *Journal of Music Theory* 26: 23–60.

———. 1986. "Music Theory, Phenomenology, and Modes of Perception." *Music Perception* 3: 327–92.

———. 1987. *Generalized Musical Intervals and Transformations.* New Haven: Yale University Press.

———. 1993. *Musical Form and Transformation: 4 Analytic Essays.* New Haven: Yale University Press.

———. 1996. "Cohn Functions." *Journal of Music Theory* 40: 181–216.

London, Justin. 2004. *Hearing in Time.* New York: Oxford University Press.

Maegaard, Jan. 1972. *Studien zur Entwicklung des dodekaphonen Satzes bei Arnold Schönberg.* Copenhagen: W. Hansen, Musik-Forlag.

Marvin, Elizabeth West. 1991. "The Perception of Rhythm in Non-Tonal Music: Rhythmic Contours in the Music of Edgard Varèse." *Music Theory Spectrum* 13: 61–78.

Marx, Adolf Bernhard. 1856. *Theory and Practice of Musical Composition.* Translated from the 3rd German ed. by Herrman S. Saroni. NY: Mason Brothers.

McCune, Mark. 1986. "Moritz Hauptmann: *Ein Haupt Mann* in Nineteenth-Century Music Theory." *Indiana Theory Review* 7, no. 2: 1–29.

Mooney, Michael Kevin. 1996. "The 'Table of Relations' and Music Psychology in Hugo Riemann's Harmonic Theory." PhD diss., Columbia University.

Morris, Robert D. 1987. *Composition with Pitch-Classes: A Theory of Compositional Design.* New Haven: Yale University Press.

Nauert, Paul. 2003. "Field Notes: A Study of Fixed-Pitch Formations." *Perspectives of New Music* 41, no. 1: 180–239.

Neumeyer, David. *Linear Analysis of Music: Conventions and Contexts.* Manuscript in preparation.

Nicholls, David Roy. 1990. *American Experimental Music, 1890–1940.* Cambridge: Cambridge University Press.

Noll, Thomas. 2005. "Musiktheorie und Mathematik." In *Musiktheorie: Handbuch der Systematischen Musikwissenschaft* Band 2, edited by Helga de la Motte-Haber and Oliver Schwab-Felisch, 409–18. Laaber: Laaber Verlag.

Oettingen, Arthur von. 1866. *Harmoniesystem in dualer Entwickelung: Studien zur Theorie der Musik.* Dorpat and Leipzig: W. Gläser.

Perle, George. 1962. *Serial Composition and Atonality.* Berkeley and Los Angeles: University of California Press.

Peterson, Ivars. 2005. "Closing the Gap on Twin Primes." *Science News Online* 168, no. 3.

Petersen, Peter. 1994. "Rhythmik und Metrik in Bartóks *Sonate fuer zwei Klaviere und Schlagzeug* und die Kritik des jungen Stockhausen an Bartók." *Musiktheorie* 9: 39–48.

Pressing, Jeffrey Lynn. 1983. "Cognitive Isomorphisms between Pitch and Rhythm in World Musics: West Africa, the Balkans, and Western Tonality." *Studies in Music* (Australia) 17: 38–61.

Rahn, Jay. 1977. "Some Recurrent Features of Scales." *In Theory Only* 2, no. 11–12: 43–52.

———. 1991. "Coordination of Interval Sizes in Seven-Tone Collections." *Journal of Music Theory* 35: 33–60.

Regener, Eric. 1974. "On Allen Forte's Theory of Chords." *Perspectives of New Music* 13, no. 1: 191–212.

Reynolds, Roger. 2002. *Form and Method: Composing Music. The Rothschild Essays,* edited by Stephen McAdams. New York and London: Routledge.

Ricci, Adam. 2002. "A Classification Scheme for Harmonic Sequences." *Theory and Practice* 27: 1–36.

Rice, Timothy. 2000. "Béla Bartók and Bulgarian Rhythm." In *Bartók Perspectives,* edited by Elliott Antokoletz, Victoria Fischer, and Benjamin Suchoff, 196–210. New York: Oxford University Press.

Riemann, Hugo. 1875. *Die Hülfsmittel der Modulation: Studie von Dr. Hugo Riemann.* Kassel (Berlin: F. Luckhardt).

———. 1877. *Musikalische Syntaxis: Grundriß einer harmonischen Satzbildungslehre.* Leipzig: Breitkopf und Härtel.

———. 1880. *Skizze einer neuen Methode der Harmonielehre.* Leipzig: Breitkopf und Härtel.

———. 1895. *Harmony Simplified; or, The Theory of the Tonal Functions of Chords,* translated by Henry Bewerunge. London: Augener.

———. 1992. "Riemann's 'Ideen zu einer Lehre von den Tonvorstellungen': An Annotated Translation," translated by Robert Wason and Elizabeth West Marvin. *Journal of Music Theory* 36: 69–117.

Roeder, John. 1989. Review of *Composition with Pitch-Classes: A Theory of Compositional Design,* by Robert D. Morris. *Music Theory Spectrum* 11: 240–51.

Santa, Matthew. 2000. "Analysing Post-Tonal Diatonic Music: A Modulo 7 Perspective." *Music Analysis* 19: 167–201.

Schachter, Carl. 1999. "Durational Reduction." In *Unfoldings: Essays in Schenkerian Theory and Analysis*, edited by Joseph N. Straus, 54–78. New York: Oxford University Press.

Schenker, Heinrich. 1954. *Harmony*, translated by Elisabeth Mann Borgese. Cambridge, MA: MIT Press.

Shimbo, Amy. 2001. "Some Transformations Among Triads and Seventh Chords." Paper presented at the Third Buffalo Symposium on Neo-Riemannian Transformations.

Shirlaw, Matthew. 1970. *The Theory of Harmony*. New York: Da Capo Press.

Silverman, Danielle, and Jim Wiseman. 2006. "Noting the Difference: Musical Scales and Permutations." *American Mathematical Monthly* 113: 648–51.

Singer, Alice. 1974. "The Metrical Structure of Macedonian Dance." *Ethnomusicology* 18: 379–404.

Smith, Charles J. 2001. "Functional Fishing with *Tonnetz*: Toward a Grammar of Transformations and Progressions." Paper presented at the Third Symposium on Neo-Riemannian Theory, Buffalo, NY.

Soderberg, Stephen. 1998. "White Note Fantasy." *Music Theory Online* 4, no. 3.

Somfai, László. 1996. *Béla Bartók: Composition, Concepts, and Autograph Sources*. Berkeley: University of California Press.

Sós, Vera T. 1958. "On the Distribution mod 1 of the Sequence nα." *Annales Universitatis Scientiarum Budapestinensis de Rolando Eötvös Nominatai Sectio Mathematica* 1: 127–34.

Stempel, Larry. 1979. Review of *Beyond Orpheus: Studies in Musical Structure*, by David Epstein. *MLA Notes* 36, no. 2: 357–58.

Sternberg, Shlomo. 1994. *Group Theory and Physics*. Cambridge: Cambridge University Press.

Stevens, Halsey. 1993. *The Life and Music of Béla Bartók*, 3rd ed. Oxford: Clarendon Press.

Stillwell, John. 2003. *Elements of Number Theory*. New York: Springer.

Strogatz, Steven H. 1994. *Nonlinear Dynamics and Chaos: With Applications to Physics, Biology, Chemistry, and Engineering*. Cambridge, MA: Westview Press.

Suchoff, Benjamin. 1992. *Béla Bartók Essays*. Lincoln, NB: University of Nebraska Press.

———. 2001. *Béla Bartók: Life and Work*. Lanham, MD: Scarecrow Press.

Tymoczko, Dmitri. 1997. "The Consecutive-Semitone Constraint on Scalar Structure: A Link between Impressionism and Jazz." *Intégral* 11: 135–79.

Weber, Gottfried. 1846. *The Theory of Musical Composition*, translated by James F. Warner. Boston: Wilkins, Carter, and Company.

Weisstein, Eric W. 2005. "String." From *MathWorld—A Wolfram Web Resource*. http://mathworld.wolfram.com/String.html.

Weitzmann, Carl Friedrich. 1853. *Das übermässige Dreiklang*. Berlin: T. Trautwein'schen.

Werckmeister, Andreas. 1698. *Die nothwendigsten Anmerkungen und Regeln, wie der Bassus continuus oder General-Baß wol könne tractiret werden*. Aschersleben: Gottlob Ernst Struntze.

Willson, Rachel Beckles. 2002. "Bulgarian Rhythm and its Disembodiment in *The Sayings of Péter Bornemisza, Op. 7.*" *Studia Musicologica* 43: 269–80.
Wilson, Paul. 1992. *The Music of Béla Bartók.* New Haven: Yale University Press.
Wooldridge, Marc C. 1992. "Rhythmic Implications of Diatonic Theory: A Study of Scott Joplin's Ragtime Piano Works." PhD diss., State University of New York at Buffalo.

Contributors

DAVID CLAMPITT is Associate Professor of Music at Yale University. His interests include scale theory, transformational theory, and Brahms and Webern studies. He holds a PhD in music theory from the University at Buffalo (SUNY) (John Clough, advisor). A recipient of the Society for Music Theory's Emerging Scholar Award (with Norman Carey) in 1999, his work has appeared in *Music Theory Spectrum, Journal of Music Theory, Perspectives of New Music, Intégral, Music Theory Online, Muzica, Gestalt Theory,* and *Notes.*

NORMAN CAREY, Associate Professor of Music Theory at CUNY Graduate Center, is also the director of the CUNY's D.M.A. program in musical performance. He holds degrees in Piano Performance from Manhattan School of Music and a Ph.D. in Music Theory from Eastman School of Music, and has been a member of the theory faculty at both Eastman and Yale. Carey has collaborated with David Clampitt on several articles on the theory of well-formed scales and was a recipient of the Society for Music Theory's Emerging Scholar Award (with Clampitt) in 1999. He has also published articles on Schenkerian topics involving issues of performance and analysis. Active as a performer, he is the pianist of the Prometheus Piano Quartet.

JOHN CLOUGH (1930–2003) taught music theory at Oberlin College, the University of Michigan, and the University at Buffalo (SUNY) where he held the distinguished Slee Professorship of Music Theory. A revolutionary thinker, he pioneered the enquiry into the nature of diatonic systems. He published many articles on diatonic sets, interval cycles and sequences, mathematical properties of diatonic systems, and scale theory in *Perspectives of New Music, Journal of Music Theory, American Mathematical Monthly, Music Theory Spectrum,* and *Muzica.* In 1993 the Society for Music Theory gave him its Outstanding Publication Award. For two decades, John Clough was unusual in his generous instinct for collaboration with colleagues and students, an instinct that led him also to organize three summer symposia that drew music theorists to Buffalo from around the country.

RICHARD COHN, Battell Professor of the Theory of Music at Yale University, is the general editor of *Oxford Studies in Music Theory*. He holds a PhD from the Eastman School of Music. He has published many articles on Bartók, atonal pitch-class theory, Schenkerian theory, and metric dissonance. He is the only two-time recipient of the Society for Music Theory's Outstanding Publication Award. His research on the group-theoretic basis of chromatic harmony benefited immeasurably from the sustained generosity of John Clough, who convened a series of seminars at the University at Buffalo to explore its implications.

JACK DOUTHETT holds two MM degrees, one in music theory and composition and the other in performance, and a PhD in mathematics from the University of New Mexico. He has published in the disciplines of mathematics, physics, acoustics, and music theory; his major area of interest is mathematical music theory. In 1993 John Clough and Jack Douthett received the Society for Music Theory's Outstanding Publication Award for their work on maximally even sets and scale theory. Douthett has taught mathematics and music at several colleges and universities, including Central New Mexico Community College, University of New Mexico, University of Chicago, and the University at Buffalo (SUNY).

NORA ENGEBRETSEN, Assistant Professor of Music Theory at Bowling Green State University, holds a Bachelor's degree from the University of Washington and a PhD in music theory from the University at Buffalo (SUNY) with the dissertation "Nascent Group-Theoretic Perspectives in German Harmonic Theory, ca. 1850–1915" (John Clough, advisor). Her research interests include transformational theory and the history of music theory. Her articles have appeared or are forthcoming in *Music Theory Spectrum*, *Theoria: Historical Aspects of Music Theory*, *Riemann Perspectives* (Oxford University Press), and *What Kind of Theory is Music Theory? Epistemological Exercises in Music Theory and Analysis* (Stockholm University Press).

JULIAN HOOK, Assistant Professor of Music Theory at Indiana University, holds advanced degrees in mathematics, architecture, and piano performance as well as in music theory. His article "Uniform Triadic Transformations," published in the *Journal of Music Theory*, earned the Emerging Scholar Award from the Society for Music Theory in 2005. Other recent publications include a survey of applications of group theory in music (in a collection of essays on mathematics published by Princeton University Press); a short "Perspective" on mathematical music theory in *Science*; and an article on the foundations of transformation theory in *Music Theory Spectrum*. He is currently Reviews Editor of the *Journal of Mathematics and Music*.

MARTHA M. HYDE, Associate Professor of Music Theory at the University at Buffalo (SUNY), holds a MMA in Piano Performance from the Yale School of

Music and a PhD in Music Theory from Yale University. She has been a faculty member at Yale University, the Luce Visiting Professor at Columbia University, and a Visiting Professor at the Eastman School of Music and the University of California, Santa Barbara. For a number of years she served as the Editor of the *Journal of Music Theory*. She received the ASCAP Deems Taylor Publication Award for her articles on Schoenberg's 12-tone music, and has published a book on Schoenberg's 12-tone sketches. She has also published a number of articles on the theory and analysis of 12-tone meter, on the music of Schoenberg and Stravinsky, on neo-classicism and on musical semiotics.

TIMOTHY A. JOHNSON, Associate Professor of Music at Ithaca College, holds a PhD in music theory from the University at Buffalo (SUNY) (John Clough, advisor), where in 2005–6 he was the Visiting Slee Professor of Music Theory. He has published two books, *Foundations of Diatonic Theory: A Mathematically Based Approach to Music Fundamentals* (Key College Publishing, 2003), which is dedicated to John Clough, and *Baseball and the Music of Charles Ives: A Proving Ground* (Scarecrow Press, 2004), which received the Sporting News–SABR (Society for American Baseball Research) Award. His articles on John Adams, Charles Ives, minimalism, and other areas have appeared in the *Journal of Music Theory, Music Theory Spectrum, Musical Quarterly*, and other journals.

JONATHAN KOCHAVI holds Bachelor's degrees in both mathematics and music from the University of Chicago, as well as a Master's degree in mathematics from the University of Wisconsin, Madison. In 2002, he completed a PhD in Music Theory at the University at Buffalo (SUNY), with the dissertation "Contextually Defined Musical Transformations" (John Clough, advisor). His areas of research include diatonic theory, transformation theory, neo-Riemannian theory, and the interconnections between music and mathematics. His work has been published in the *Journal of Music Theory* and *Music Theory Spectrum*. He is currently on the music faculty at Swarthmore College, and he has previously taught at Temple University, the University of Pennsylvania, and the University at Buffalo.

DAVID LEWIN (1933–2003) taught composition at the University of California, Berkeley, and at the State University of New York, Stony Brook, and later taught music theory at Yale and Harvard Universities. His three books, *Generalized Musical Intervals and Transformations* (1987), *Musical Form and Transformation* (1993), and *Studies in Music with Text* (2006), have recently been published or reprinted by Oxford University Press. He was the recipient of honorary degrees from the University of Chicago, New England Conservatory of Music, and the University of Strasbourg for his distinguished and unparalleled work in music theory.

Charles J. Smith, Associate Professor of Music Theory at the University at Buffalo (SUNY), is a specialist in tonal harmony who has written most often about chromaticism in several of its manifestations. His lengthy article in *Music Analysis* on Schenker's *Formenlehre* won the Society for Music Theory's Outstanding Publication Award in 1999. Before coming to Buffalo in 1987, he was on the faculty at the University of Connecticut. Since his move to Buffalo, he has been the coordinator of the Music Theory area and is now serving as chair of the Music Department.

Stephen Soderberg is Senior Specialist for Contemporary Music in the Music Division of the Library of Congress. He is also Project Coordinator for the Music Division's website Content and currently serves on the boards of directors for the Contemporary Music Forum in Washington, DC and the Garth Newel Music Center in Warm Springs, VA. He has published articles in the field of mathematical music theory in the *Journal of Music Theory* and *Music Theory Online*. He has delivered papers in this field at various conferences, including the second Buffalo symposium in 1997, "Neo-Riemannian Transformations: Mathematics and Applications."

Index

An italicized page number indicates a figure or table.

Eastman Studies in Music

The Poetic Debussy: A Collection of His Song Texts and Selected Letters
(Revised Second Edition)
Edited by Margaret G. Cobb

Concert Music, Rock, and Jazz since 1945: Essays and Analytical Studies
Edited by Elizabeth West Marvin and Richard Hermann

Music and the Occult: French Musical Philosophies, 1750–1950
Joscelyn Godwin

"Wanderjahre of a Revolutionist" and Other Essays on American Music
Arthur Farwell, edited by Thomas Stoner

French Organ Music from the Revolution to Franck and Widor
Edited by Lawrence Archbold and William J. Peterson

Musical Creativity in Twentieth-Century China: Abing, His Music, and Its Changing Meanings
(includes CD)
Jonathan P. J. Stock

Elliott Carter: Collected Essays and Lectures, 1937–1995
Edited by Jonathan W. Bernard

Music Theory in Concept and Practice
Edited by James M. Baker, David W. Beach, and Jonathan W. Bernard

Music and Musicians in the Escorial Liturgy under the Habsburgs, 1563–1700
Michael J. Noone

Analyzing Wagner's Operas: Alfred Lorenz and German Nationalist Ideology
Stephen McClatchie

The Gardano Music Printing Firms, 1569–1611
Richard J. Agee

"The Broadway Sound": The Autobiography and Selected Essays of Robert Russell Bennett
Edited by George J. Ferencz

Theories of Fugue from the Age of Josquin to the Age of Bach
Paul Mark Walker

The Chansons of Orlando di Lasso and Their Protestant Listeners: Music, Piety, and Print in Sixteenth-Century France
Richard Freedman

Berlioz's Semi-Operas: Roméo et Juliette *and* La damnation de Faust
Daniel Albright

The Gamelan Digul and the Prison-Camp Musician Who Built It: An Australian Link with the Indonesian Revolution
Margaret J. Kartomi

"The Music of American Folk Song" and Selected Other Writings on American Folk Music
Ruth Crawford Seeger, edited by Larry Polansky and Judith Tick

Portrait of Percy Grainger
Edited by Malcolm Gillies and David Pear

Berlioz: Past, Present, Future
Edited by Peter Bloom

The Musical Madhouse
(Les Grotesques de la musique)
Hector Berlioz
Translated and edited by Alastair Bruce
Introduction by Hugh Macdonald

The Music of Luigi Dallapiccola
Raymond Fearn

Music's Modern Muse: A Life of Winnaretta Singer, Princesse de Polignac
Sylvia Kahan

The Sea on Fire: Jean Barraqué
Paul Griffiths

"Claude Debussy As I Knew Him" and Other Writings of Arthur Hartmann
Edited by Samuel Hsu, Sidney Grolnic, and Mark Peters
Foreword by David Grayson

Schumann's Piano Cycles and the Novels of Jean Paul
Erika Reiman

Bach and the Pedal Clavichord: An Organist's Guide
Joel Speerstra

Historical Musicology: Sources, Methods, Interpretations
Edited by Stephen A. Crist and Roberta Montemorra Marvin

The Pleasure of Modernist Music: Listening, Meaning, Intention, Ideology
Edited by Arved Ashby

Debussy's Letters to Inghelbrecht: The Story of a Musical Friendship
Annotated by Margaret G. Cobb

Explaining Tonality: Schenkerian Theory and Beyond
Matthew Brown

The Substance of Things Heard: Writings about Music
Paul Griffiths

Musical Encounters at the 1889 Paris World's Fair
Annegret Fauser

Aspects of Unity in J. S. Bach's Partitas and Suites: An Analytical Study
David W. Beach

Letters I Never Mailed: Clues to a Life
Alec Wilder
Annotated by David Demsey
Foreword by Marian McPartland

Wagner and Wagnerism in Nineteenth-Century Sweden, Finland, and the Baltic Provinces: Reception, Enthusiasm, Cult
Hannu Salmi

Bach's Changing World: Voices in the Community
Edited by Carol K. Baron

CageTalk: Dialogues with and about John Cage
Edited by Peter Dickinson

European Music and Musicians in New York City, 1840–1900
Edited by John Graziano

Schubert in the European Imagination, Volume 1: The Romantic and Victorian Eras
Scott Messing

Opera and Ideology in Prague: Polemics and Practice at the National Theater, 1900–1938
Brian S. Locke

Ruth Crawford Seeger's Worlds Innovation and Tradition in Twentieth-Century American Music
Edited by Ray Allen and Ellie M. Hisama

Schubert in the European Imagination, Volume 2: Fin-de-Siècle Vienna
Scott Messing

Mendelssohn, Goethe, and the Walpurgis Night: The Heathen Muse in European Culture, 1700–1850
John Michael Cooper

Dieterich Buxtehude: Organist in Lübeck
Kerala J. Snyder

Musicking Shakespeare: A Conflict of Theatres
Daniel Albright

Pentatonicism from the Eighteenth Century to Debussy
Jeremy Day-O'Connell

Maurice Duruflé: The Man and His Music
James E. Frazier

Representing Non-Western Music in Nineteenth-Century Britain
Bennett Zon

The Music of the Moravian Church in America
Edited by Nola Reed Knouse

Music Theory and Mathematics: Chords, Collections, and Transformations
Edited by Jack Douthett, Martha M. Hyde, and Charles J. Smith

The essays in *Music Theory and Mathematics: Chords, Collections, and Transformations* define the state of mathematically oriented music theory at the beginning of the twenty-first century. The volume includes essays in diatonic set theory, transformation theory, and neo-Riemannian theory-the newest and most exciting fields in music theory today.

The essays constitute a close-knit body of work-a family in the sense of tracing their descent from a few key breakthroughs by John Clough, David Lewin, and Richard Cohn in the 1980s and 1990s. They are integrated by the ongoing dialogue they conduct with one another.

The editors are Jack Douthett, a mathematician and music theorist who collaborated extensively with Clough; Martha M. Hyde, a distinguished scholar of twentieth-century music; and Charles J. Smith, a recognized expert in tonal theory. The contributors are all prominent scholars, teaching at institutions such as Harvard, Yale, Indiana University, and the University at Buffalo. Six of them (Clampitt, Clough, Cohn, Douthett, Hook, and Smith) have received the Society for Music Theory's prestigious Publication Award, and one (Hyde) has received the ASCAP Deems Taylor Award. The collection includes the last paper written by Clough before his death, as well as the last paper written by David Lewin, an important music theorist also recently deceased.

Contributors: David Clampitt, John Clough, Richard Cohn, Jack Douthett, Nora Engebretsen, Julian Hook, Martha M. Hyde, Timothy A. Johnson, Jonathan Kochavi, David Lewin, Charles J. Smith, and Stephen Soderberg.

"These essays, by leading American music theorists, continue the development of some of the most important research of the last twenty years into mathematical models of basic musical structures. These models are elegant in the abstract, but they are also shown to have many practical applications in explaining a wide range of art music. Several of the contributions are bound to be classics in this literature."

—John Roeder, professor of music theory,
University of British Columbia

"*Music Theory and Mathematics* is a fitting memorial to John Clough, one of music theory's great pioneers. Clough was among the first scholars to introduce non-trivial mathematics into what has emerged as diatonic set theory or scale theory. This volume consists of essays by important theorists on a variety of topics ranging from scale and Riemannian theory to analysis of works by Bartók and Schoenberg. Building on Clough's research, *Music Theory and Mathematics* poses new questions and approaches to what are perhaps the most exciting directions in music theory today."

—Robert Morris, professor of composition,
Eastman School of Music (University of Rochester)

Printed and bound by CPI Group (UK) Ltd, Croydon, CR0 4YY

15/09/2026

14962416-0002